LABORATORY MANUAL

Bert Atsma

Union County College

Human Biology

Concepts and Current Issues

EIGHTH EDITION

Michael D. Johnson

PEARSON

Editor in Chief: Beth Wilbur
Senior Acquisitions Editor: Star MacKenzie Burruto
Executive Marketing Manager: Lauren Harp
Program Management Team Lead: Michael Early
Project Management Team Lead: David Zielonka
Program Manager: Anna Amato
Project Manager: Brett Coker
Rights & Permissions Project Manager: Donna Kalal
Production Service and Composition: Lumina Datamatics, Inc.
Illustrations: Imagineering
Cover Production: Seventeenth Street Studios
Manufacturing Buyer: Stacey Weinberger
Cover Image: Alfred Pasieka / Science Photo Library / Getty Images

ISBN 10: 0-134-28381-3
ISBN 13: 978-0-134-28381-4

6 17

 www.pearsonhighered.com

CONTENTS

Preface   iv
To the Student   vi

EXERCISE 1   Scientific Method and Laboratory Protocol   1 14/14

EXERCISE 2   The Microscope   15   14

EXERCISE 3   The Anatomy and Diversity of Cells   27   14

EXERCISE 4   Cell Physiology   35   13/14

EXERCISE 5   Tissues   45

EXERCISE 6   Orientations to the Human Body   57   14

EXERCISE 7   The Integumentary System   71

EXERCISE 8   The Skeletal System   85   —7 no Review

EXERCISE 9   The Muscular System   103 0 not done

EXERCISE 10   The Nervous System I: Organization, Neurons, 14
Nervous Tissue, and Spinal Reflexes   115

EXERCISE 11   The Nervous System II: The Spinal Cord, Brain, and
Autonomic Nervous System   125   14

EXERCISE 12   The Senses   137   14

EXERCISE 13   The Endocrine System   153

EXERCISE 14   The Cardiovascular System I: Blood   165   14

EXERCISE 15   The Cardiovascular System II: Heart and Blood Vessels   177 14

EXERCISE 16   The Respiratory System   193   14

EXERCISE 17   The Digestive System and Nutrition   209

EXERCISE 18   The Urinary System   225

EXERCISE 19   The Reproductive System   235

EXERCISE 20   Genetics   253   14/14

EXERCISE 21   DNA Technology and Genetic Engineering   267

EXERCISE 22   Evolution   279

EXERCISE 23   Human Ecology   293

APPENDIX   The Metric System   305

Art, Text, and Photo Credits   306
Index   308

PREFACE

This laboratory manual was written for use with *Human Biology: Concepts and Current Issues*, Eighth Edition, by Michael Johnson, a wonderful textbook written for students who may not have a strong background in science and, indeed, may not be science majors at all. Similarly, the goal of my laboratory book is to provide a meaningful laboratory experience for students, regardless of their level of preparation for a human biology course. I firmly believe that providing the nonmajor with a solid foundation in the science of human biology is an appropriate goal and that a laboratory experience is an important part of that foundation. For this edition, improvements were made to the art, introductory information, and activities for several exercises.

Features

Although it is possible that students with little or no laboratory experience may sometimes find the science laboratory an intimidating environment, much can be done to make students more comfortable with laboratory learning. This laboratory manual contributes to a more comfortable learning environment for the nonmajor by using a simple, consistent format.

- **Objectives** are listed first in each exercise so that students immediately understand what they are supposed to get out of the laboratory experience.
- **Materials** needed to perform the laboratory activities are set forth, both at the beginning of the exercise and often at the start of each individual activity. This placement is convenient for the instructor or staff in preparing the laboratory and for the student in executing the activity.
- A short **introduction** is provided for each topic so that the student can learn the information relevant to the activity he or she will be performing, regardless of whether he or she has read the matching textbook section or covered it in the lecture portion of the course.

- **Activities** are clearly called out with distinct headings so that students are alerted to places where they should be performing lab work. The activities are set forth in a clear, step-by-step format and presented in a way that assumes the student is doing this kind of work for the first time.
- **Embedded questions** are asked frequently throughout each exercise to keep students focused on what they should be learning while they are doing their lab work.
- **Critical thinking questions** at the end of each exercise may be used by students to practice and reinforce the concepts they have covered, and these questions may also be used by instructors to evaluate student learning.
- **Answers** to the questions in this book are available online in the instructor resources area at http://www.pearsonmylabandmastering. com. Instructors should contact their local Pearson sales representative for a username and password.

Organization

The initial exercises of this lab manual introduce students to the lab environment. Exercise 1 introduces the scientific method and lab protocols. Exercise 2 covers the use and care of the microscope. The remaining exercises closely follow the organization of the Johnson textbook, *Human Biology*. Exercises 3 and 4 concern cell anatomy and physiology, and Exercise 5 provides the link between cells, tissues, and organs. In Exercise 6, students are introduced to the terminology used to characterize and discuss anatomical structures and regions. Exercises 7–19 cover the 10 body systems, while Exercise 20 provides a fun introduction to genetics. Exercise 21 introduces students to the fascinating field of DNA Technology. Finally, Exercise 22 covers evolution, and Exercise 23 invites students to address ecological issues, such as the impact of various lifestyles and population growth upon the environment and human interaction.

Approach

There are so many fascinating things to be learned about the human body that it is not surprising that human biology course instructors often have varying approaches to its presentation. As with many things in life, there is more than one right way to organize a human biology course, and my somewhat ambitious table of contents reflects an awareness of this. It is expected that most instructors, with 2 to 3 hours of lab time per week, will pick and choose from these numerous exercises and activities to match their goals and methods of presentation. Some of the shorter exercises were written with the idea in mind that in courses with longer laboratory periods, they would be combined with all or parts of other exercises.

Some instructors disagree with the use of dissection work for nonmajors, and others feel that dissection is an integral part of learning how organs and structures look *and* feel. I tend to believe that a certain amount of dissection work is meaningful, even for the nonmajor, in many places. For example, the fetal pig does much to convey the flimsy, saclike nature of the gallbladder, and the sheep brain demonstrates the toughness of the dura mater around the brain. A few minutes with an actual specimen makes a lasting impression, perhaps in more ways than one. Due to the time it takes to carefully open the fetal pig, dissection of the fetal pig is introduced in an exercise that is light in dissection work (the endocrine system).

There are certainly legitimate reasons why instructors do not wish to include dissection in their human biology courses. Understanding this preference, I provided an opportunity in most exercises that call for dissection to learn much of the same material by using widely available models.

Acknowledgments

Many people contributed their efforts to this book. First, I thank Kay Ueno, whose unique talents and incredibly hard work made the first edition of this book possible. Kay, I couldn't have done it without you. I also wish to thank Frank Rugirello, Daryl Fox, Marie Beaugureau, Deirdre Espinoza, Alex Streczyn-Woods, Becky Ruden, Leslie Allen, Ziki Dekel, Alison Rodal, Emily Portwood, and Sandra Hsu for their contributions to the success of this book.

For this Eighth Edition, I thank our talented editorial team: Star MacKenzie Burruto, Beth Wilbur, Ginnie Simione Jutson, and Maja Sidzinska. Their leadership, advice, and guidance made the substantial work on this edition go smoothly. On the production side, I thank Michael Early, Anna Amato, David Zielonka, Donna Kalal, Brett Coker, Anju Joshi, and Andrea Stefanowicz for their hard work bringing this book across the finish line. Thank you all for your help.

Finally, I thank my past students for teaching *me* what works best through their questions and efforts in the laboratory. And of course, thanks to Zac, Sandy, Amanda, and Carol, for their cooperation while dad was writing.

BERT ATSMA, Union County College

TO THE STUDENT

Many students dread a science course, particularly its laboratory component. You have had reading, writing, and arithmetic, even a little history, and possibly a full year or two of high school science, but your classwork may not have prepared you for a science laboratory. For most nonmajors, college-level science in general, and biology specifically, may be unfamiliar territory to explore. But exploring new territory can be exciting, especially with good maps and a good guide. With the aid of your instructor, I hope that this manual helps you successfully navigate the laboratory portion of your human biology course.

This laboratory manual was designed to prepare you for a positive laboratory experience, regardless of your level of preparation for a course in human biology. I have used a simple, consistent format to make laboratory learning a more comfortable and enjoyable experience. The **objectives** alert you to what you are supposed to learn from each exercise. Wherever I thought it would be helpful, the **materials** required are listed both at the beginning of each exercise and each activity. A short **introduction** begins each exercise providing the background information necessary to fully appreciate the activities you will perform. The **activities** are spelled out in a clear, step-by-step format, helping you to more easily follow the directions, even if this is your first laboratory experience. **Questions** are asked frequently throughout the activities and at the end of each exercise to ensure that you are "learning while doing" your work.

To get the most out of your laboratory experience and perform your best, I recommend that you read the exercise you will be performing *before* you come to class. This really does work.

The introductory information will prepare you for new terms and concepts that you will encounter in the lab. Being familiar with the procedures will help you perform the activities more smoothly and more safely.

Speaking of safety, you will find safety guidelines printed on the inside front cover of this lab manual. Please take the precautionary statements in these guidelines seriously, as well as any additional safety information provided by your instructor. I have made an effort to reduce the risk of personal injury by recommending the use of the safest chemicals and procedures appropriate to each activity. Careless use of chemicals and equipment is something that only you can prevent. Your instructor will advise you of any protective equipment that you should use prior to each activity. In particular, I feel it is a good idea to wear eye protection when handling any chemicals, especially if you wear contact lenses. Remember, safety first!

Finally, be prepared for your instructor to modify some of the activities in this laboratory manual, as there is more than one way to learn about the marvels of the human body. If you wish to make comments on this lab manual or offer suggestions for improvements in future editions, please use the student response form at the back of this book. I sincerely hope that your human biology laboratory experience will be positive and memorable, and I wish you great success!

BERT ATSMA
Pearson
Biology
1301 Sansome Street
San Francisco, CA 94111

EXERCISE
1

Scientific Method and Laboratory Protocol

Objectives

After completing this exercise, you should be able to

1. Describe the process of scientific discovery and information gathering.
2. Explain the concept of scientific method and its steps.
3. Discuss the metric system, and calculate metric conversions.
4. Convert percentages, decimals, and fractions.
5. Demonstrate proper dissection and experimental protocols.
6. Describe proper lab safety protocols.

Materials for Lab Preparation

Safety Equipment
- ○ Hood
- ○ Safety shower
- ○ Eyewash
- ○ Fire extinguisher
- ○ Fire blanket
- ○ Gloves
- ○ Goggles

Dissection Tools
- ○ Scalpel or knife
- ○ Probe
- ○ Forceps
- ○ Scissors
- ○ Dissection tray (waxed or unwaxed)
- ○ Dissection pins or twine

Supplies
- ○ Coin
- ○ Cleaning disinfectant (e.g., Sanisol in spray bottle)
- ○ Autoclave bag for biologically hazardous material
- ○ Bleach jar for soaking glassware
- ○ Timer

Introduction

All activities in this exercise are intended to serve as an introduction to the scientific laboratory. They also may be useful to you as a reference when embarking on later activities in this laboratory manual.

The first three activities will acquaint you with the scientific process designed for testing ideas and making discoveries. The remaining activities introduce you to some of the scientific tools of the trade: the use of metric measurements and commonly used dissection tools. Last, but not least, you will review safety protocol in the laboratory.

Scientific Method of Discovery

Even before the concept of "science" was around, people observed the world around them, formulated questions, tested their ideas, and communicated their conclusions to their friends and neighbors. The tried and tested ideas eventually became part of our general body of knowledge. Today, whether in our homes, in school, or in business, we continue to add or make adjustments to that body of knowledge. Without realizing that we are doing so, we employ some of the same elements of the discovery process, albeit informally or unconsciously employed by the scientist in the laboratory.

However, informal attempts at applying what may *seem* to be the scientific process can have mixed results to say the least. Sometimes they add to our understanding of the world, and sometimes they do a disservice to our understanding (Figure 1.1). Many stereotypes and prejudices people hold to are the result of such "pseudoscience." To ensure valid scientific discovery, scientists are trained observers, and follow a more formal approach to asking and answering questions about the world around us.

FIGURE 1.1 **The film industry hasn't always presented scientists and their work accurately.** However, *Medicine* Man was a fairly good portrayal of the work of field biologists and their counterparts in the science laboratory.

Science seeks to learn what we can prove by careful observation, prediction, and experimentation, which is accomplished through the step-by-step process of the **scientific method.** Other branches of learning can be quite important and useful, but they are not considered science if they do not follow the scientific method. That is an important distinction to keep in mind as we go through the course.

ACTIVITY 1

Developing the Scientific Method

There is more than one right way to describe the scientific method. For example, the process of hypothesis building can be rather complex, and the way it is typically simplified for grade school and nonmajor college students can vary depending upon the instructor's viewpoint. Nonetheless, there are several key conceptual elements to the scientific method. In the following activity, you will learn each step of the scientific method and practice performing that step yourself.

Step 1: Observation and Collecting Objectively Acquired Information

Scientists are typically trained observers, noticing and recording even the smallest details. An often overlooked and underappreciated part of the process of scientific observation is reducing the possibility of error and bias.

Amusing examples of error/bias are some of the things attributed to Murphy's law(s) about things *supposedly* going wrong much more often than they go right. Do they really? Does the pencil you drop nearly always bounce and roll to a hard-to-reach place under the desk? It certainly seems so, but what really happens? Most people (scientists included) tend to notice an event more when it has a negative or uncomfortable effect. The many times the pencil dropped at your feet or on your lap may have been far less memorable.

The scientific method requires making note of *every* spot the pencil dropped, not just the annoying ones. Many widely accepted but mostly false notions are the result of such **selective perception.** Individually or with a partner, come up with your own example of selective perception, and discuss the things you could do to properly observe the occurrence in a way that eliminates bias. Summarize on the following page.

Observations are underappreciated in science. Typically bias and error are the issues.

Another amusing example of the bias often applied to experience is the **self-fulfilling prophecy.** Imagine someone becoming annoyed at a person with blond hair, and recalling the so-called dumb blond stereotype. This person may go through the next few days (or years) noticing every blond-haired person who annoys him or her, but not noticing the many nice, intelligent blonds encountered. Furthermore, he or she may tend to *not* recall this "theory" when irritated by dark-haired persons. So despite having encountered no more annoyingly dim-witted blonds than brunettes over time, the person may remain convinced that experience has shown this "theory" to be true. Unfortunately, many commonly held prejudices follow this model (Figure 1.2).

Individually or with a partner, come up with your own example of a self-fulfilling prophecy, and discuss the things you could do to properly observe these things in a way that eliminates bias. Summarize below.

Filmmaking is the self-fulfilling prophecy. People believe the work is perfect.

More Practice Identifying Error/Bias

Ralph decides to observe/record snowfall depth on his side of town, which is on the north side of a mountain that runs through the middle of the township. He looks for the deepest snowdrift on the side of his house facing the wind. He then pushes a yardstick into the snow as far as it will go, writes "7 inches," and goes back inside. He does this after every snowstorm. Are these scientific observations? *Yes*
Why or why not? *Only if he is going to use the data for good.*

Step 2: Proposing a Question or Generalization Based upon the Information

This question or generalization is called a **hypothesis.** A critical point here is that consistent trends or patterns *in the information/observations* should be the basis for the hypothesis, and a scientist should *not* start off with a personally preferred idea. In the previous example, Ralph apparently wanted to prove that he got the most snow on his side of town, and that biased his observations (as well as the conclusions he would draw from them).

At the end of the winter, Ralph compares his data for the north side of town with Frank's data from the other side of the mountain. Frank, a successful biology student, reports average data that is between 1 and 2 inches less than Ralph's data in snow depth for three of the storms. However, there are two

FIGURE 1.2 Bias is everywhere! Although a miniature golf course is an unlikely place for bias related to science, this is an amusing (even though extreme) example of how subtle bias can be, and why scientists must be trained to be vigilant about recognizing it. A scientist would probably disagree with Mr. Blue Baseball Cap here, as sometimes a miniature golf windmill is just a windmill. (Hillary B. Price, Rhymes with Orange, July 17, 2010)

storms for which the depth was about the same and one for which Frank reported a snow depth 2 inches *higher* than Ralph's. Ralph looks at this information and states that the north side of town gets more snow per storm. Is this a scientific hypothesis? __Yes__
Why or why not?

It has data from both sides to back up his hypothesis
Bias?

Step 3: Test the Hypothesis

A frequent misunderstanding is that the tests of a hypothesis or theory must be laboratory experiments. This is not true, because science has many ways of testing ideas. Tests can include lab experiments, predictions followed by further observation, consistency with future discoveries, models, simulations, and so on. The one thing the test *must* do is allow for possible outcomes that support or reject the hypothesis.

Imagine that you and your lab partner just happen to live on the north and south sides of the same town as Ralph and Frank. Describe the experimental procedure you would use to test Ralph's hypothesis next winter.

You would measure the depth of snow in the North, South, East and West of town to prove/disprove Ralph's hypothesis

Step 4: Reject, Revise, or Confirm the Hypothesis Based upon the Test Results

Let us assume that you designed a test procedure whereby you sampled the depth of snow in many places and on all sides of the house. Your data is compiled in Box 1.1.

Based upon this data, does your test confirm, reject, or force revision of Ralph's hypothesis? __Yes__
Explain.

It rejects his hypothesis as depths are equal

Note the last column. This is why *detailed* observations, including other possible factors, are important in science. Although there is no difference in depth overall, careful examination of Box 1.1 may allow you to see a pattern when matching the storm direction variable with depth data. What *specific* hypothesis would you propose as an alternative?

Different parts of town get more snow depending on snow direction

Step 5: Some Hypotheses Become Theories

This step only happens with more general, encompassing hypotheses, when they have been successfully tested many times, and by many different scientists in the scientific community. The revised hypothesis, that snowfall over time is about the same, but the side of a mountain facing the storm (windward) gets more than the other (leeward) side, can be tested in a number of ways. Brainstorm (no pun intended) some ideas with

Box 1.1 **Depth of Snow Samples**			
Storm Date	**Average Depth (N)**	**Average Depth (S)**	**Storm Direction**
Dec. 24	4	5	SW
Jan. 5	5	4	NE
Jan. 11	3	4	SW
Feb. 9	9	10	SW
Feb. 25	9	8	NE
Mar. 28	8	7	NE
Total	**38**	**38**	

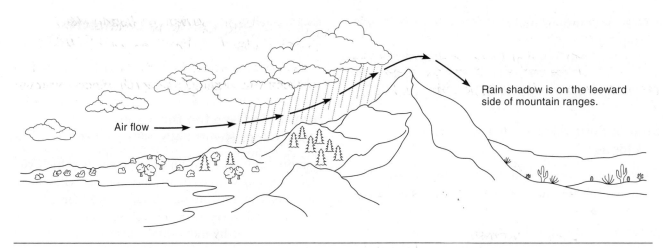

FIGURE 1.3 The rain shadow phenomenon.

a lab partner for how this could be tested enough to become a theory, and summarize your ideas below.

Do the same test as before but with more locations

Step 6: Some Theories Become Laws

This step may follow when theories are tested enough to make the scientific community confident that no new testing or information is likely to substantially alter them. However, this is sometimes modified when a thoroughly and successfully tested theory becomes so useful in studying other aspects of science that it is accepted as a **core principle** or even a **branch of science.** For example, genetics and evolution have been so thoroughly accepted and useful that they stand alone as scientific disciplines within biology. They have become so thoroughly incorporated in the science of biology that it is unnecessary to refer to them as "laws." In the preceding, you were discovering something more basic—the principle of the "rain shadow" that frequently affects rain and snow patterns around mountains (Figure 1.3). ■

ACTIVITY 2

Theory Building: The Common Cold

Throughout history, many ideas have been tossed around about the cause(s) of the common cold. In the following activity, you will apply the steps of the scientific method to the common cold.

Step 1: Observe and Generalize

Observe different characteristics related to causation of the common cold, and then generalize about it. Take into consideration what is already known on the subject, varying opinions, past mistakes, and new aspects that you may not have foreseen. Summarize below:

Observation 1: The common cold is one of the most common and well known sicknesses.

Observation 2: People get the cold from germs, not cold air and water.

Observation 3: The cold is easily fought with modern medicines.

Generalization: Germs are the main cause of the cold and can be prevented with vitamins

Step 2: Formulate a Hypothesis

Formulate a hypothesis by making an explicit statement about the cause(s) of the common cold based on the observations and generalizations you made in Step 1.

Hypothesis: The cold is caused by common germs and the lack of vitamins.

Step 3: Formulate Valid Test Criteria and Ideas

Consider how you would test your hypothesis. Remember, the test must allow for showing the hypothesis to be right or wrong. Formulate a test idea.

Not exactly movale, but expose different groups of people to germs with or without vitamins in controlled environment

Step 4: Design an Experiment

An experiment is a carefully planned and executed manipulation of the natural world that has been designed to test your prediction. You must design an experiment that accounts for all factors that may vary during the course of the experiment, or **variables,** except for the **controlled variable,** which is the factor or element under study.

Let's design an experiment to study how colds may be transmitted. In this case, the controlled variable is the common cold. We will have two groups: the **control group,** which will not be exposed to the controlled variable (the common cold), and the **experimental group,** which will be exposed to the controlled variable. Propose an experimental design for this test:

Experiment Step 1: Take one group of people in a controlled environment and give them vitamin (with controlled diet.

Experiment Step 2: Allow them to be exposed for a while then observe each condition.

Experiment Step 3: Allow them to "reset" their system. Then give no vitamins in the same conditions.

Experiment Step 4: Allow them to be exposed to the germs then observe their conditions and compa.

Step 5: Analyze the Data and Modify Your Hypothesis

Imagine that you ran an experiment in which you placed people with colds in close contact with seemingly healthy individuals (Group A), and fortunately remembered to create a control group (Group B) of seemingly healthy individuals, kept in close contact with other healthy individuals during the same time interval. You also remembered to keep the groups in isolation for the duration of a two-week period so that other variables (other exposures) would not affect the outcome. At the end of the test, Group A reports 15 out of 100 people developed colds, and Group B reports 3 people out of 100 developed colds. Does this support, reject, or require modification of the hypothesis that colds are caused by something that is transmitted from person to person through close contact?

Yes, since you observe the effects of truly healthy exposed to cold causing germs.

Consider that three people in Group B developed colds, even though they were not kept in close contact with people with cold symptoms. If your hypothesis stated that people develop colds only after being in close contact with cold sufferers, you would need to modify your hypothesis in light of this data. This is where the scientist applies knowledge from other areas or other tests to speculate reasons for test results that may not fully fit the hypothesis. Those reasons are used to modify the hypothesis.

For example, we know that there are **subclinical** cases of diseases like colds, in which a person may occasionally have and transmit the virus without showing symptoms. For another, there is an **incubation period** for many infectious diseases like colds, in which a person shows no symptoms during the time the virus is multiplying but has not yet reached sufficient numbers to cause disease symptoms. For colds, this incubation period is typically several days.

Describe below how you might use the examples just provided to explain the Group B results, and how you might modify your hypothesis and/or experimental procedure for future tests.

I could try exposing healthy people to sick people in controlled environments.

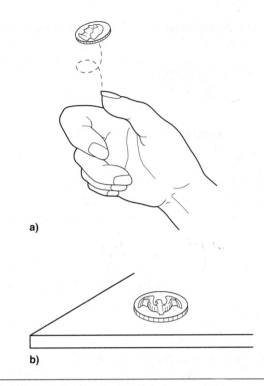

a)

b)

FIGURE 1.4 **Coin-toss procedure.**

Math and Experimental Error

Generally speaking, **experimental error** is anything in either the design or execution of observations and experiments that can lead to collection of invalid data or information. Biased observation and uncontrolled variables that can affect the outcome of an experiment are examples we have already learned about. Another fundamental source of error is related to statistics and sample size. There are statistical devices that scientists can use to determine if the sample size or numbers of repetitions in their observations/experiments are enough for a valid study. Fortunately, for a nonmajors biology course, we do not need to dwell on those details, and a more "common sense" example may suffice.

Most coins with different engravings on their obverse and reverse, more commonly called the heads and tails sides, will drop about half of the time on one side, and half on the other. If you were to test this by flipping the coin only twice, you could easily enough flip two tails in a row. A coin has no memory and feels no obligation to land on the opposite side if the first flip was tails up. So, after the first coin toss, there is still the same 50–50 chance of landing on either side. If you got two tails in a row, would you conclude that a flipped coin always lands tails up? Or, if you tossed the coin four times and got three heads and one tails, would you conclude that a coin lands heads up 75% of the time? Probably not, as "common sense" would tell you that you did not flip the coin enough times to do a fair test and observe a reliable pattern.

Scientists often investigate things that are much less straightforward than the coin toss probability discussed earlier, and must use statistics to determine if they have a valid sample size. But hopefully, the following activity will impress upon you the importance of using a valid number of subjects or repetitions in a study.

ACTIVITY 3

Size and Experimental Error

Work individually or in groups of two, but make sure there are at least 10 different experiments (100 total tosses) performed in the classroom. Plan your procedure for tossing a coin such that it will turn end over end before being caught and viewed. How you do this is not important, but it is important that you use the *same method* each time (Figure 1.4). Otherwise, you are actually letting another variable (method of toss) into your experiment. Toss the coin 10 times and record your data in Box 1.2.

Box 1.2 **Coin-Toss Data**											
Toss 1	**Toss 2**	**Toss 3**	**Toss 4**	**Toss 5**	**Toss 6**	**Toss 7**	**Toss 8**	**Toss 9**	**Toss 10**	**Total H**	**Total T**
H	H	T	H	H	T	H	T	H	T	6	4

If you obtained exactly five heads and five tails, you were pretty "lucky," as this sample size is too small for expected results to match actual results in most cases. It would take only one toss randomly "defying the odds" and not behaving as expected to end up with ⅗ (or 60%) of one and ⅖ (or 40%) of the other instead of the expected ½ to ½. Aside from the occasional lucky individual, most people in the class probably obtained data something other than exactly five heads and five tails.

Now, pool your data with at least nine other groups and calculate your percentage of each:

Heads: _57%_; Tails: _43%_ .

In this case, results of 42–58% or closer to "50–50" would be considered statistically close enough to support a hypothesis of the two events happening ½ of the time for each possible outcome. Did your class data from 100 tosses fall within this range? _Yes_ If not, repeat the experiment and pool your class data from 200 or more tosses. Did your class data from 200 tosses fall within this range? _~~~~_ Speculate as to what might contribute to obtaining unexpected results, if more than 100 tosses would be desirable.

If you test more, the data will be more precise and accurate.

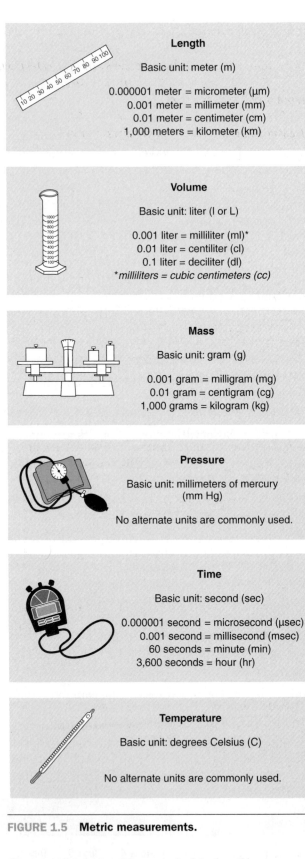

FIGURE 1.5   Metric measurements.

Length

Basic unit: meter (m)

0.000001 meter = micrometer (μm)
0.001 meter = millimeter (mm)
0.01 meter = centimeter (cm)
1,000 meters = kilometer (km)

Volume

Basic unit: liter (l or L)

0.001 liter = milliliter (ml)*
0.01 liter = centiliter (cl)
0.1 liter = deciliter (dl)
milliliters = cubic centimeters (cc)

Mass

Basic unit: gram (g)

0.001 gram = milligram (mg)
0.01 gram = centigram (cg)
1,000 grams = kilogram (kg)

Pressure

Basic unit: millimeters of mercury (mm Hg)

No alternate units are commonly used.

Time

Basic unit: second (sec)

0.000001 second = microsecond (μsec)
0.001 second = millisecond (msec)
60 seconds = minute (min)
3,600 seconds = hour (hr)

Temperature

Basic unit: degrees Celsius (C)

No alternate units are commonly used.

Metric Measurements

In nearly all fields of science, the physical dimensions of a phenomenon must be measured. At this time, all nations use the **metric system** for scientifically measuring length, mass, volume, and temperature. The metric system has two advantages:

1. Standard references. This allows measurements to be checked independent of the language of any single counting system.
2. Conversion. Measurements can be converted between different units in a universally agreed-upon manner (Figure 1.5).

Standard References

Without a standard reference, such as the metric system, we could not know that a gram (g) in Japan weighs exactly the same as a gram in the United States, which would severely hinder the process of scientific discovery. In both countries, the weight of a gram is measured by weighing the mass of 1 cubic centimeter of water at the temperature of its maximum density. Water, of course, behaves the same in all countries, so by basing the standard reference for mass on water, all countries will achieve the same standards for a gram.

The standard reference for a meter is more complicated. A meter (m) represents a unit of length equal to 1,650,763.73 wavelengths in a vacuum of the orange-red line of the spectrum of krypton 86. You would almost certainly not be expected to memorize that reference number, but it should impress upon you the precision and universality of this kind of measurement standard.

One degree Celsius (C) represents one-hundredth of the temperature scale, of which 0° is the freezing point of water and 100° is the boiling point. As you can see, the Celsius scale is a more compressed scale than the Fahrenheit temperature scale, more commonly used in the United States. That is, each rise in one degree Celsius is nearly the same as a two-degree increase on the Fahrenheit scale.

You should memorize the basic units for each physical dimension (see Figure 1.5), as well as the most common metric units, for example, centi- (one-hundredth), milli- (one-thousandth), micro- (one-millionth), and nano- (one-billionth). Note that there is no period after a metric unit. For example, 1 m (no period) is equivalent to writing and saying "one meter."

Conversions

Making conversions easily is the second advantage of using the metric system. Metric conversions are based on powers of 10 (Table 1.1). Thus, different orders of magnitude are easily converted by moving the decimal point to the right or left. For example, a common question is: How many cm are there in 5 m? The answer is 500 cm, which was determined by moving the decimal two places to the right. Use the acronym **LSR,** which stands for **Large Small Right,** to help you remember this rule. You can also rely on common sense to determine the answer, if you keep in mind, in this instance, that a centimeter (cm) is smaller than a meter (m). So when you are trying to convert a larger unit (m) to a smaller unit (cm), the equivalent numerical value you seek will be bigger.

Let's calculate some metric conversions. Suppose that you want to convert 0.4 m to millimeters:

$$0.4 \text{ m} \times 1{,}000 \text{ mm/m} = 400 \text{ mm}$$

Table 1.1 Metric Prefixes and Conversions

Metric Prefixes				
Prefix	**Symbol**	**Term**	**Exponent**	**Multiplication Factor**
deci	d	one-tenth	10^{-1}	$\times 0.1$
centi	c	one-hundredth	10^{-2}	$\times 0.01$
milli	m	one-thousandth	10^{-3}	$\times 0.001$
micro	m	one-millionth	10^{-6}	$\times 0.000001$
nano	n	one-billionth	10^{-9}	$\times 0.000000001$

Metric Conversions			
To Convert from	**To**	**Move Decimal Point**	**Factor**
meter (liter, gram)	decimeter	1 place to right	$\times 10$
meter (liter, gram)	centimeter	2 places to right	$\times 100$
meter (liter, gram)	millimeter	3 places to right	$\times 1{,}000$
meter (liter, gram)	micrometer	6 places to right	$\times 1{,}000{,}000$
meter (liter, gram)	nanometer	9 places to right	$\times 1{,}000{,}000{,}000$
decimeter	meter (liter, gram)	1 place to left	$\div 10$
centimeter	meter (liter, gram)	2 places to left	$\div 100$
millimeter	meter (liter, gram)	3 places to left	$\div 1{,}000$
micrometer	meter (liter, gram)	6 places to left	$\div 1{,}000{,}000$
nanometer	meter (liter, gram)	9 places to left	$\div 1{,}000{,}000{,}000$

Or, because we are moving from larger to smaller units, let's apply the LSR rule and move the decimal to the *right* three places:

$$0.4 \text{ mm} \rightarrow 400 \text{ mm}$$

Let's convert 5 µg to grams:

$$5 \text{ µg} \times 1 \text{ g}/1{,}000{,}000 \text{ µg} = 0.000005 \text{ g}$$

The opposite of the LSR rule applies because we are moving from smaller to larger units. Let's move the decimal to the *left* six places.

$$5 \text{ µg} \rightarrow 0.000005 \text{ g}$$

Fortunately, there are some lucky similarities that make it a little easier to develop a sense of English–metric conversions. A meter is just a little more than a yard (3.37 inches more). A kilogram is a little over two pounds, and a liter is a little more than a quart. Keeping these rough equivalents in mind may help you do a common sense reality check on your math. For example, if you asked for a 2-liter bottle of a soft drink and were handed a pint-sized bottle instead of one nearly a half-gallon, you'd know you were getting ripped off. Or, if an average-sized person claimed he or she calculated his or her weight as 300 kilograms you'd suspect a math error was made (e.g., perhaps he or she *multiplied* his or her weight in pounds by 2.2 instead of dividing).

For a complete list of metric conversions, including temperature conversions between Celsius and Fahrenheit, refer to Appendix A on page 305.

Converting Percentages to Fractions

Many of our "eyeball estimates" involve percentages or fractions. For example, let's talk about length. Your fingers are longer than your toes by about $2\frac{1}{2}$ times, or 250%, or by a factor of 2.5. Another way of saying the same thing is the length of your toes is about 40%, or two-fifths, of your fingers. Converting between numbers in decimals, percentages, and fractions is a useful skill, not only in biology, but in everyday life.

Let's calculate the percentage represented by the fraction $\frac{2}{5}$:

Step 1: $2 \div 5 = 0.4$

Step 2: $0.4 \times 100 = 40$

Step 3:   Attach the %.

Answer: 40%

Let's calculate the fraction represented by the percentage 40%:

Step 1:   Set up a fraction: $\dfrac{40}{100}$

Remember, in percentages the denominator is always 100.

Step 2:   Divide the numerator and denominator by the largest possible number.

In this case, divide both numbers by 20:

$$\frac{40}{100} \div \frac{20}{20} = \frac{2}{5}$$

It is recommended that you make a table of the most common fractions and percentages. Then treat it like the multiplication table. Review it again and again until you know it by heart so that you will "automatically" know the answer. Doing so will give you a head start in the following lab report. Remember that if you're not sure, you can always calculate the answer.

ACTIVITY 4

Review of Dissection Protocols

In human biology, the lab exercises will include conducting anatomical dissections and physiological experiments. Some animals are similar enough in structure or function to humans for dissection to provide a useful means for understanding how your own body works. Proper dissection requires patience, development of basic dissection skills, and an understanding of commonly used dissection tools. To acquaint you with the dissection tools that you will encounter in future exercises, look at Table 1.2 and

Table 1.2   Dissection Tools and Their Purposes

Dissection Tool	Purpose
Scalpel or knife	Cutting quickly through large areas
Probe	Lifting, separating, and pointing to parts of a specimen
Forceps	Pulling and/or removing parts of a specimen
Scissors	Cutting carefully through skin, muscles, and smaller bones
Dissection tray (waxed or unwaxed)	Holding specimen; containing fluids, debris, etc.
Dissection pins and twine)	Marketing or separating small parts of a specimen

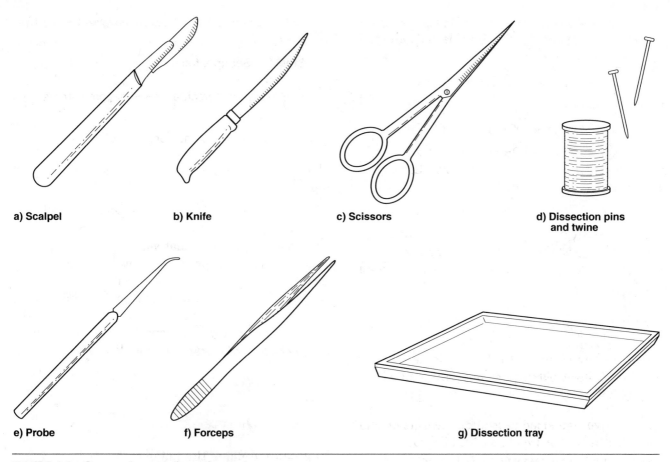

a) Scalpel   b) Knife   c) Scissors   d) Dissection pins and twine

e) Probe   f) Forceps   g) Dissection tray

FIGURE 1.6 **Commonly used dissection tools.**

Figure 1.6. Examine any dissecting equipment your instructor may have made available in the lab.

Review of Lab Safety Protocols

Conducting an experiment involving glassware, flames, acids, and other people also requires skill and patience. The most important thing is for you to be fully prepared. Always read ahead carefully before beginning an experiment, visualize the steps, write down your questions, and clear them with the instructor before the lab begins. Think about safety issues as you review the experiment, and consider how you might address them.

Every laboratory contains some safety equipment. Although the equipment maintained from one laboratory to another may vary according to needs, Figure 1.7 illustrates the equipment most commonly found in biology labs. Remember the following:

- To avoid injury or contamination, always dispose of hazardous material properly.

- Place used glassware into a jar filled with a bleach solution.

- When handling preserved specimens, wearing gloves may reduce the possibility of skin irritation.

- Place biological material into an autoclave bag.

- Clean, dry, and return all equipment to their proper locations. This is especially important for microscopes and small machines such as the centrifuge, calorimeter, and pH meter.

- After you complete your experiment, always wipe down the counter with a disinfectant so that the next student has a clean surface to use.

- If any glassware or equipment is broken, notify your instructor immediately.

- Wash your hands thoroughly upon entering and leaving a biological laboratory.

Human biology can be a fascinating course. We hope you have an enjoyable learning experience!

negative

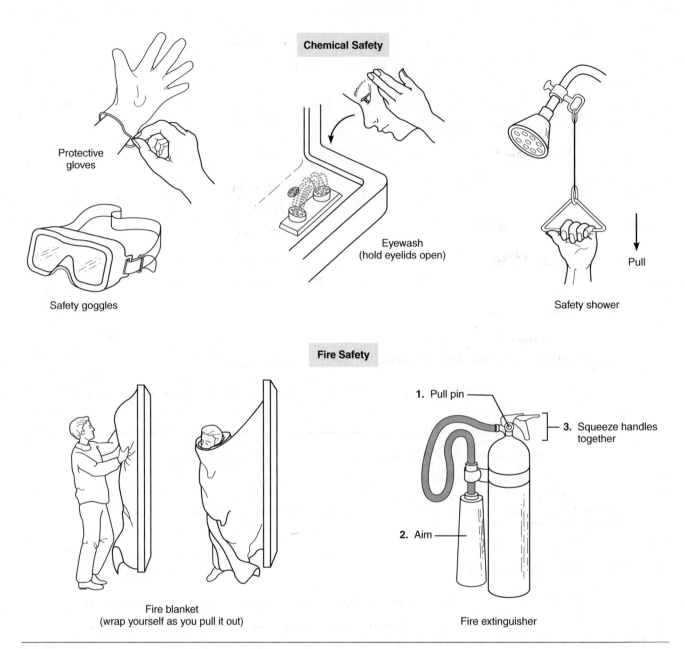

Chemical Safety

Protective gloves

Safety goggles

Eyewash (hold eyelids open)

Safety shower

Pull

Fire Safety

Fire blanket (wrap yourself as you pull it out)

1. Pull pin
2. Aim
3. Squeeze handles together

Fire extinguisher

FIGURE 1.7   **Laboratory safety equipment.**

SCIENTIFIC METHOD AND LABORATORY PROTOCOL

Critical Thinking and Review Questions

1. In your own words, describe the purpose and process of the scientific method.
   Observe, question, hypothesis, predict, experiment, repeat, and conclude. This process is used to find an answer for a question.

2. According to the scientific method, if the results of an experiment contradict the predicted results, is the hypothesis invalid? What is the next step?
   No since the results were similar to the predicted results.

3. Show all the calculations for the metric conversions below. Indicate if it is not possible to solve the problem.

   a. 6.5 g to mg: _6500_

   b. 4,300 ml to l: _4,3_

   c. 43°C to °F: _109.4_

   d. 9 m to cm: _900_

   e. 5 l to ml: _5000_

   f. 2.5 min to msec: _150000_

   g. 6,000 sec to hr: _100_

4. Using Table 1.1 and Appendix A (page 305) as needed, complete the missing parts in the table below.

15,000 mm	_150_ cm	_15_ m	_16.4_ yd	_5006_ in.
_162934_ mm	_162934_ cm	_1609.3_ m	5280 ft	_63360_ in.
_____ mg	_1000000_ g	100 kgs	_250.3_ lbs	_3524_ oz
_3790_ cc	_3790_ ml	3.79 L	_4.00_ qt	_1.0_ gal
_75708_ cc	_75708_ ml	_7.37_ L	8 qt	_2_ gal
37°C	_98.6_ °F		_21.1_ °C	70°F

13

5. Show all the calculations for the conversion of percentages, decimals, and fractions.

   a. 4.5% to fractions: _9/200_

   b. 30% to decimals: _.3_

   c. ½ to %: _50%_

   d. ⅓ to %: _33.3%_

   e. ¼ to %: _25%_

   f. ⅕ to %: _20%_

   g. ⅙ to %: _16.6%_

   h. 1/7 to %: _14.28%_

   i. ⅛ to %: _12.5%_

   j. 1/9 to %: _11.1%_

   k. 1/10 to %: _10%_

6. Describe the difference in the use of the forceps and blunt probe.

   A probe can lift or point out something of a specimen.
   Forceps can pull apart pieces of a specimen.

7. Describe the difference in the use of the scalpel and scissors.

   A scalpel cuts large areas quickly. Scissors are used
   to carefully cut skin, muscle, and bone.

8. What would you do if you cut your finger while in the lab?

   You should seek help from your instructor to get first aid
   attention.

9. What would you do if an open flame was out of control in the lab?

   Have a student call the instructor, and use the
   fire extinguisher to put the flame out.

10. What would you do if a corrosive chemical squirted into your eye?

    As a student gets the instructor, use the eyewash
    unit to flush out the chemical. You should keep your
    eyes open in the water until instruction is given.

EXERCISE 2

The Microscope

Objectives

After completing this exercise, you should be able to

1. Care properly for microscopes.
2. Recognize magnification issues and calculate the total magnification.
3. Identify the parts of the microscope and list their functions.
4. Focus a specimen with a microscope.
5. Compare and describe the effects of scanning-, low-, and high-power lenses.
6. Determine the depth of focus and field of view.
7. Prepare a wet mount slide.
8. Properly clean up and store the equipment.

Materials for Lab Preparation

Equipment and Supplies

- ○ Compound light microscopes
- ○ Millimeter ruler (plastic)
- ○ Lens paper and cleaner
- ○ Autoclave bag
- ○ Jar of 10% bleach for soaking glassware
- ○ Disinfectant

Prepared Slides

- ○ Letter "e"
- ○ Three colored threads
- ○ Whitefish blastula
- ○ Ox spinal cord cross section

Supplies for Wet Mount Slide Preparation

- ○ Clean slides
- ○ Clean coverslips
- ○ Flat toothpicks
- ○ Paper towels
- ○ Methylene blue or Wright's stain in dropper bottle

Introduction

In a human biology laboratory, the microscope is probably the most widely used instrument. Since the seventeenth century, it has enabled scientists to extend their senses into the microscopic world, to literally see beyond the naked eye. Blood looks like red ink to the naked eye, but under the microscope it becomes an amber fluid filled with hundreds of thousands of tiny red pancakes and tiny clear bags containing funny-shaped purple structures. A hangnail, under a microscope, becomes a cylindrical specimen composed of flat cells arranged in a whorl-like pattern, while a drop of clear pond water may become a large body of water alive with small, swimming organisms covered with yellow and green flecks.

This exercise will introduce you to the use and care of the standard compound light microscope.

Care of Microscopes

Remember that the microscope is a delicate and expensive instrument. A good microscope should last many, many years. Accidents and damage can be avoided by using the following guidelines:

1. Carry a microscope with two hands close to your body.
2. Clean the lenses only with lens paper and cleaner. *Do not use paper towels!*
3. Do not force anything—lenses, knobs, levers. Ask for help if something seems stuck.
4. Never leave a slide on the microscope stage once you have completed your observations and are preparing the microscope for storage.
5. Store the microscope with the lowest objective lens in place.
6. Store the microscope with the lamp cord around the arm, not the objective lens.
7. Report microscope problems, then use another microscope. *Do not store damaged microscopes!*

Microscopes and Magnifications

The microscopes commonly used in this course are compound light microscopes. Light is produced from a lamp, passed through the condenser, iris diaphragm, stage, and finally through the specimen. The image is magnified as it passes through two lenses—first, the objective lens, and then, the ocular lens. The image that you see, depending on the objective lens you are using to view your specimen, will have been magnified from 40 to 400 times, which is far greater than the typical magnifying glass.

You might think that the more magnification, the better. This is only correct to a certain extent. There are three reasons that magnification in a light microscope cannot or should not be increased indefinitely:

1. The maximum magnification for a light microscope is approximately 1,000 times. Beyond this, the image is magnified but no longer clear.
2. The more you magnify, the smaller the area you are observing. In other words, you're seeing a smaller and smaller part of the specimen, being blown up more and more.
3. It's more important (for beginning students) to see the whole specimen than to see a very small part of the specimen well. This is comparable to "seeing the forest, rather than the leaves." Remember that whenever you work with microscopes, you will eventually need to correlate all of these separate images and relate them back to the original whole specimen.

Because not all microscope models are alike, your objective lenses may differ. In this exercise, the power of a lens will be referred to as follows:

Scanning power (about 4×)

- Used to locate specimen; low power should be used if the microscope does not have scanning power

Low power (about 10×)

- Used to see the whole or large portions of the specimen

High power (about 40×)

- Also called the high-dry lens, as no oil is required for clear viewing
- Used to see small, detailed parts of the specimen

Oil immersion (about 100×)

- Used to see very small specimens (e.g., bacteria)
- Some microscopes require a drop of oil on the slide to improve clarity of the image
- Usually not used in beginning biology labs

The job of the microscope is to magnify small images. The compound light microscope accomplishes this task with two sets of lenses: the ocular lenses, which provide a constant magnification factor, and various objective lenses, which provide a variable magnification factor. It is important to know, and to be able to communicate, the total magnification of an image. In biology books, you will notice that photographs of slides usually indicate how much

magnification has occurred. For example, 106×
means the image has been magnified 106 times.

Total magnification (TM) is determined by
multiplying the power of the ocular lens (OcL) by
the power of the objective lens (ObL).

$$TM = OcL \times ObL$$

For example, if your objective lens is 4×, and
your ocular lens is 10×, total magnification is
40×. Work through Activity 1 to acquaint yourself with your microscope, then calculate the total
magnification of the lens combinations of your
microscope and record the results in Box 2.1.

ACTIVITY 1

Identifying the Parts of the Microscope

As you read the description of the parts of the microscope, begin to label the structures shown in Figure
2.1 on page 18. We will start at the top and generally move downward. Identify all bold face items.

Ocular Lens

- Also known as the eyepiece
- May be one only or a set of two
- Lens closest to your eye, usually the highest part
  of the microscope
- Includes a magnification factor engraved on
  the barrel of the ocular lens; for example, 10×
  indicates the image is magnified 10 times
- Should be cleaned *only* with lens cleaner and lens
  paper

Nosepiece

- Revolving circular mechanism that holds the
  different objective lenses

- Rotation of the nosepiece (*not* pushing on the
  barrel of the objective lens) is used to change the
  objective lens

Objective Lenses

- Individual lenses attached to the nosepiece
- Includes a magnification factor engraved on the
  barrel of each objective lens; for example, 4×, 10×,
  40×, or 100× indicates the magnification
- Should be changed by rotating the nosepiece,
  not by pushing the barrel of the objective lens
  directly

Arm

- Upright bar that connects the ocular lenses to the
  rest of the microscope
- Provides the safest way to hold a microscope: Use
  two hands—one hand on the arm, the other under the base of the microscope

Stage

- Also called the mechanical stage
- A surface that supports and secures the **slide,**
  with the help of stage clips

Stage Controls

- Usually located on the side or on top of the stage
- Front/back controls: move the slide front to back,
  and vice versa
- Side/side controls: move the slide from side to side

Condenser

- Located under the stage

Box 2.1 **Calculating Total Magnification for the Microscope**				
	Scanning	**Low**	**High**	**Oil Immersion**
Objective lens (ObL)	4x	10x	40x	λ
Ocular lens (OcL)	+	0	—	χ
Total magnification (TM)	4x	10x	40x	χ

- Focuses the light through the hole in the stage and onto the specimen
- Adjusts the quality and amount of light passing through the specimen
- May be raised or lowered with the **condenser-adjustment knob**

Iris Diaphragm

- Located under the condenser
- Adjusts the intensity of light passing through the specimen
- Opens or closes as the **iris diaphragm wheel** is manipulated

Coarse-Adjustment Knob

- Large knob located on the arm
- Adjusts, in large increments, the distance between the stage and objective lens
- Used initially to bring the specimen into general focus; it is dangerous to use this knob when the objective lens is already near the slide
- Should be turned very slowly to avoid breaking the slide or damaging the objective lens

Fine-Adjustment Knob

- Small knob attached to the coarse-adjustment knob, on the arm

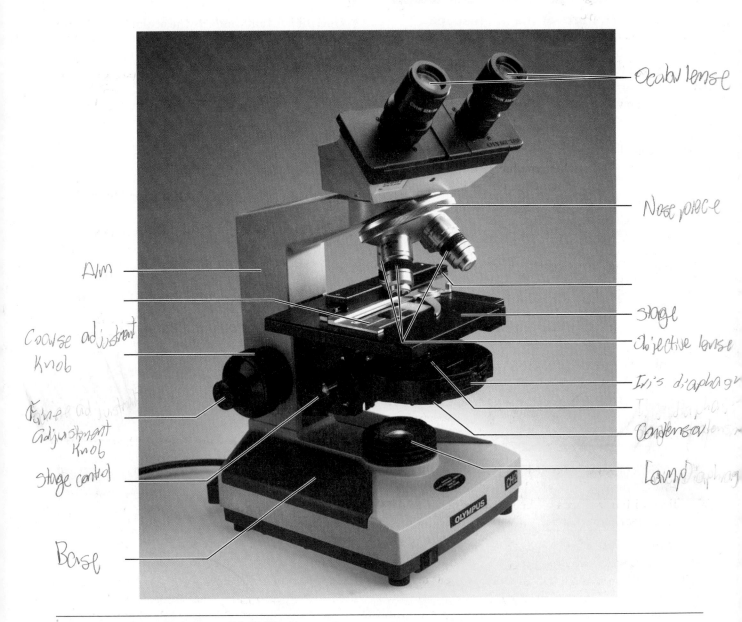

Ocular lense

Nose piece

stage

Objective lense

Iris diaphragm

Condenser

Lamp diaphragm

Arm

Coarse adjustment knob

Fine adjustment knob

stage control

Base

FIGURE 2.1   Parts of a compound light microscope.

- Adjusts, in small increments, the distance between the stage and objective lens
- Typically used after the objective lens is already near the slide and the specimen is almost in focus
- Should be turned very slowly to avoid breaking the slide

Lamp

- Small light source located under the condenser
- May be turned on or off with a switch located on the base

Base

- Square or horseshoe-shaped support for the whole microscope
- Usually quite heavy to prevent tilting ■

ACTIVITY 2

Cleaning and Caring for the Microscope

One of the signs that your microscope needs routine cleaning occurs when you have done everything right to focus your slide but the image is still blurry or otherwise hard to see. Even the tiny amount of oil on your eyebrows can rub off on the ocular lens and make it hard to see through. Because these lenses are delicate, you should use the special lens paper and lens cleaner provided in your laboratory, and definitely not paper towels or other materials as they may damage the lens surface.

1. Dab a drop of lens cleaner near one corner of the lens paper sheet.
2. Rub the wet area of the lens paper on the tip of each objective lens and then on the top of the ocular as well. Quickly use a dry section of the lens paper to wipe away the lens cleaner from any surfaces you have cleaned.
3. Also examine the microscope slide to see if dirt or oil appears to be obscuring a clear view of the specimen. If so, rub both sides of the slide, first with a wet area of the lens paper sheet, and then with a dry portion.

When you are finished with your microscope, the recommended way to prepare it for storage is to wrap the cord around your hand and place it around the ocular tube. Then place the dust cover over the top, and carry it to the storage location by holding the arm and base of the microscope. ■

ACTIVITY 3

Focusing with the Microscope

Materials for This Activity

Prepared slide of the letter "e"

Obtain a letter "e" slide, and place it on the **stage.** Looking from the side of the microscope, use the **stage controls** to center the letter "e" over the hole in the center of the stage. Secure the slide with the **stage clips.** Rotate the **nosepiece** to bring the **scanning-power lens** (or the lowest power objective lens if your microscope does not have a scanning-power lens) into place; listen for the "click," which indicates the objective lens is in position. To prevent damage to the lens, remember to look at the objective lens from the side while turning the nosepiece. Do not look through the ocular lens yet.

Again put your head to the side, and watch while you slowly turn the **coarse-adjustment knob** forward (clockwise), until the stage is ½ to 1 inch from the objective lens. Do not touch the slide. Observe how the stage moves upward, bringing the slide closer to the objective lens. Now put your eye to the **ocular lens.** You should see the blurry shape of the letter "e."

Sharpen the focus further in two stages: Begin by slowly using the **coarse-adjustment knob.** Finish by slowly using the **fine-adjustment knob.** Notice that only a few turns of the fine-adjustment knob are necessary.

Light control is another important factor to focusing an image. For example, when a slide specimen is quite thick, you may need to increase the amount of light. This may be the case if the letter "e" was printed on thick paper. If the specimen is very thin, you may find that reducing the light will produce a better contrast, and thus a better focus.

Put your head to the side and find the **iris diaphragm** wheel, which is usually located on the condenser, underneath the stage. Place your finger on the wheel and look at the slide. See what happens as you move the wheel. In one direction, light will flood through the slide; in the opposite direction, the light will begin to dim. Later, when viewing biological specimens, you will have to play with the iris diaphragm wheel to determine the amount of light needed to obtain a clear focus.

Draw the letter "e" in the circle of Box 2.2 as you observe it through the eyepiece.

Box 2.2 Drawing of Letter "e" Using Scanning (or Low) Power

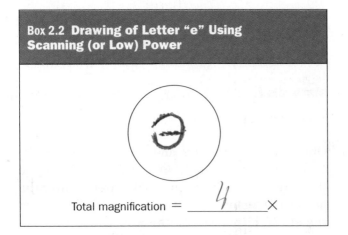

Total magnification = ____4____ ×

1. Compare the image of the letter "e" as it is placed on the stage to its image as seen through the microscope. How does the orientation differ? Is the top side still on the top? Is the right side still on the right?

   It is upside down, also you can see all ink splotches instead of a fine lined "e"

2. If the light was too bright or too dim, how did you adjust it?

   We used the slider to brighten or darken, or more IRIS

3. Let's say that you never focused on the letter "e." You just saw streaks and blurs. What are the possible reasons for this? What can you do differently?

   We could focus to fix blurs.

Now switch to the low-power lens, and then to the high-power lens. Most microscopes are **parfocal,** which means that the image remains focused when you change objective lenses. Some minor adjustment may be necessary using the fine-adjustment knob. Putting your head to the side, observe that the distance between the objective lens and the slide gets closer and closer as the power of the lens increases. When the image is focused, the **working distance** between the slide and the scanning-power lens is about ½ to 1 inch. With the low-power lens, it's about ¼ to ½ inch, and with the high-power lens, it's about ⅛ inch. In other words, at the highest magnification, there is very little distance between the lens and the slide. Slides are easily broken at this stage. *Never adjust the coarse focus when using high power!*

Box 2.3 Letter "e" at Different Magnifications

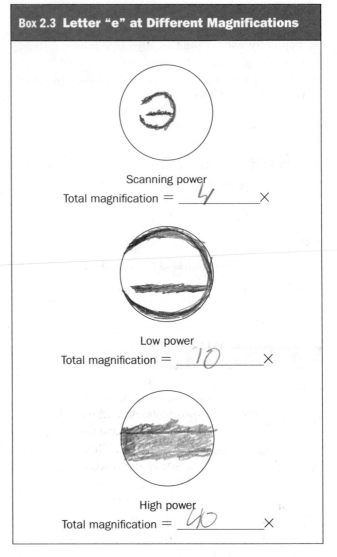

Scanning power
Total magnification = ____4____ ×

Low power
Total magnification = ____10____ ×

High power
Total magnification = ____40____ ×

Sketch the entire field in Box 2.3, as observed using the three powers. Try to indicate the actual proportions of the letter "e" in relationship to the whole field. Observe how much of the letter "e" is visible at each magnification. As much as possible, draw in details like the blotchiness of the ink, the increasing amount of space around the ink, the jagged edges of the letters, and the strands and small objects of the paper itself.

Compare the three drawings.

1. At what total magnification was the ink thickest and most uniform?

   4x

2. At what total magnification was the ink splotchy and inconsistent?

   40x

3. Which power provided the clearest image?

   4x

4. Was the orientation of the letter "e" the same at all powers?

yes

5. Did you have to adjust the light at each objective lens? Were substantial adjustments necessary?

no ■

ACTIVITY 4

Determining the Depth of Focus

Prepared Slide

Three colored threads

It is possible to focus on different depths in the same specimen. Put your head to the side of the microscope, and observe what happens as you turn the coarse-adjustment knob forward, or clockwise. Generally, the stage moves upward, which brings the slide closer to the objective lens (Figure 2.2). This means that if you have focused the lenses on the top level of the specimen, turning the adjustment knob forward, or clockwise, enables you to focus on successively lower sections of the specimen until you have finally focused on the bottom surface of the slide. The opposite occurs when you turn the adjustment knob backward, or counterclockwise. The stage moves downward, which moves the slide away from the objective lens. This means that if you have focused on the top level of the specimen, turning the adjustment knob backward will bring the focus to successively higher levels of the slide, until you are finally focused on the top surface of the slide (i.e., the glass of the coverslip).

Obtain a prepared slide of three crossed strands of colored thread, usually red, blue, and yellow. All three strands cross at or near the same point. Center and secure the slide, and focus with the scanning-power lens.

Turn the fine-adjustment knob backward, or counterclockwise, and note the order of the colored threads as they come into focus. When the specimen goes out of focus, you are at the top surface of the slide. Now, turn the fine-adjustment knob forward, or clockwise, and note the threads' order of appearance. The order should be reversed. Record the sequence and repeat the process.

Try these steps using the low-power lens, then the high-power lens. Note the additional sharpness of the colors and individual strands of the thread.

1. What is the order of the threads, from top to bottom?

Red, blue, yellow

2. Which lens allowed you to most easily distinguish the order of appearance of the threads?

4x

3. If the high-power lens was not the most effective, propose a reason why not.

Pretty effective to view one color ■

FIGURE 2.2 The importance of "coarse" adjustments! Although this is a telescope rather than a microscope, the same issue applies. The reason you were instructed to look from the side of the microscope when setting up the slide is to make sure you have it lined up somewhat close to the specimen. You might otherwise have the microscope lens centered over the label, coverslip glue, or something else inappropriate. Glue on the slide magnified 50 times might look almost as alarming as Kermit the alien! (Hillary B. Price, Rhymes with Orange, June 19, 2010)

ACTIVITY 5

Understanding Size of Field

Obtain a clear plastic metric ruler and place it on the microscope stage such that the metric calibration is *centered* across the field of view. Remember, you *always* start with your lowest-power objective lens. Move the ruler so that one of the mm lines is at the left edge of the field of view, and count the number of mm lines across to the right edge. If a mm line does not appear precisely at the right end, approximate how much of that last mm is in your field of view (e.g., ½, ¼, etc.). Also make a mental note of the relative brightness of the field under each of these objective lenses. Record the diameter of your field for your 4× objective: __4__ mm.

Repeat this for your 10× and high-power ("high-dry") objectives, and record: 10× __2__ mm; high power __25__ mm.

1. Did the field become brighter or darker as you increased magnification?

   __No__

2. What did you (or should you) do to correct this if it was too dark to count the lines?

   __Raise the brightness.__

3. Based upon your observations, describe a general rule regarding the relationship between increasing magnification and diameter of field.

   __Diameter Goes down__

   _____

4. From your observations, describe a general rule regarding the relationship between increasing magnification and brightness of field.

   __Brightness go up__

   _____ ■

ACTIVITY 6

Applying Magnification and Field Size to Biology

Prepared Slides

Whitefish blastula
Ox spinal cord cross section

Part 1: Whitefish Blastula

The 40× total magnification obtained with the low-power objective may seem like a lot on the scale of viewing everyday objects; however, cells and most of the things you will view in this course will be easily visible only under higher magnification. It is a common mistake to assume that a circular or rectangular object viewed under low power is a single cell, when it is almost certainly a tissue, a group of cells, or perhaps even an air bubble. Only unusually large cells such as some egg cells and neurons are discernable as cells under low power. This activity will serve as an example.

View a slide of whitefish blastula under low power, and note the large circular objects that are easily visible. These are not single cells, but a blastula (embryo) made of many small cells. Use your 10× and 40× objective lenses to view the interior of the blastula and note the numerous tiny cells contained therein. At which power were you best able to discern the individual cells and their contents? __40x__

Make a sketch of a few adjoining cells in the space provided next.

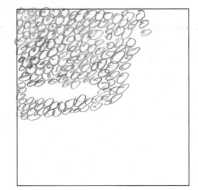

Part 2: Ox Spinal Cord

Place a slide of ox spinal cord cross section on your microscope, and focus when using the low-power objective lens. Can you see the entire specimen in your field? __No__

Move the slide around to create a mental picture and describe the appearance of the specimen as a whole.

__As a whole, it looks like a big gooey dop that had a line around it.__

Check your mental picture and description of the spinal cord with the photomicrograph in Figure 11.5 on page 128 of this lab manual.

Next, center one of the bulbous structures that you see toward either the left or right side of the rounded spinal cord. Switch to your 10× objective, and look for the large cells contained in this structure. These are the large cell bodies of sensory neurons ("nerve cells")—some of the largest cells found in animals. Select an area where you clearly see several of these cells and draw them in the space provided next.

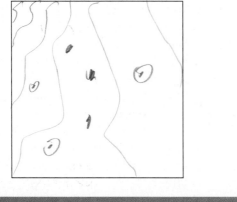

ACTIVITY 7

Preparing a Wet Mount Slide

Materials for This Activity

Paper towel

Clean slide and coverslip

Methylene blue stain or Wright's stain in dropper bottle

Flat toothpick

1. Place a small drop of methylene blue or Wright's stain on the clean slide (Figure 2.3a).
2. Gently scrape the inner surface of your cheek with the flat part of the toothpick. (Be careful not to draw blood.) Transfer the cells to the slide by carefully stirring the end of the toothpick with the cheek cells in the drop of stain (Figure 2.3b). The stain serves to darken the nuclei and edges of the cheek cells, thus substantially improving the contrast between different parts of the cheek cells. Be careful to use no more than a small drop of stain, or the cells will appear too dark.
3. Immediately dispose of the toothpick in the autoclave bag.

4. Hold the coverslip by the sides to avoid getting fingerprints on it. Place one edge of the coverslip next to one edge of the smear, then lower it slowly into the smear (Figure 2.3c). By lowering it slowly you will avoid forming air bubbles. Absorb any excess fluid around the edges of the coverslip and on the slide with a twist of paper towel. Dispose of the paper towel in the autoclave bag.
5. Place the cheek smear slide onto the stage of your microscope. Focus first with the scanning-power lens. Scan for some individual cheek cells with the scanning-power lens. Look for cells that are separated from the other cells and are not folded at the edges. Notice that each cell has a distinctive dark center called the nucleus. Draw one of these cells in Box 2.4, under *Individual Cheek Cells*, in the circle labeled *Scanning power*. Be sure to include the proportion of the cell to the whole visual field, including details like the shape of the cell and nucleus and if the cytoplasm appears clear, cloudy, or grainy.
6. Change to the low-power lens, and draw the cell in the appropriate circle in Box 2.4. Remember this is a parfocal microscope, and the lens change should not involve any major adjustments of the microscope. Finally, move on to the high-power lens, and draw the cell again in the appropriate circle.
7. Return to the scanning-power lens, and scan for a cluster of cheek cells. Again, look for a cluster that is separate from the other clusters and that is not crumpled or squeezed at the edges. The nucleus of each cell should be approximately in the center. Draw the cluster of cheek cells in Box 2.4, under *Grouped Cheek Cells* and the appropriate magnification.
8. Repeat this process for the low- and high-power lenses.

After completing the box on cheek cells, study it for a moment. Compare the view and details at

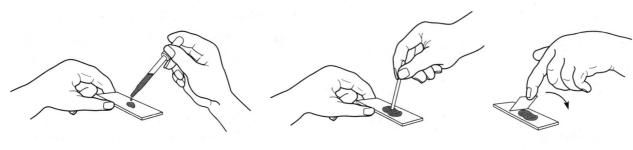

| a) **Add drop of stain to the slide.** | b) **Stir cheek cells into the stain.** | c) **Lower coverslip onto the smear.** |

FIGURE 2.3   **Preparation of a wet mount slide.**

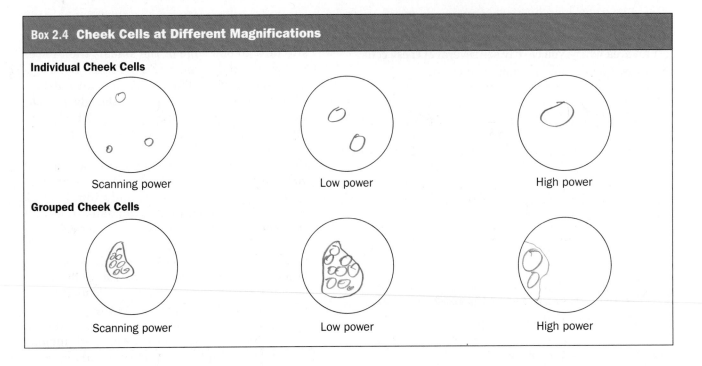

Box 2.4 Cheek Cells at Different Magnifications

Individual Cheek Cells

Scanning power   Low power   High power

Grouped Cheek Cells

Scanning power   Low power   High power

the different magnifications. What are the changes? Look at the drawings of the grouped cells. How do they hold together? At increased magnifications, is this question answered?

Imagine how the cheek cells are arranged inside your mouth. Does it make sense that they form the mouth lining or that they can be scraped off so easily with a toothpick? When you chew an apple or a piece of lean meat, how do the cheek cells protect the mouth?

1. What is the shape of the individual cheek cells?

   Circle

2. Do you think cheek cells are thick or thin?

   I think thin

3. What does the thickness of a cheek cell tell you about its function?

   Used to line your mouth.

4. Was the nucleus centrally or peripherally located?

   Central

5. When the cells are spread out, how many layers did you observe?

   One layer

6. Do you think the natural cheek structure contains one or more layers of cells? Give a reason for your answer.

   Yes to keep lined and developer

Cleanup and Storage of Equipment

After you are satisfied with your drawings, begin to clean up your work area. The toothpicks and soiled paper towels should have been placed into the autoclave bag at the instructor station. The glassware (slides and coverslips) now should be placed into a jar containing a bleach solution.

Look at the microscope, and clean, if necessary, as directed in Activity 2, and store as directed in Box 2.5. Wipe down your counter with a disinfectant and paper towels in order to leave a clean work surface for the next student.

Box 2.5 Checklist for Proper Microscope Storage

❑ Store the microscope with the lowest objective lens in place.

❑ Store the microscope with the lamp cord around the arm, not the objective lens.

❑ Report microscope problems, then use another microscope.

   DO NOT STORE DAMAGED MICROSCOPES!

THE MICROSCOPE

Critical Thinking and Review Questions

1. Match the part of the microscope to the description on the right.

   _e_ Condenser   a. move the slide

   _c_ Coarse-adjustment knob   b. used for precise focusing

   _a_ Stage controls   c. used for general focusing

   _b_ Fine-adjustment knob   d. have different magnifications

   _d_ Objective lenses   e. focuses light on the specimen

2. What is the total magnification if the ocular lens is 10× and the objective lens is 100×? Show the calculations.

   _110%, 10x + 100x = 110x_

3. Why should you scan a specimen under scanning power before using higher magnifications?

   To get a general view of the specimen.

4. After focusing on the top of the specimen, what happens to the depth of focus as you turn the fine adjustment knob backward?

   It expands to better focus. - can focus on diff layers

5. How is poor light quality affected by opening the iris diaphragm?

   It brightens and allows a clear view.

6. Summarize the procedure for preparing a wet mount slide.

   Put drop of stain, get cheek cells, mix cells in, cover then view

7. Briefly describe how to put away and store your microscope.

   Clean, then cover the microscope and store safely

The Anatomy and Diversity of Cells

Objectives

After completing this exercise, you should be able to

1. Explain why cells are the basic unit of life.
2. Identify (on a model or in a diagram) the major cellular organelles and describe their functions.
3. Explain why cellular diversity and cellular specialization are necessary for the human body.

Materials for Lab Preparation

Equipment and Supplies

○   Compound light microscope

Prepared Slides

○   Kidney
○   Sperm
○   Skeletal muscle
○   Multipolar neuron
○   Adrenal gland

Introduction

The basic units of life are the **cells,** which separate the often disordered, random chemistry of *nonliving* things from the ordered biochemistry of *living* things. Although many primitive cells such as bacteria do not have a large central **nucleus,** all cells have a **plasma membrane** to encase their special biochemistry. Cell membranes keep their complex chemistry discrete from contamination by the chemistry of the nonliving world.

Organelles

Within the plasma membrane are various **cytoplasmic organelles** that are responsible for particular life functions (Figure 3.1). **Organelle** is the general name given to cellular structures made of complex macromolecules. Each has a job essential to the life of complex cells, such as producing energy or directing cell division. The basic study of cells involves the study of membranes and organelles.

There are two basic categories of organelles. **Membranous organelles,** as the name implies, are cellular structures wrapped in phospholipid bilayers. The **nonmembranous organelles** are typically made of complex proteins. See Table 3.1 for a summary of the most important organelles.

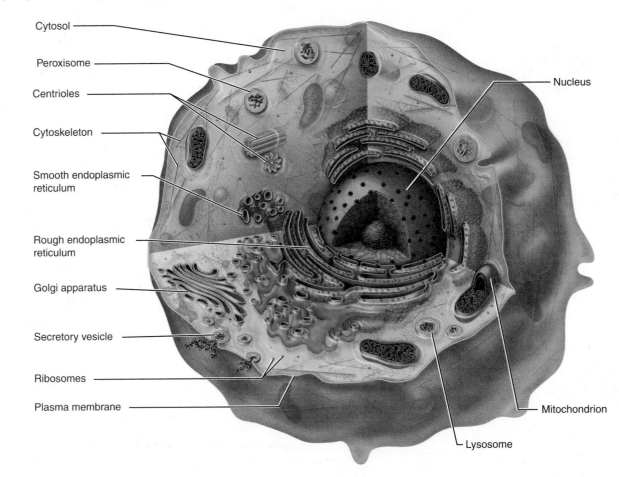

Cytosol
Peroxisome
Centrioles
Cytoskeleton
Smooth endoplasmic reticulum
Rough endoplasmic reticulum
Golgi apparatus
Secretory vesicle
Ribosomes
Plasma membrane
Nucleus
Mitochondrion
Lysosome

FIGURE 3.1   Generalized animal cell.

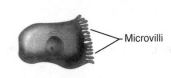

Microvilli

FIGURE 3.2a   Diversity in human cells. A single cell with its outer membrane of microvilli.

Flagellum
Head

FIGURE 3.2b   A mature sperm.

Table 3.1   **Important Cell Structures and Organelles**

Organelle	Description
Membranous	
Nucleus	Usually a spherical membrane shell that houses DNA and makes RNA for use in directing protein synthesis. By controlling what proteins are made, the nucleus controls most cellular functions.
Endoplasmic reticulum (ER)	"Mazelike" network of membrane often spanning out from the nucleus. The rough ER is dotted with ribosomes and is often an industrial complex for protein production. The smooth ER is a site where some of the proteins produced do their jobs and also where detoxification, materials processing, and lipid synthesis occur.
Golgi apparatus	Site for packaging and special processing of molecules (mainly proteins) for export out of the cell.
Mitochondria	Nicknamed "the powerhouse of the cell," these membranous energy transducers convert molecular energy from one molecule to another (usually from sugars, lipids, and amino acids to ATP).
Vacuole	Typically a large bubble of membrane used for storage inside a cell.
Vesicle	Small membrane spheres with various storage or transport functions.
Lysosome/peroxisome	Vesicle or small vacuole containing digestive enzymes.
Nonmembranous	
Cytoskeleton	Network of protein fibers and tubules that supports and moves the cell.
Flagella and cilia	Whip-like projections composed of contractile proteins; important for moving a cell (e.g., sperm) or moving other substances in a multicellular organism (e.g., ciliated epithelium of the human respiratory tract).
Ribosomes	Mixture of ribosomal RNA (rRNA) and proteins; worksites for protein synthesis.
Centrioles	Part of the larger centrosome, responsible for directing many aspects of cell division in most animal cells.
Microvilli	Folds in the plasma membrane to increase surface area (e.g., for absorption, transport).

ACTIVITY 1

Identifying Organelles

Using Figures 3.1 and 3.2 and Table 3.1, match the following organelles to their descriptions.

_b_   Mitochondria
_a_   Lysosome
_d_   Golgi apparatus
_c_   Centrosome
_h_   Smooth ER
_g_   Rough ER
_e_   Flagella
_f_   Cytoskeleton

a. vesicle containing digestive enzymes
b. converts molecular energy from sugars to ATP
c. contains centrioles; directs parts of cell division
d. packages materials for export out of the cell
e. composed of contractile proteins that move a cell
f. network of protein fibers/tubules that supports the cell
g. membranous organelle associated with protein synthesis
h. membranous organelle associated with lipid synthesis ■

Diversity of Human Cells

Although all human cells come from the original, generic fertilized egg, human development eventually produces a wide variety of cells. Oocytes are very large, and sperm are very small. Muscle cells are long and slender, and neurons are highly branched.

The reason why each of the cells you will be viewing is so different from the others is that each cell has been modified to its own specialized task. For example, sperm cells need to swim and wouldn't be able to do so unless modified to include a structure like a tail or flagellum. Neurons need to communicate information over great distances and need to be modified with long, slender projections that can reach distant parts of the body. You will learn more about these cells and what they do in later exercises.

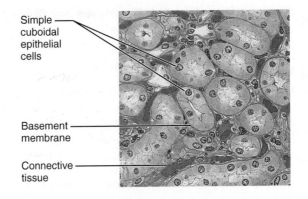

FIGURE 3.3 **Simple cuboidal epithelium in kidney tubules (270×).**

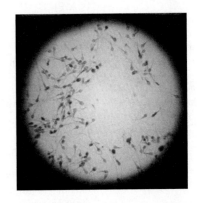

FIGURE 3.4 **Sperm cells from a sperm smear (430×).**

ACTIVITY 2

Observing Cellular Diversity

Materials for This Activity

Compound light microscope

Prepared Slides

Kidney
Sperm
Skeletal muscle
Multipolar neurons

1. Obtain microscope slides of the kidney, sperm, skeletal muscle, and multipolar neurons.
2. View each slide using the high-power (high-dry) magnification on your microscope (see Figures 3.3 through 3.6).
3. Sketch a few of the representative cells from each slide that you view in Box 3.1. Remember to record the magnification used to view the slide in the space provided.
4. For each slide you observed, note in the appropriate section of Box 3.1 a description of it, for example, its shape, size, arrangement, or other characteristics.

Keeping in mind the notion that "form follows function," choose each cell's job from the following list and enter it in the function column of Box 3.1.

Movement through fluid
Contraction
Communication
Absorption and secretion

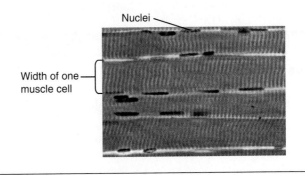

FIGURE 3.5 **Skeletal muscle (~100×).**

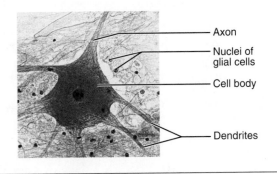

FIGURE 3.6 **Multipolar neuron (170×).**

Box 3.1 Cellular Diversity

Cell	Sketch of Representative Cells	Brief Description of Characteristics	Function (see suggested list of functions on previous page)
Kidney	Magnification __4__ ×	Very liquid like and "blobby", Many "circles" of cells. Purplish	Absorb and secrete
Sperm	Magnification __40__ ×	Looks like lines and pretty clear.	Used for swimming.
Skeletal muscle	Magnification __4__ ×	Looks like black streaks.	Used to contraction skeletal's
Neuron	Magnification __10__ ×	Purple, looks stretches, has gaps, and gapped	Used for communication

ACTIVITY 3

Observing Cellular Diversity Within a Single Organ

Materials for This Activity

Compound light microscope

Prepared Slide

Adrenal gland

Obtain a prepared microscope slide of the adrenal gland.* View this slide with your 10× and high-power (high-dry) objectives. Scan over the entire slide and look for cells that are of different shapes and sizes.

1. How many different sized/shaped cells could you observe? _____

2. Briefly describe the different cells you observed.

   _____

   _____

   _____

3. Based upon what you have learned about cellular diversity, propose a hypothesis about the reasons for the different kinds of cells you observed in the adrenal gland.

   _____

   _____

   _____

   _____

4. Use your textbook and at least one other source (medical dictionary, Mayo Clinic website, etc.) to check your hypothesis about adrenal gland cells, and summarize your findings below.

   _____

   _____

   _____

   _____

*Note: Keep in mind that the purpose of this activity is not to memorize the functions of the adrenal gland (that will be covered much later) but rather to gain further appreciation of cellular specialization. ▪

THE ANATOMY AND DIVERSITY OF CELLS

Critical Thinking and Review Questions

1. Why is the plasma membrane so important to cells and to life in general?

   The membrane encases the cell's biochemistry. This keeps everything in tact

2. Name and describe four membranous organelles found in cells.

   The nucleus oversees function
   Mitochondria creates ATP
   Vacuole is the storage
   Vesicle is storage and transport

3. Name and describe four nonmembranous organelles found in cells.

   Cytoskeleton keeps structure
   Ribosomes control

4. Name and describe four organelles that produce or process materials for the cell.

   The nucleus helps hold DNA and produce RNA
   Golgi Apparatus processes protein for export
   Lysosome helps digestive enzymes
   Ribosome, used for protein synthesis

5. Do you think all cells have roughly the same number of mitochondria? Explain your answer.

   I think that all cells do have a similar amount since all alive cells need to have the same amount of energy.
   diff cells need diff amount of energy

6. Why is cellular diversity important in building complex organisms such as the human body?

Cellular diversity helps build complex organisms since each cell has its own job. Some cells help with structure, growing, thinking, etc.

Cells can specialize

7. Choose two of the cells you examined in this lab, and describe how each cell's shape and/or size fits its function.

The sperm is able to "swim" its way to eggs for reproduction. A multipolar neuron can spread out and communicate over stretched areas.

8. Label the following diagram.

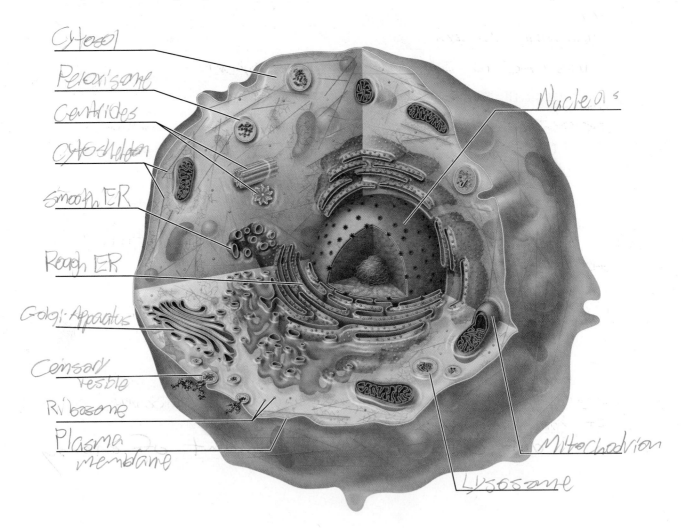

Cytosol
Peroxisome
Centrioles
Cytoskeleton
Smooth ER
Rough ER
Golgi Apparatus
Sensory Vesicle
Ribosome
Plasma membrane
Nucleus
Mitochondrion
Lysosome

EXERCISE

4

Cell Physiology

Objectives

After completing this exercise, you should be able to

1. Describe the basic functions of cell membranes.
2. Explain how substances move across membranes.
3. Explain the concept of diffusion.
4. Explain the concept of osmosis and its relationship to body fluids.

Materials for Lab Preparation

Supplies

- ❍ Petri dish with agar gel
- ❍ Cork borer or sturdy drinking straw
- ❍ China marker
- ❍ Ruler
- ❍ Precut, presoaked dialysis tubing (4-inch segments)
- ❍ Thread
- ❍ 200-ml beakers
- ❍ Clinistix or other suitable sugar testing strips
- ❍ Small test tube

- ❍ Fresh potato
- ❍ Knife
- ❍ Sodium chloride
- ❍ Paper towels
- ❍ Sample of whole blood (ox blood recommended)
- ❍ Clean microscope slides and coverslips
- ❍ Medicine dropper
- ❍ Dropper bottle
- ❍ Biohazard container

Equipment

- ❍ Triple-beam balance or other scale with sensitivity of at least 0.1 g
- ❍ Compound light microscope

Solutions

- ❍ Dropper bottle of methylene blue solution
- ❍ Dropper bottle of potassium permanganate solution
- ❍ Mixture of 5% glucose and 1% boiled starch solution in squeeze bottle
- ❍ Iodine solution in dropper bottle
- ❍ Distilled or deionized water
- ❍ 2.5% saline
- ❍ Isotonic (0.9%) saline

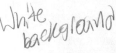

Introduction

As you learned in the previous exercise, cells are the basic units of life, and their **plasma membranes** are the structures that keep their complex chemistry discrete from contamination by the chemistry of the nonliving world. But there are times when substances need to move in and out of cells, and the plasma membrane regulates that process as well.

The Plasma Membrane

The plasma membrane is the most essential component of a cell because it encases and protects the complex chemistry of life inside the cell. Its main component is a molecule called a **phospholipid.** Phospholipids are unusual "hybrid molecules" that are made up of a polar "head" and nonpolar "tails." The polar head is attracted to water, while the neutral tails are repelled by water. In the plasma membrane, these molecules form two layers, where the polar heads form the outside layers and the hydrophobic tails make up the interior. This structure, which allows certain molecules to pass through while keeping others out, is the **phospholipid bilayer.** As such, the plasma membrane is **selectively permeable.**

Cells can modify their plasma membranes by making protein channels that span the membrane. These protein channels allow for the movement, or **diffusion,** of specific items, such as ions and polar molecules, from areas of high concentration to low concentration. Without these channels, the ions and polar molecules would not be able to pass across the plasma membrane (Figure 4.1).

Passive transport is the name for the process of simple diffusion across the plasma membrane. Lipid-soluble items pass right through the phospholipid bilayer, while items that are not lipid soluble (ions and polar molecules) must utilize the protein channels or carriers (if present) to pass through. Molecules of different sizes tend to diffuse at different rates, as you will see in the following activity.

ACTIVITY 1

Measuring Diffusion Rates of Dye Through an Agar Gel

Materials for This Activity

Petri dish with agar gel

Dropper bottle of methylene blue solution

Dropper bottle of potassium permanganate solution

Cork borer or sturdy drinking straw

China marker

Ruler

1. Obtain a petri dish containing a layer of agar gel. Create a "well" in the agar near one side of the petri dish using a cork borer or sturdy drinking straw. Repeat near the opposite side (Figure 4.2).
2. Carefully place a drop of methylene blue solution in one well and a drop of potassium permanganate solution in the opposite well. Try not to overfill the well to the point that it overflows.

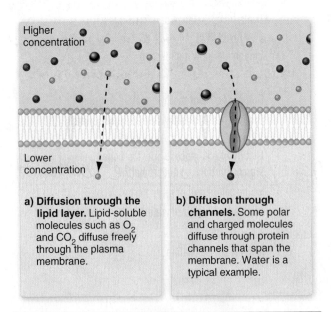

FIGURE 4.1 Two forms of passive transport. Both involve transport down a concentration gradient without the expenditure of additional energy.

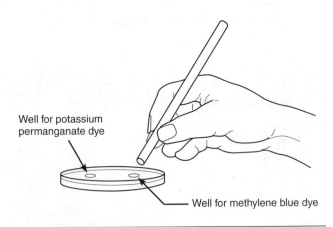

FIGURE 4.2 Setup for measuring diffusion rates using methylene blue and potassium permanganate dyes through an agar gel.

3. If you overflow the well, use a china marker to put a circle on the underside of the petri dish (without turning the petri dish upside down) to mark the border of your placement of these materials.

4. At 15-minute intervals, measure the distance the dye has moved from the edge of the well (or your marked circle). Enter the data in Box 4.1.

Which dye diffused through the agar at a faster rate?

Meth Blue

Compare the diffusion rates. What is the approximate diffusion distance ratio of methylene blue to potassium permanganate (e.g., 2:1, 3:1, 2:3)?

2:1

Given that potassium permanganate has a molecular weight approximately half that of methylene blue, what is the connection between molecular size/weight and diffusion rate?

Had 2:1 ratio due to weight

Selectively Permeable Membranes

Now that you understand diffusion, it should be clear that only the wall of a solid container stops most molecules once they head toward the area of low concentration. But what if a porous wall or selectively permeable membrane were placed in the way of these molecules? Would they pass through the barriers or bounce off them? Common sense suggests that passing through will depend at least partly upon the size (which is proportional to the molecular weight) of the molecule versus the size of the pores in the membrane. Visualize straining cooked spaghetti. Water passes through the holes in the strainer/colander, but the spaghetti noodles are trapped on one side. The colander is like a selectively permeable membrane.

There are factors other than size (such as the polarity of the molecules) that determine whether molecules move through a membrane. But because those factors are more difficult to test in the laboratory, we will focus on the molecular size issue for the following experiment.

ACTIVITY 2

Observing Movement Through a Selectively Permeable Membrane

Materials for This Activity

Precut, presoaked dialysis tubing (4-inch segments)

Thread

200-ml beaker

Mixture of 5% glucose and 1% boiled starch solution in squeeze bottle

Clinistix or other suitable sugar testing strip

Iodine solution in dropper bottle

Small test tube

1. Obtain a piece of presoaked dialysis tubing about 4 inches long. Fold about 1 inch of the bottom end of the tube. Use thread to tie the middle of the folded area. Tie it as tightly as you can, and then tie this same area with another piece of thread. This duplication will ensure that you have made your tube into a bag that will not leak.

Box 4.1 **Distance Dye Moved Over Time**		
Time	**Methylene Blue**	**Potassium Permanganate**
15 minutes	2.5 mm	1 mm
30 minutes	3 mm	2 mm
45 minutes	4 mm	3 mm
60 minutes	4 mm	3 mm

2. Open the top of the bag and insert the spout of a squeeze bottle of glucose/starch solution into the open end of your bag. Add the solution to the bag until it is about two-thirds full. Twist the bag closed about 1 inch from the top of the bag, and tie it with thread as you did for the bottom. You may find that a helping hand from a partner will be useful during this step.

3. Rinse the outside of the bag with tap water, and gently blot it dry on a paper towel. Place the bag in a 200-ml beaker that is about half full of tap water. Figure 4.3 shows how your setup should look.

4. Wait 40 minutes, and then test the water using the following procedures:

   Sugar test: Obtain a Clinistix strip or other suitable testing strip for sugar. Dip the strip into the water according to the directions on the product's container.

   Is there sugar in the water? _Yes_

   Starch test: Pour some of the water from the beaker into a clean small test tube. About 1 inch of water in the test tube should be sufficient. Obtain a dropper bottle of iodine solution, and add several drops to the water in the test tube. If the liquid in the test tube is yellow-orange, there is no starch in the water (a negative starch test). If the liquid turns a dark color, usually navy blue or black, there is starch in the water (a positive starch test).

   Is there starch in the water? _No_

   Which molecules, sugar or starch, were small enough to pass through the membrane (the dialysis tubing bag)? Which molecules were too large to pass through the membrane?

   Starch

   _____

   _____ ■

Osmosis—Part One

Water easily passes through the pores in a selectively permeable membrane. However, as you have already discovered, other molecules do not always do so. Could these other molecules, which are trapped on one side of the membrane, affect the movement of water molecules across the membrane? The net diffusion of water across a selectively permeable membrane, known as **osmosis,** is driven by solutes affecting the concentration of water. (Consult your textbook for a more complete discussion of osmosis.) Where there is more solute, there is less water available. Like all molecules, water will move toward its area of lower concentration.

Tonicity is the term used to describe the solute concentration in the fluid surrounding a cell. An **isotonic** fluid has the same solute concentration as the cell. **Hypotonic** fluids have lower solute concentrations, and **hypertonic** fluids have higher solute concentrations. Keeping in mind the general rule that "more solute means less water," which fluid would have more water, the hypertonic or hypotonic solution?

_____

| ACTIVITY 3 |

Observing Osmosis in Potato Cubes

Materials for This Activity

Fresh potato

Knife

Distilled or deionized water

200-ml beakers

Forceps

China marker

Triple-beam balance or other scale with sensitivity of at least 0.1 g

Sodium chloride

Paper towels

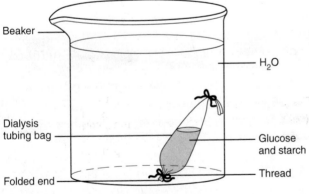

FIGURE 4.3 **Setup for observing diffusion through a nonliving selectively permeable membrane.**

Solutions

2.5% saline

Isotonic saline

1. Peel and cut the fresh potato into three cubes that are approximately 1 inch in size.
2. Obtain three 200-ml beakers, and label each with a china marker as follows: the first as 0.9% saline (isotonic), the second as distilled water (hypotonic), and the third as 2.5% saline (hypertonic).
3. Weigh 1.5 g of sodium chloride and add it to the beaker labeled 2.5% saline. Weigh 0.54 g of sodium chloride and add it to the beaker labeled 0.9% saline. Add 60 ml of distilled water to each beaker. Gently stir the salt solutions until all salt is dissolved.

   Check your work: Recalling that 1 ml of distilled water weighs 1 g, what is the percent solution of 1.5 g sodium chloride in 60 ml water? _____ What is the percent solution of 0.54 g of solute in 60 ml water? _____ Which solution do you think most closely corresponds to most human body fluids? (*Hint:* The answer is implied in the tonicity terminology section earlier.) _____
   Note: Your lab instructor may have these solutions prepared in advance in order to save time during the lab period. If so, just add about 1½ inches of solution to each appropriate beaker.
4. With your forceps, dip one potato cube into the first beaker, remove promptly, pat dry with a paper towel, and weigh as accurately as possible. Record this weight as the initial weight in Box 4.2.

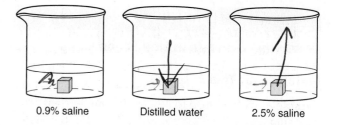

FIGURE 4.4   **Osmosis in potato cubes.**

6. Place the potato cubes back in their respective beakers for one hour. Remove, pat dry, and weigh them as accurately as possible. Record this weight in Box 4.2.

   Which potato cube gained weight? _All_

   Which potato cube lost weight? _None_

   Which potato cube remained closest to its original weight? _2.5_

7. In Figure 4.4, draw an arrow indicating the direction in which diffusion of water is occurring: into the cube, out of the cube, or equally in both directions. Explain your results below.

   all cubes gained weight in the experiment

Osmosis—Part Two

In the previous experiment, you established that water was moving either into or out of the cells in the potato cube depending upon the solute concentration in the water that surrounded it. But what of human cells? It is true that plant cells respond to the changes driven by osmosis differently than animal cells. For example, the cell wall of a plant cell prevents the cell from bursting when excess water enters from the surrounding environment, but animal cells have no cell walls and would burst. The same overall process occurs in all cells. Imagine what would happen to trillions of cells in your body if your body fluids were permitted to become too dilute.

Keeping cells from shriveling up or bursting is a major reason behind the mechanisms the human body has in place for carefully controlling water and solute level (e.g., kidney function). In the next activity, we will demonstrate this principle with red blood cells.

Box 4.2 **Weight of Potato Cubes**		
Beaker Solution	**Initial Weight**	**Final Weight**
0.9% saline (isotonic)	.674	2.3
Distilled water (hypotonic)	2.496	2.61
2.5% saline (hypertonic)	2.865	2.9

5. Repeat Step 4 for the other two cubes and beakers. *Note:* This step is to reduce experimental error regarding dry and wet cubes.

ACTIVITY 4

Observing Osmosis in Red Blood Cells

Materials for This Activity

Sample of whole blood (ox blood recommended)

Distilled water

Clean microscope slides and coverslips

China marker

Medicine dropper

Compound light microscope

Biohazard container or other suitable disposal vessel for the slides

Solutions

2.5% saline

Isotonic saline

1. Set three clean, dry slides on your work area, and label them near one edge as "A," "B," and "C," respectively, with a china marker.
2. On slide "A," place a drop of isotonic saline solution. On slide "B," place a drop of distilled water. On slide "C," place a drop of hypertonic saline solution.
3. Using the medicine dropper, add a small drop of blood to each slide, and rock each slide gently to mix the solutions. Place a coverslip over the liquid on each slide. Allow an immersion time of 5 minutes before examining slides "A," "B," and "C" using your microscope.
4. Now use your microscope to look at slide "A," and carefully focus the image of the red blood cells using your high-power objective lens. The cells should have a normal "biconcave" shape, like a jelly doughnut slightly pushed in at the center (Figure 4.5a). Summarize the results in Box 4.3.

Review what you did in the potato cube experiment when you placed the potato cube in the isotonic saline solution. Look at the arrow(s) you were asked to draw in Figure 4.4 for the potato cube in the isotonic saline solution. Keeping in mind that the same process is occurring here, briefly describe what is happening to the cells on slide "A."

_____

_____

_____

5. Next, use your microscope to look at slide "B," and carefully focus the image of the red blood cells using your high-power objective lens. The cells should have taken on a somewhat swollen shape (Figure 4.5b). The bursting of these cells that normally follows is referred to as **hemolysis.** Let the slide sit for another 15 minutes, and then look at it a second time. Compare the two views in terms of estimating the percentage of size change, and describe the shape change. Summarize the results in Box 4.3.

Review what you did in the potato cube experiment when you placed the potato cube in distilled water. Look at the arrow(s) you were asked to draw in Figure 4.4 for the potato cube in distilled water. Keeping in mind that the same process is occurring here, briefly describe what is happening to the cells on slide "B."

_____

_____

6. While waiting to look at slide "B" a second time, look at slide "C" using your microscope, and

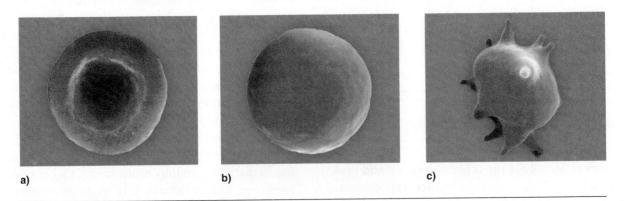

a)　　　　　b)　　　　　c)

FIGURE 4.5　Red blood cells. a) Normal. **b)** Swollen. **c)** Crenated.

carefully focus the image of the red blood cells using your high-power objective lens. Now let this slide sit for another 15 minutes, and then look at it a second time. Compare the two views in terms of estimating the percentage of size change and describing the shape change. The cells should have taken on a somewhat wrinkled shape (Figure 4.5c). This wrinkling is referred to as **crenation.** Summarize the results in Box 4.3.

Review what you did in the potato cube experiment when you placed the potato cube in the hypertonic saline solution. Look at the arrow(s) you were asked to draw in Figure 4.4 for the potato cube in the hypertonic saline solution. Once again, keeping in mind that the same process is occurring here, briefly describe what is happening to the cells on this slide.

_____

_____

Box 4.3 **Results of Osmosis in Red Blood Cells**		
Solution Added	**Appearance of Cells**	**Net Movement of Water**
0.9% saline (isotonic)		
Distilled water (hypotonic)		
2.5% saline (hypertonic)		

CELL PHYSIOLOGY

Critical Thinking and Review Questions

1. What is a selectively permeable membrane, and why is it so important that the plasma membrane is selectively permeable?

   Allows certain molecules into the cell. The plasma membrane is important so that the cell isn't contaminated

2. Define *diffusion*.

   Spreading something widely

3. Do all molecules diffuse at the same rate? Explain your answer based on the experiment you did in this lab exercise.

   Some molecules diffuse at a different rates due to the area surrounding the cells

4. What are two important factors that determine whether or not molecules pass through a selectively permeable membrane? Explain your answer.

5. Define *osmosis*.

   Movement of molecules through membranes

6. Explain how the experiments you performed demonstrated the principle of osmosis.

7. A potato cube is placed in an unknown solution and weighs significantly more when removed and weighed after one hour. Is the unknown solution hypertonic, isotonic, or hypotonic? Explain your answer.

_____

_____

_____

_____

Tissues

Objectives

After completing this exercise, you should be able to

1. Explain why tissues are a crucial link between cells and organs.
2. List the four basic tissue types and describe them.
3. List the major epithelial tissues and describe them.
4. List the major connective tissues and describe them.
5. List the three kinds of muscle tissue and describe them.
6. List the two general categories of nervous tissue and describe them.

Materials for Lab Preparation

Equipment and Supplies

- ○ Compound light microscope

Prepared Slides

- ○ Simple squamous epithelium/lung
- ○ Stratified squamous epithelium/esophagus
- ○ Simple cuboidal epithelium/kidney
- ○ Simple columnar epithelium/small intestine
- ○ Pseudostratified ciliated columnar epithelium/trachea
- ○ Areolar connective tissue
- ○ Dense white fibrous connective tissue
- ○ Hyaline cartilage
- ○ Adipose tissue
- ○ Skeletal muscle
- ○ Cardiac muscle
- ○ Smooth muscle
- ○ Multipolar neuron smear

Introduction

As you learned in the cell exercise, there is quite a bit of specialization and diversity among the cells of the human body. But one lonely cell cannot do much for an organism the size of the average human. **Tissues,** or groups of similar cells working together to accomplish a basic function, form a crucial link between cells and organs.

There are four primary types of tissues in the human body. **Epithelial tissues** form linings. **Connective tissues** connect, support, protect, and store things for the human body. **Muscle tissue** contracts and produces movement. Finally, **nervous tissue** conducts electrochemical impulses important for communication.

ACTIVITY 1

Making the Link Between Cells, Tissues, and Organs

In Box 5.1, choose any six organs and list them, one in each box, under *Organ*. Under *Tissues,* list two tissues found in each organ. Under *Function of Tissue,* briefly describe what that tissue does for the organ and system to which it belongs. You may find your textbook or later chapters in this lab manual helpful as references for this exercise, but first try to reason it out based upon the description in the introduction. ■

Epithelial Tissues

Epithelial tissues form the linings of the human body. This is why we typically find them on the inside and/or outside of an organ. They are made of tightly packed cells, seldom having room for blood vessels, and are described as **avascular** (without blood vessels). Epithelial cells are attached to the tissue they are lining via a thin layer of connective tissue called the **basement membrane.**

The names of epithelial tissues may seem daunting, but the naming system is fairly simple. They are first named for their degree of layering: one layer = **simple,** two or more layers = **stratified.** Then they are named for the shape of the cell: flat = **squamous,** cube-shaped to spherical = **cuboidal,** and tall = **columnar.** So, for example, an epithelial tissue made of a single layer of flat cells would be called simple squamous epithelium.

Box 5.1 Organs, Tissues, and Functions		
Organ	**Tissue**	**Function of Tissue**
	1. 2.	1. 2.
	1. 2.	1. 2.
	1. 2.	1. 2.
	1. 2.	1. 2.
	1. 2.	1. 2.
	1. 2.	1. 2.

ACTIVITY 2

Viewing and Learning About Selected Epithelial Tissues

Materials for This Activity

Compound light microscope

Prepared Slides

Simple squamous epithelium/lung

Stratified squamous epithelium/esophagus

Simple cuboidal epithelium/kidney

Simple columnar epithelium/small intestine

Pseudostratified ciliated columnar epithelium/ trachea

View the various epithelial tissues described in the following paragraphs. While examining each tissue carefully, be on the lookout for characteristics that may be clues to the tissue's function.

It is important to understand that, for most slides, you will need to do a little prep work before choosing an area to sketch. Some slides will be easy—they'll contain almost nothing but the tissue you want. In those cases, your main prep work will be ensuring you are using appropriate magnification and focusing skills. Other slides will be made from thin slices of appropriate *organs,* as it is too difficult to remove just the one tissue that the technician would like to place on the slide. Recall that organs are made up of two or more (and usually three or more) tissues. In such cases, you will need to move the slide around until you locate an area resembling the descriptions and pictures provided.

Also, consider that, just as the faces of your classmates all look different (unless you have twins attending) yet are clearly recognizable as human, a tissue may look slightly different from one slide to the next (or even different parts of the same slide!). Therefore, taking the time to patiently move the slide around and carefully select a good representation of the tissue is a good investment in properly learning about that tissue type.

Simple Squamous Epithelium

Simple squamous epithelium is made from a single layer of flat cells (Figure 5.1). As such, it provides the thinnest possible layer for diffusion and filtration. It is important for the air sacs of the lungs and blood capillaries to have this tissue so that molecules such as oxygen can easily diffuse through this layer of cells.

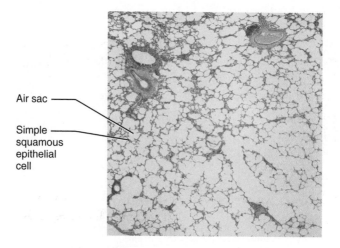

Air sac ——

Simple —— squamous epithelial cell

FIGURE 5.1   Simple squamous epithelium in normal lung alveoli (400×).

View the slide of this tissue, using your microscope, and sketch the image in Box 5.2. Remember to record the magnification. How do your observations compare with the photomicrograph in the figure?

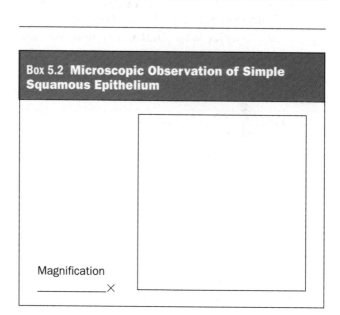

Box 5.2 Microscopic Observation of Simple Squamous Epithelium

Magnification

————————×

Stratified Squamous Epithelium

Stratified squamous epithelium is made from many layers of flat cells, although the mitotic cells near the basement membrane may appear more rounded (Figure 5.2). This tissue provides protection from friction and is generally a good barrier. That is why this tissue is found lining the mouth, esophagus, vagina, and epidermis of the skin.

View a slide of this tissue with your microscope, and sketch the image in Box 5.3. Remember to record

Nuclei

Stratified squamous epithelium

Basement membrane

Connective tissue

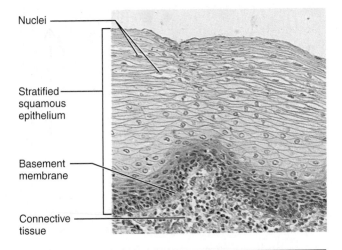

FIGURE 5.2   Stratified squamous epithelium in lining of esophagus (151×).

Simple cuboidal epithelial cells

Basement membrane

Connective tissue

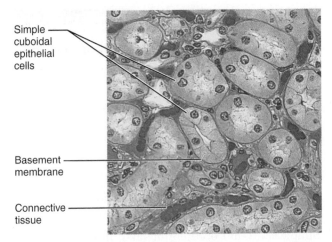

FIGURE 5.3   Simple cuboidal epithelium in kidney tubules (270×).

Box 5.3 Microscopic Observation of Stratified Squamous Epithelium

Magnification

_____×

Box 5.4 Microscopic Observation of Simple Cuboidal Epithelium

Magnification

_____×

---

the magnification. How do your observations compare with the photomicrograph in the figure?

_____

_____

Simple Cuboidal Epithelium

Simple cuboidal epithelium is made from one layer of rounded or cube-shaped cells (Figure 5.3). Although poorly stained membranes sometimes make it difficult to determine the exact shape of the cell, the mostly round nuclei of the cells in this tissue are easy to spot. This tissue provides an absorptive or secretory layer in organs such as the kidney and is also found in many of the glands of the human body, such as the thyroid.

View a slide of this tissue with your microscope, and sketch the image in Box 5.4. Remember to record the magnification. How do your observations compare to the photomicrograph in the figure?

_____

_____

Simple Columnar Epithelium

Simple columnar epithelium is made from a single layer of tall cells with oval nuclei (Figure 5.4). This

Simple columnar epithelial cell

Basement membrane

Connective tissue

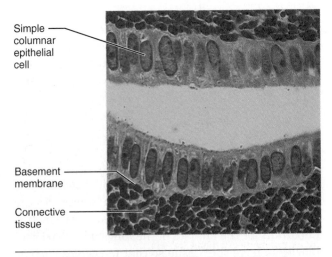

FIGURE 5.4   Simple columnar epithelium of stomach mucosa (500×).

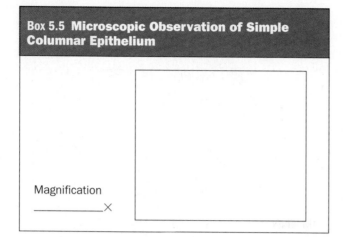

Box 5.5 Microscopic Observation of Simple Columnar Epithelium

Magnification
_____×

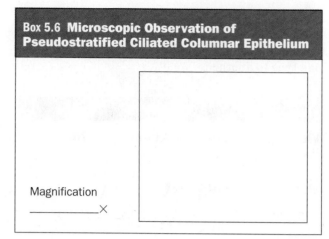

Box 5.6 Microscopic Observation of Pseudostratified Ciliated Columnar Epithelium

Magnification
_____×

tissue performs absorption, secretion, and sometimes glandular functions for the human body. The stomach, intestines, and a few miscellaneous ducts are places where this tissue is located.

View a slide of this tissue with your microscope, and sketch the image in Box 5.5. Remember to record the magnification. How do your observations compare to the micrograph in the figure?

This epithelium is a specialized tissue found throughout most of the respiratory tract, where it protects the lungs by removing dirt particles and mucus.

View a slide of this tissue with your microscope, and sketch the image in Box 5.6. Remember to record the magnification. How do your observations compare to the photomicrograph in the figure?

Pseudostratified Ciliated Columnar Epithelium

Pseudostratified ciliated columnar epithelium has the appearance of several layers of tall cells, but all of these cells touch the basement membrane (Figure 5.5). So, it is technically one layer of taller and shorter cells, thus the name *pseudostratified*.

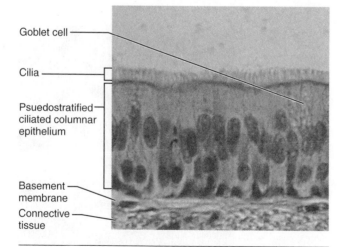

Goblet cell
Cilia
Psuedostratified ciliated columnar epithelium
Basement membrane
Connective tissue

FIGURE 5.5 **Pseudostratified ciliated columnar epithelium lining the human trachea (800×).**

Connective Tissues

Connective tissues perform a wide variety of functions for the human body. They connect, protect, store, and provide the substrate in which some structures in the body may be embedded (such as sweat glands embedded in the dermis of the skin). In many ways, connective tissues are the opposite of epithelial tissues. That is, they tend to contain comparatively few cells; instead they have a great deal of nonliving material called the **matrix** that is secreted by the cells. There are two components in the matrix: ground substance (which varies in composition from very hard to liquid) and protein fibers, such as collagen or elastin. Collagen gives strength and elastin gives flexibility.

Although many connective tissues contain protein fibers, others, such as adipose tissue, qualify for this tissue category because of their protective role as a tissue that cushions. Connective tissues are the largest and the most diverse of the four tissue groups. In the following activity, you will explore selected connective tissues. Bone, which is a classic example of a connective tissue because of its protective role, will be covered in the skeletal system. Blood, which chemically connects all parts of the body, will be covered in the circulatory system.

When viewing the different kinds of connective tissue listed in the following activity, note the diverse matrix elements found as you progress from one type to another.

ACTIVITY 3

Viewing and Learning About Selected Connective Tissues

Materials for This Activity

Compound light microscope

Prepared Slides

Areolar connective tissue
Dense white fibrous connective tissue
Hyaline cartilage
Adipose tissue

Areolar Connective Tissue

Areolar connective tissue is made of an irregular arrangement of fibers with **fibroblasts** (fiber-producing cells) and other cells sprinkled in between them (Figure 5.6). As such, it permits flexible attachment and makes a good, loose packing material. It is found below the skin, between muscles, and around various organs.

View a slide of this tissue with your microscope, and sketch the image in Box 5.7. Remember to record the magnification. How do your observations compare to the photomicrograph in the figure?

_____

_____

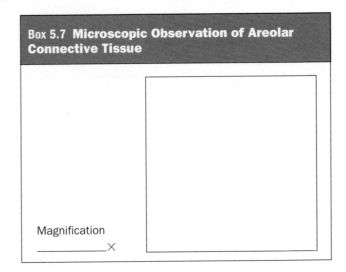

Box 5.7 **Microscopic Observation of Areolar Connective Tissue**

Magnification
_____×

Dense White Fibrous Connective Tissue

Dense white fibrous connective tissue is made of densely packed fibers, running mostly in the same direction, with fibroblasts wedged in between the fibers (Figure 5.7). This tissue provides strength of attachment, especially when the mechanical stress will be along a predictable direction. This is the case with ligaments, tendons, and capsules around various organs, where we find dense fibrous connective tissue.

View a slide of this tissue with your microscope, and sketch the image in Box 5.8. Remember to record the magnification. How do your observations compare to the photomicrograph in the figure?

_____

_____

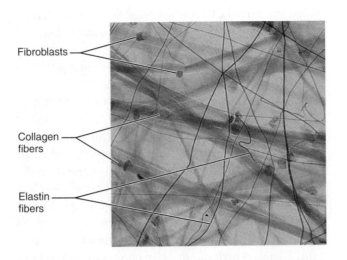

FIGURE 5.6   **Loose areolar connective tissue (160×).**

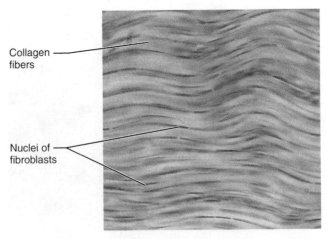

FIGURE 5.7   **Dense white fibrous connective tissue (160×).**

Box 5.8 Microscopic Observation of Dense White Fibreous Connective Tissue

Magnification
_____×

Box 5.9 Microscopic Observation of Hyaline Cartilage

Magnification
_____×

Hyaline Cartilage

Hyaline cartilage is made of thick, tough collagen fibers "glued" and packed so densely that the matrix appears smooth (Figure 5.8). These fibers produce a tissue that is strong yet somewhat flexible. Sites where the ribs attached to the sternum and the ends of bones attach to movable joints are examples of places where hyaline cartilage is found. Instead of fibroblasts, hyaline cartilage has **chondrocytes** (mature cartilage-producing cells) in little holes called **lacunae.**

View a slide of this tissue with your microscope, and sketch the image in Box 5.9. Remember to record the magnification. How do your observations compare to the photomicrograph in the figure?

_____

_____

Adipose Tissue

Adipose tissue is made of large, rounded irregular cells that look empty but are really filled with a large fat vacuole (Figure 5.9). These **adipocytes** do more than store energy. They also form tissue that cushions and insulates. That is why adipose tissue is located below the skin and around various organs.

View a slide of this tissue with your microscope, and sketch the image in Box 5.10. Remember to record the magnification. How do your observations compare to the photomicrograph in the figure?

_____

_____

_____

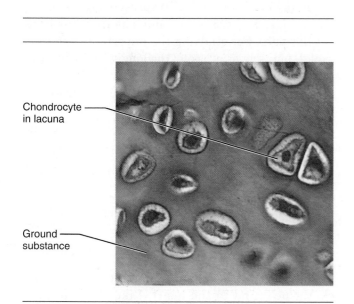

Chondrocyte in lacuna

Ground substance

FIGURE 5.8   **Hyaline cartilage from the trachea (300×).**

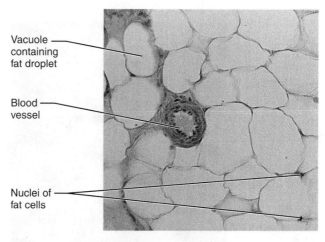

Vacuole containing fat droplet

Blood vessel

Nuclei of fat cells

FIGURE 5.9   **Adipose tissue from the subcutaneous layer under skin (140×).**

Box 5.10 Microscopic Observation of Adipose Tissue

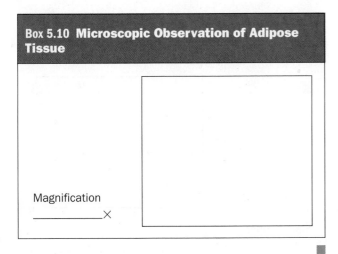

Magnification
_____ ×

Muscle Tissue

Muscle tissue is unique in its ability to contract and cause movement. Like nervous tissue, it is also conductive. The **contractile proteins** (actin and myosin), and the structure of the cell, are modified into three different categories of muscle tissue: skeletal, cardiac, and smooth muscle.

ACTIVITY 4

Viewing and Learning About Selected Muscle Tissues

Materials for This Activity

Compound light microscope

Prepared Slides

Skeletal muscle
Cardiac muscle
Smooth muscle

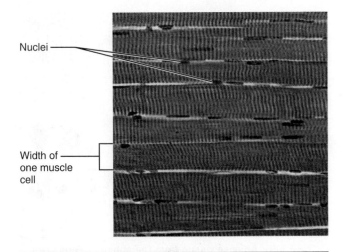

Nuclei

Width of one muscle cell

FIGURE 5.10   Skeletal muscle (100×).

Box 5.11 Microscopic Observation of Skeletal Muscle

Magnification
_____ ×

Skeletal Muscle

Skeletal muscle, as the name implies, is the muscle typically attached to the skeleton for mostly voluntary movement. The contractile proteins are packed so tightly into alternating patterns of light and dark areas that the cells take on the appearance of having cross-stripes, or **striations** (Figure 5.10). If this tissue looks familiar, it is because the multinucleated skeletal muscle cell slide was the one chosen in the cellular diversity activity in Exercise 3.

Depending on the slides available in the laboratory, you may come across an "orientation" issue that will be covered in the next chapter. When skeletal muscle is prepared for mounting on a slide by cutting it *longitudinally,* it will look as shown and described on this page. However, if it is cut in cross section (like cutting a banana into many small pieces for adding to cereal), you will instead see many circles and dots (representing the cells, wrappings, and occasional nucleus that has been cut through). For studying the important aspects of skeletal muscle as a tissue, the longitudinal orientation as pictured here is the one you want.

View a slide of this tissue with your microscope, and sketch the image in Box 5.11. Remember to record the magnification. How do your observations compare to the photomicrograph in the figure?

_____

_____

_____

Cardiac Muscle

Cardiac muscle is a highly fatigue-resistant muscle tissue found in the heart. Like skeletal muscle, cardiac muscle is also striated. Its branched appearance and dark areas of connection between cells, called **intercalated discs,** help to differentiate this tissue from skeletal muscle (Figure 5.11).

View a slide of this tissue with your microscope, and sketch the image in Box 5.12. Remember to record the magnification. How do your observations compare to the photomicrograph in the figure?

_____

_____

_____

View a slide of this tissue, and make a sketch in Box 5.13. Remember to record the magnification. How do your observations compare to the photomicrograph in the figure?

_____

_____

_____

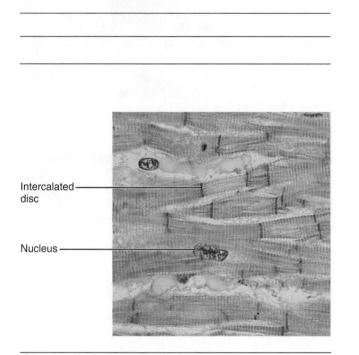

FIGURE 5.11   Cardiac muscle (225×).

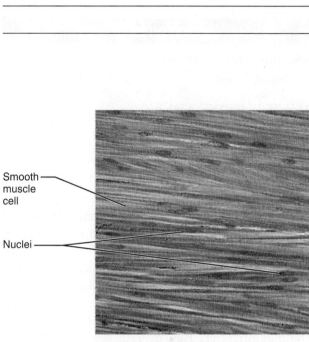

FIGURE 5.12   Smooth muscle (250×).

Smooth Muscle

Smooth muscle is the involuntary muscle found in many internal organs such as the stomach. The cells lack striations and are shorter than those of skeletal muscle (Figure 5.12).

Box 5.12 Microscopic Observation of Cardiac Muscle

Magnification _____×

Box 5.13 Microscopic Observation of Smooth Muscle

Magnification _____×

Nervous Tissue

Nervous, or neural, tissue falls into two categories: **neurons,** which are the conductive cells, and **neuroglia,** which are the nonconductive supporting cells of the nervous system. For this exercise, we will examine only the neuron.

ACTIVITY 5

Viewing and Learning About Selected Nervous Tissues

Materials for This Activity

Compound light microscope

Prepared Slide

Multipolar neuron smear

Neurons are usually highly branched cells with a large spherical area containing the nucleus (Figure 5.13). The long branches pick up or carry information to other neurons (and other cells such as muscle cells), sometimes over long distances.

View a slide of a multipolar neuron smear, and sketch the image in Box 5.14. Remember to record the magnification. How do your observations compare to the photomicrograph in the figure?

_____

_____

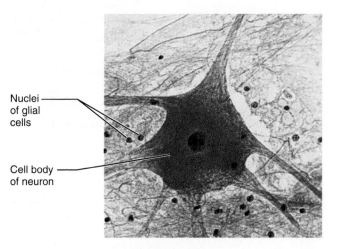

Nuclei of glial cells

Cell body of neuron

FIGURE 5.13   Multipolar neuron (170×).

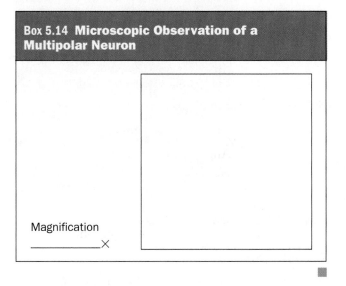

Box 5.14 **Microscopic Observation of a Multipolar Neuron**

Magnification

_____×

TISSUES

Critical Thinking and Review Questions

1. Review Box 5.1 in Activity 1, *Making the Link Between Cells, Tissues, and Organs*. For each organ you selected, do the two tissues you listed perform similar or different functions for that organ? Explain, using at least two examples.

_____

_____

_____

_____

2. Match the tissue type to its basic description.

   _____ Simple squamous epithelium   a. single layer of cube-shaped to spherical cells

   _____ Simple cuboidal epithelium   b. several layers of flat cells

   _____ Stratified squamous epithelium   c. single layer of flat cells

   _____ Areolar connective tissue   d. branched cells with intercalated discs

   _____ Cardiac muscle   e. loose, irregular array of protein fibers

3. Match the tissue type to an example of its location.

   _____ Simple columnar epithelium   a. ligaments and tendons

   _____ Adipose tissue   b. between ribs and sternum

   _____ Dense fibrous connective tissue   c. lining of the small intestine

   _____ Hyaline cartilage   d. below the skin

   _____ Smooth muscle   e. the wall of the stomach

4. Fill in Box 5.15, listing three tissue types that provide the human body with some kind of protective benefit, and briefly explain what these tissues do.

Box 5.15 Tissue Types with a Protective Benefit	
Tissue	**What They Do to Protect an Organ**

5. Fill in Box 5.16, listing three tissue types that either secrete, absorb, or store things for the human body, and explain what these tissues do.

Box 5.16 **Tissue Types That Secrete, Absorb, or Store**	
Tissue	**What Does It Secrete, Absorb, or Store?**

6. Speculate as to why the tissue level of organization is necessary (why can't one big cell take the place of a tissue?). Note that there is more than one right way to answer this question.

_____

_____

_____

_____

Orientations
to the Human Body

Objectives

After completing this exercise, you should be able to

1. Understand and use anatomical terminology.
2. Describe the standard anatomical position (SAP).
3. Understand and use directional terms.
4. Identify anterior and posterior surface regions.
5. Describe and give examples of how the body and its organs would be cut by the three anatomical planes.
6. Identify and describe the human body cavities.

Materials for Lab Preparation

All Human Models Available in the Lab

○ Torsos, heads, organs
○ Dissectible models

Introduction

Let's say that you're going to a party in a city nearby. You've never been there before, but no problem—you have a map and directions! There are things you automatically know and do when following road trip directions that you probably take for granted. You know your right from your left, and you already understand the meaning of terms like *stop sign, intersection, traffic light,* and so on. You understand that landmarks are often used to alert you to important steps—like "turn right just after crossing the railroad tracks."

Similarly, landmarks and terminology used to describe the human body are useful in following directions about how to find organs, bones, and other

things you will need to learn. For example, it will be quite easy to find the thyroid gland when you understand that the directions to find it, *inferior to the larynx in the cervical region,* simply translate as "below the Adam's apple in the neck." This laboratory exercise will teach you how to use the same terminology used by scientists and medical professionals to successfully navigate the human body.

Standard Anatomical Position

In biology courses and in clinical settings, precise terminology is used to describe body parts and position. For that terminology to be meaningful, there must

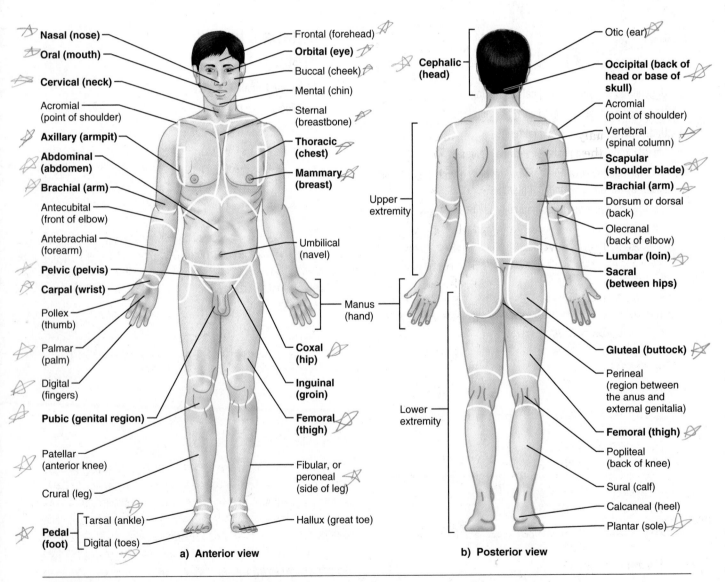

FIGURE 6.1 **The standard anatomical position with surface landmarks.**

be a universally agreed-upon reference point. From which perspective should we view the body? How should this body be positioned? Clearly, perspective and position have great impact on how we describe parts of the body.

All anatomical terminologies assume an anatomical reference point, which is the human body in the **standard anatomical position (SAP)** (Figure 6.1). The standard anatomical position includes both the **anterior** (front) and **posterior** (back) views of the human body. Whether you are describing exterior landmarks or internal ones, always assume you're referring to the SAP.

ACTIVITY 1

Anatomical Landmarks and Reference Points

Standing face to face with a lab partner, raise your right hand, and ask your partner to do the same. From *your* perspective, your partner will have raised his or her left hand. However, custom and experience teaches us very early to adapt to this apparent anomaly and to understand in situations like this that the person we are facing indeed has raised his or her right hand. What we have done is simply shift our perspective from our own to theirs.

1. Stand facing your lab partner. Visualize a long imaginary arrow piercing through both of your bodies starting at your right shoulder area and passing through the upper left part of your partner's body as shown in Figure 6.2. Using the terms *anterior* or *posterior*, *right* or *left*, and the appropriate surface landmark term from Figure 6.1, describe the entry and exit points of the arrow. *Note:* The first entry point is given next as an example.

First entry point (A):

Posterior, right scapular area.

First exit point (B):

Anterior, left thoracic

Second entry point (C):

Anterior, right thoracic

Second exit point (D):

Posterior, left scapular

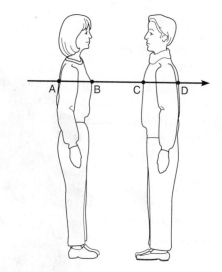

FIGURE 6.2 **Two facing bodies pierced with an arrow.** A, B, C, and D indicate entry and exit points.

The point of this exercise is to make you aware of how important perspective can be when trying to accurately describe body parts and position. In anatomy, it is *not* the observer that is the reference point, but rather it is the body (or a lab model or a photograph of a body) being viewed.

2. Test your understanding of perspective and reference by filling in the blank lines in Figure 6.3, using the terms *right* or *left*. Remember, the heart is positioned as you would see it in a person you are facing. ■

There are three useful orientation tools: **directional terms, surface regions** (also called landmarks or body regions), and **planes.**

Directional Terms

Just as you use directional terms in your everyday conversations—such as *north, south, up, down, outside, inside, front, and back*—anatomists use corresponding terms that are more precise and universally agreed upon to describe body parts. Where we might say, "The eyes are above the mouth," the anatomist would say, "The eyes are superior and lateral to the mouth."

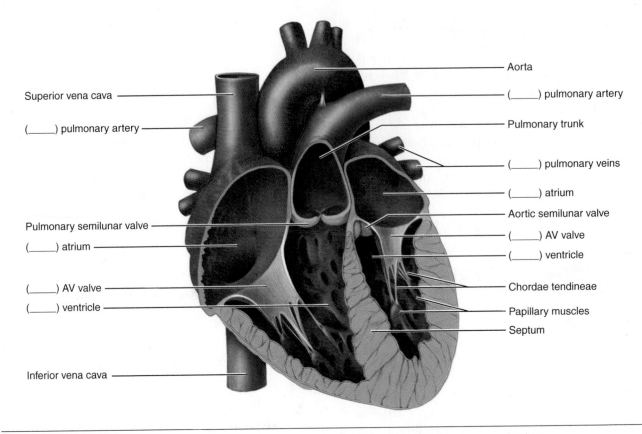

FIGURE 6.3 **Gross anatomy of heart.** Frontal section, showing major blood vessels, chambers, and valves.

ACTIVITY 2

Using Directional Terms

Review the directional terms and definitions in Table 6.1. Note the direction of the arrows, and imagine how you would describe the relationship between common features of your body—for example, the position of the head relative to the abdomen. In the far right column, you will see examples of the use of each term.

1. Describe the relationship between the following parts of your body:

Shoulders to neck

Lateral

Lips to teeth

Superficial

Ears to shoulders

Superior

Feet to knees

Distal

Belly button to buttocks

Anterior

2. Refer to Figure 6.3. Use the directional terms from Table 6.1 to describe the relationships of the following structures of the heart.

Superior vena cava to pulmonary trunk

Interior vena cava to aorta

Aorta to right atrium

Pulmonary trunk to aorta

Table 6.1   **Orientation and Directional Terms**

Term	Definition		Example
Superior (cranial) *Toward head*	Toward the head end or upper part of a structure or the body; above		The head is superior to the abdomen
Inferior (caudal) *Away from head*	Away from the head end or toward the lower part of a structure or the body; below		The navel is inferior to the chin
Anterior (ventral)* *Front*	Toward or at the front of the body; in front of		The breastbone is anterior to the spine
Posterior (dorsal)* *Back*	Toward or at the back of the body; behind		The heart is posterior to the breastbone
Medial *middle*	Toward or at the midline of the body; on the inner side of		The heart is medial to the arm
Lateral *Away from middle*	Away from the midline of the body; on the outer side of		The arms are lateral to the chest
Intermediate *In between*	Between a more medial and a more lateral structure		The collarbone is intermediate between the breastbone and shoulder
Proximal *Towards point of attachment*	Closer to the origin of the body part or the point of attachment of a limb to the body trunk		The elbow is proximal to the wrist
Distal *Away from point of attachment*	Farther from the origin of a body part or the point of attachment of a limb to the body trunk		The knee is distal to the thigh
Superficial (external) *Towards surface*	Toward or at the body surface		The skin is superficial to the skeletal muscles
Deep (internal) *Inside*	Away from the body surface; more internal		The lungs are deep to the skin

*The terms *ventral* and *anterior* are synonymous in humans with reference to the torso. This is not the case in four-legged animals. *Ventral* specifically refers to the "belly" or inferior surface of four-legged animals. Likewise, the terms *dorsal* and *posterior* are the same in humans with reference to the torso, but *dorsal* specifically refers to the "back" or superior surface of four-legged animals.

Surface Regions

Remember our discussion on the importance of local landmarks when following directions to a party? In anatomy, recognizing landmarks is equally crucial, whether looking at the surface of the body, at bones, or at internal organs.

On the human body, these landmarks are called **surface** or **body regions.** We use surface regions to represent specific locations on the surface of the human body, but eventually we will also use them to represent the underlying anatomical parts—for example, the bones, muscles, nerves, and blood vessels.

As you study the anatomy, notice that the names of specific anatomical parts reflect the surface region. For example, the femoral region represents the thigh portion of the leg. In that area, the bone is called the femur, the major blood vessels are called the femoral artery and vein, the major nerve is called the femoral nerve, and the two major muscles are called the biceps femoris (one of the hamstrings) and rectus femoris. Making this connection between the names of surface regions and names of organs will help you learn and retain the memory of the many regions, vessels, and organs you will encounter.

Let's look at the front and back surface regions of the human body, which in anatomy are described as the **anterior** and **posterior** surface regions.

Anterior and Posterior Surface Regions

Figure 6.1 illustrates the human figure in SAP, both anterior and posterior. Notice that the surface (or body) regions are labeled and include, in parentheses, the common, everyday term for each region.

With a lab partner, study these terms and quiz each other. As you read the labels, pronounce the words (aloud or silently) as best as you can. You will find that if you can "hear" each word (aloud or silently), your brain will register and remember it much more than if you "look" at the letters of each word. In fact, after "looking" at about five new terms, you will not be able to retain any more. What happens if you mispronounce the word? You will still remember it better than if you do not pronounce it at all. Some brain traces are still better than none! Remember the purpose of this exercise is for you to see, understand, and retain the material, not simply to turn in correct answers.

Identifying Anterior and Posterior Surface Regions

1. Using a pencil and Figure 6.4a and b, fill in the blanks (without referring to Figure 6.1). Your instructor will specify which terms you should label (e.g., all or just the bold print terms from Figure 6.1). Then check your work against Figure 6.1. Noting those regions you identified incorrectly, discuss ideas with your lab partner(s) for techniques to help remember these regions. Now take a second look at your work. If you were to cover up the labels, could you do it again without referring to the definitions?

2. Draw the anterior and posterior views of any upright organism in SAP. Label the surface regions. Add in your personal touches, such as hair, skin ornamentation, cartoon figures, or alien figures. Make it your work of art; have some fun with this!

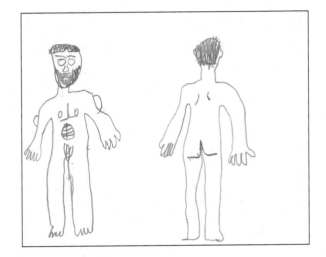

3. Identify the following surface regions, indicate if they are anterior or posterior terms, and indicate if there are any terms that apply to both the anterior and posterior surfaces:

Knee

Patellar anterior

Breastbone

Thoracic anterior

Arm

Brachial posterior

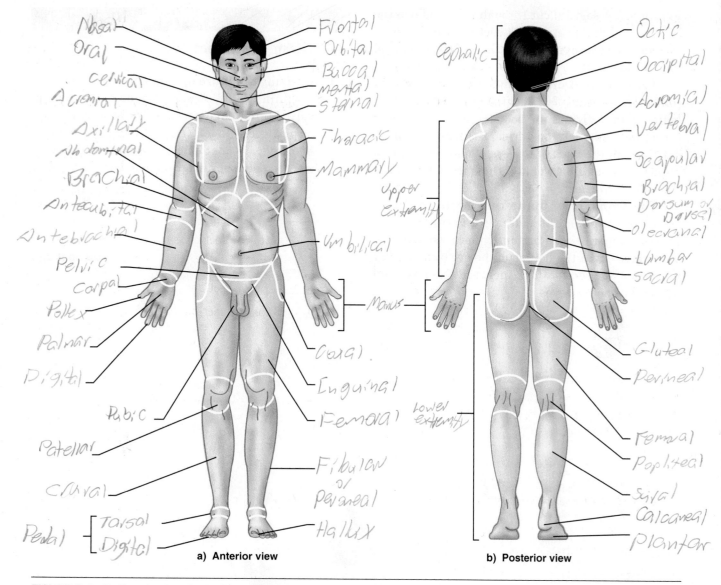

a) Anterior view

b) Posterior view

FIGURE 6.4 **Surface landmarks for identification.**

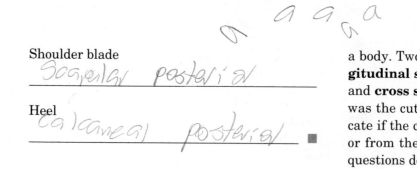

Shoulder blade

Scapular _posterior_

Heel

Calcaneal _posterior_ ■

Planes

It is sometimes useful to show and discuss the sections that result from a straight cut of the body or through an organ. A **plane** is a flat surface that conceptually cuts through a tissue, organ, or part of a body. Two traditional terms for sections are **longitudinal section** (ls), indicating a lengthwise cut, and **cross section** (cs) for a crosswise cut. But how was the cut performed? How does lengthwise indicate if the cut was made from the front to the back or from the top to the bottom? In anatomy, these questions do not arise because three basic anatomical planes are used to describe various cuts and sections made to a body.

A **sagittal plane** (not shown in Figure 6.5) runs longitudinally (top to bottom), dividing the body into right and left sections. When it divides the body or an organ, top to bottom, into *exactly equal* sections, it is called a **median (midsagittal) plane,** as shown in

Figure 6.5. A sagittal cut through the middle of the head would result in a right section, which would include the right eye, the right half of the nose, and half of the mouth.

The **frontal plane,** sometimes called a **coronal plane,** involves another lengthwise cut from top to bottom, but it divides the body (or organ) into front (anterior) and back (posterior) sections. A frontal cut through the back of the eye would result in an anterior section that would include the face but not the back of the head.

The **transverse plane** runs horizontally (or crosswise from side to side). It divides the body into top (superior) and bottom (inferior) sections. A transverse cut at the bottom of the eye would result in an inferior section that included the nose and mouth, but not the eyes.

ACTIVITY 4

Identifying Body Sections

1. Referring to Figure 6.5, imagine that you need to obtain the following sections of the human body. Match each section below with the appropriate cut (a–c). (*Hint:* Always remember, the body is in anatomical position, regardless of how the person or diagram may be positioned.)

_____ Right section	a. Sagittal cut
_____ Left section	b. Frontal cut
_____ Anterior section	c. Transverse cut
_____ Posterior section	
_____ Inferior section	
_____ Superior section	

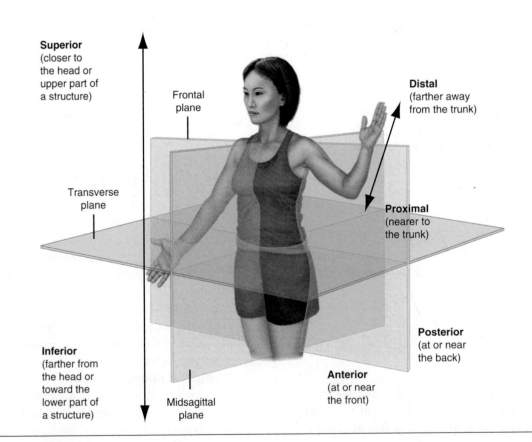

Superior
(closer to the head or upper part of a structure)

Frontal plane

Transverse plane

Distal
(farther away from the trunk)

Proximal
(nearer to the trunk)

Inferior
(farther from the head or toward the lower part of a structure)

Midsagittal plane

Anterior
(at or near the front)

Posterior
(at or near the back)

FIGURE 6.5   Anatomical planes.

Body Cavities

The hollow body cavities of the human body provide a secure place to hold and protect many important organs. The two main body cavities are the **posterior** and **anterior** cavities (Figure 6.6). The posterior cavity (also called the dorsal cavity) contains the brain and spinal cord. The portion that contains the brain is called the **cranial cavity,** and the portion that houses the spinal cord is called the **vertebral canal.**

The anterior cavity (also called the ventral cavity) contains many cavities of its own, and is home to many of the organs you may already be familiar with. The two main divisions of the anterior cavity are the **thoracic,** which is above the diaphragm, and the **abdominopelvic** below the diaphragm. The thoracic cavity is further divided into the two pleural cavities holding the lungs, and the pericardial cavity holding the heart.

There is no physical separation between the abdominal and pelvic parts of the abdominopelvic cavity, and the area near the top of the pelvis is an imaginary separator between the two. The abdominal cavity contains mainly digestive organs, and the pelvic cavity contains mainly organs associated with the urinary and reproductive systems.

ACTIVITY 5

Orientation to the Body Cavities

Use a torso model and any labeled dissection specimens provided in the lab to locate the major structures associated with the human body cavities. (See also Figure 6.7.)

1. Locate the thin, muscular diaphragm, which separates the thoracic and abdominopelvic cavities. Find the heart and lungs within the thoracic cavity.

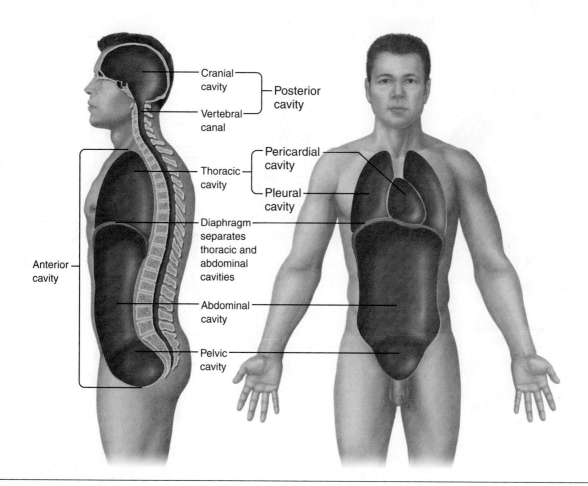

FIGURE 6.6 Body cavities.

Do you think the thoracic cavity changes size and shape during breathing, or remains fixed in size and shape? *Yes*

Why or why not? *The lungs need space to expand for breathing*

2. Look below the diaphragm in the abdominal cavity, and find the liver and spleen (on the left side). Next, look below the liver to find the gallbladder, stomach, and intestines. *Note:* If a dissected rat is used for this activity, you will not find a gallbladder. Lift up the stomach and intestines to find the pancreas, kidneys, and several large blood vessels. In a dissection specimen, you will find the kidneys are behind a membrane and stuck into the posterior body wall. For this reason, the kidneys are usually classified as being retroperitoneal, and technically not in the abdominopelvic cavity. As you

look at the blood vessels attached to the kidneys, speculate as to why the "kidney punch" is illegal in boxing.

There are main arteries connected to the kidney vessels

3. In the pelvic cavity (below the top border of the hips), find the lower portions of the intestines, the urinary bladder, and if your specimen or model is female, the uterus and vagina. When the uterus expands to hold a full-term fetus, will it remain entirely within the pelvic cavity?

Yes

What organs would an expanded uterus restrict or compress?

Intestines

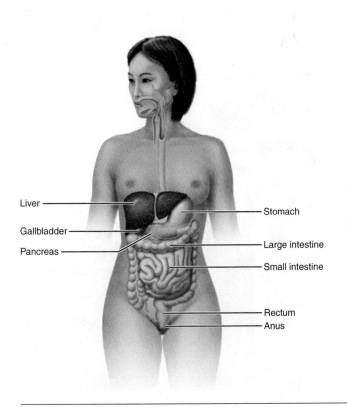

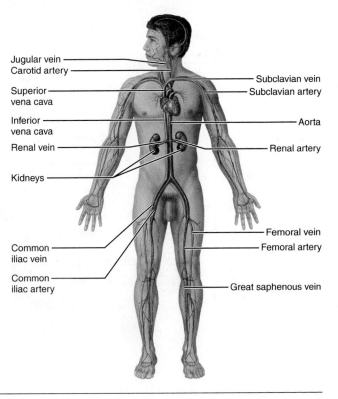

FIGURE 6.7 **Major organs and vessels of the human body.**

ORIENTATIONS TO THE HUMAN BODY

Critical Thinking and Review Questions

Directional Terms

Fill in the directional term for the questions below. More than one term may be acceptable.

1. Where are the shoulders relative to these parts?
   a. Throat _Inferior and lateral_
   b. Hip _Superior_
   c. Ears _Inferior_
   d. Hands _Superior_
   e. Eyes _Inferior_

2. Where is the nose relative to the same parts?
   a. Throat _Superior_
   b. Hip _Superior_
   c. Ears _Medial_
   d. Hands _Superior_
   e. Eyes _Medial_

Surface Regions

1. Fill in the anatomical terms for these anterior surface regions:
   a. Shoulder _Acromial_
   b. Armpit _Axillary_
   c. Wrist _Carpal_
   d. Thigh _Femoral_
   e. Forearm _Antebrachial_
   f. Side of leg _Fibular_
   g. Foot _Pedal_
   h. Chest _Thoracic_
   i. Navel _Umbilical_
   j. Hip _Coxal_

2. Fill in the anatomical terms for these posterior surface regions:

   a. Ear — Octic
   b. Thigh — Femoral
   c. Spine — Vertebral
   d. Heel — Calcaneal
   e. Elbow — Olecranal
   f. Calf — Sural
   g. Head — Cephalic
   h. Sole — Plantar
   i. Arm — Brachial
   j. Back — Upper extremity

3. Imagine that you are rock climbing with your best friend. Your friend slips down a rock about 50 feet and lands awkwardly on his right side. The following figures show the location and types of your friend's injuries. Using your knowledge of surface regions and directional terms you learned in this exercise, you call for help and need to describe the injuries as accurately as possible. How would you describe each injury to the emergency dispatcher assisting you?

   His orbitals are unfocused. Superior to that, his nasal is bleeding. Also bleeding on his brachial, and his crural just superior to tarsal. Also bleeding on the occipital, and acromial. Back to his anterior, he has protruding bones on his thoracic, carpal, and tarsal. On his posterior, he's bruised on his upper brachial, and sural.

   Unfocused   Bleeding   Bruise

   Protruding bone

Planes

4. Draw and label a picture of a football (assuming the laces represent the belly side) cut in the following ways:

    a. Transverse plane

    b. Frontal plane

    c. Sagittal plane

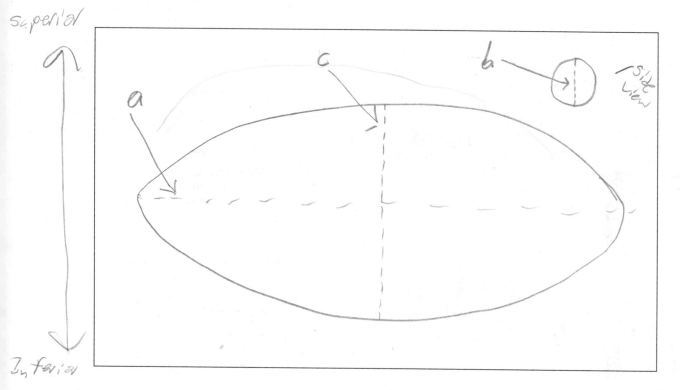

superior

Inferior

Body Cavities

5. Match the following organs and body cavities:

    _C_ Liver and spleen   a. dorsal cavity

    _b_ Heart   b. thoracic cavity

    _d_ Urinary bladder   c. abdominal cavity

    _a_ Brain and spinal cord   d. pelvic cavity

    _e_ Lungs   e. pleural cavities

    _d_ Uterus

    _C_ Gallbladder

The Integumentary System

Objectives

After completing this exercise, you should be able to

1. Describe the functions of the integumentary system.
2. Explain the basic structure and organization of the skin.
3. Name the five epidermal layers and their characteristics.
4. Compare thick and thin skin.
5. Describe the dermis and its skin derivatives.
6. Identify the two layers of human skin under the microscope.
7. Map the cold and touch sensations of the hand.

Materials for Lab Preparation

Equipment and Supplies

- ○ Compound light microscope
- ○ Magnifying glass or hand lens
- ○ Ice cubes
- ○ Blunt probe

Prepared Slides

- ○ Human skin (with and without hair follicles)

Human Skin Models (three-dimensional if possible)

- ○ Regular skin models (scalp and palm omitted)
- ○ Skin models of scalp, body skin, and palm or sole
- ○ Skin model showing epidermal layers
- ○ Human skin charts

Introduction

If we conduct an informal street survey and ask people to name the largest organ in the body, few would say skin; some might say lungs, while others might choose the heart. Recall, however, what you learned about how the body is organized: Cells form tissues; tissues form organs; two or more organs that work together to carry out a general function important to the survival of the whole organism are the basis of an organ system. Now think again about your body, both externally and internally, and you quickly will realize that the **skin** (or **integument,** from the Latin *integere,* meaning "to cover") is the organ with the largest surface area. The skin and its derivatives, such as hair, nails, sweat and oil glands, form the **integumentary system.**

Functions of the Skin

What functions are performed by the skin? The most obvious function is to provide protection from injury, such as abrasion, but there are other equally important ones. They include the following:

- protection from dehydration
- defense against bacteria and viruses
- regulation of body temperature
- synthesis of vitamin D
- receipt of sensation—touch, vibration, pain, and temperature
- provision of a blood reserve

Structure of the Skin

Skin is a tissue membrane, and it contains two basic regions, the epidermis and dermis, which are organized as shown in Figure 7.1.

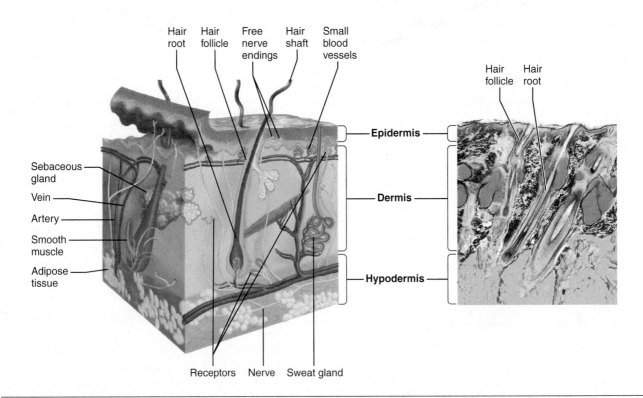

FIGURE 7.1 **The skin.** The two layers of skin (epidermis and dermis) rest on a supportive layer (hypodermis). Although not part of the skin, the hypodermis serves important functions of cushioning and insulation.

The thin, outer region of the skin is called the **epidermis.** The epidermis is composed mainly of stratified squamous epithelial tissue, which you were introduced to in Exercise 5, Tissues. There are five layers of cells in the epidermis, which we will examine in more detail later in this exercise. For now, keep in mind that the outermost cells of the epidermis are dead and continually slough off. They are replaced by cells from the deeper epidermal layers. This allows the epidermis to provide protection from everyday wear and tear on underlying body tissues.

The **dermis** is the thicker region of the skin, and it is composed primarily of connective tissue and a variety of fibers: reticular, collagen, and elastic. This combination of tissue and fibers gives the dermis its strength and flexibility. There are abundant capillaries in the superficial dermal region, which provide nutrients to the epidermal layers above.

The uneven, wavelike protuberances on the superior surface of the dermis attach it to the epidermis above. This "fit" is analogous to the tongue and groove joint found in finished lumber, which is created by shaping pieces of wood so that glue has more surface area to hold the pieces together. These protuberances force the inferior surface of the epidermis to fold, and this arrangement is the source of the unique patterns we call fingerprints.

A deeper layer called the **hypodermis** (*hypo* means "under"), or superficial fascia, is composed primarily of loose connective tissue containing fat cells and a large number of blood vessels. This layer is thickest on the palms and soles, areas that are subject to a great deal of wear and tear, and it is thinner on the eyelid, which undergoes considerably less wear.

The hypodermis anchors the dermis to muscle, while being flexible enough to allow the skin to move and bend. Its fat cells provide insulation to protect the body from excessive heat loss and absorb shock.

ACTIVITY 1

Identifying Skin Layers and Organization

Materials for This Activity

Skin models

Skin charts

1. Using the skin models or charts available in your lab, examine the structure of each layer and compare their differences and similarities. Notice that there are no blood vessels in the epidermis. How does the lack of blood vessels affect this layer?

   _____

   _____

   In a first-degree burn, only the epidermis is damaged. This is not a serious burn, and it tends to heal in a few days without special treatment. Can you think of a common example of a first-degree burn?

   _____

2. Look at the thickness of the dermis. The dermis appears to be a network of thick-to-thin fibers and connective tissue holding together many structures of the dermis, such as hair cells, sweat glands, and oil glands. The vertical blood vessels of the dermis connect two horizontal blood vessel supplies, one at the top of the dermis and one at the top of the hypodermis. If the dermis were damaged, would it affect the epidermis at all? If so, why?

   _____

   _____

   Notice the smooth muscle attached to the base of the hair follicle. What is its effect? (Think about frightened dogs or cats and what happens to their hair.)

   _____

   _____

3. Look at the main features of the hypodermis: adipose tissue (or fat cells), nerves, and blood vessels. The hypodermis, like the dermis, is a supportive structure. A network of thin fibers and connective tissue is also found here and anchors the skin to the underlying skeletal muscles. Notice the adipose tissue is located in this layer. What is the function of the adipose tissue? Of the three layers, what are the advantages for the location of adipose tissue in this layer?

   _____

   _____

In a second-degree burn, the epidermis and part of the dermis are damaged. This condition is more serious than a first-degree burn. With proper care, skin regeneration may take place in three to four weeks, and scarring may be avoided. What are the effects on skin functions if the blood vessels, sweat glands, and oil glands are seriously damaged?

_____

_____

A third-degree burn penetrates past the dermis; a first-degree burn affects only the epidermis. What additional skin functions are lost from a third-degree burn as compared to a first-degree burn?

_____

_____ ■

Structure of the Epidermis

The epidermis is composed of two major cell types. **Keratinocytes** are cells that produce the protein keratin, which is a strong gluelike material that also helps prevent water loss through our skin. **Melanocytes** are cells that produce melanin, a brown pigment that protects the nucleus of the skin cells from the ultraviolet rays of the sun.

The epidermis consists of four to five layers, called strata, that vary in thickness. There are five strata in **thick skin,** which covers the palms, the fingertips, and the soles of the feet. **Thin skin,** which covers the remainder of the body, consists of only four strata, which are thinner than those found in thick skin. The five strata of epidermis, from the outermost stratum to the deepest, are as follows:

1. **Stratum corneum:** most superficial portion, 25–30 layers of cells in thick skin; considerably fewer in thin skin.
   - This is also called the cornified layer.
   - It consists of dead, dried-out keratinocytes.
   - No nucleus is evident.
2. **Stratum lucidum:** 3–5 layers of cells.
   - This is also called the translucent layer.
   - It consists of dying keratinocytes.
   - No nucleus is evident.
   - This layer, which is not found as a distinct layer in thin skin, is mainly found on palms and soles and provides additional protection for these heavy-traffic skin areas.

3. **Stratum granulosum:** 3–5 layers of cells.
   - This is also called the granular layer.
   - It consists of living keratinocytes, which are becoming flatter.
   - The nucleus is barely evident.
4. **Stratum spinosum:** 8–15 layers of cells.
   - This is also called the spiny layer.
   - It consists of living keratinocytes, which are still rounded or cuboidal.
   - The nuclei are evident.
   - The cells are protected by melanin granules produced by melanocytes and concentrated in the layer above.
5. **Stratum basale:** deepest portion, 1 layer of cells only.
   - This is also called the base layer.
   - It consists of living keratinocytes and melanocytes.
   - The nuclei are evident, and cell division is in progress.
   - A basement membrane attaches this layer to the dermis.

ACTIVITY 2

Identifying the Layers of the Epidermis

Materials for This Activity

Skin models with epidermal layers
Skin models of scalp, body skin, and palm or sole
Magnifying glass or hand lens

1. Obtain a skin model with epidermal layers, and compare it to Figure 7.2. Analyze the cells that make up each stratum in the epidermis. For each stratum, compare these points: number of layers, shape of the cells, shape of the nucleus, and any special features of the stratum:

   stratum corneum _____

   _____

   stratum lucidum _____

   _____

   stratum granulosum _____

   _____

   stratum spinosum _____

   _____

   stratum basale _____

   _____

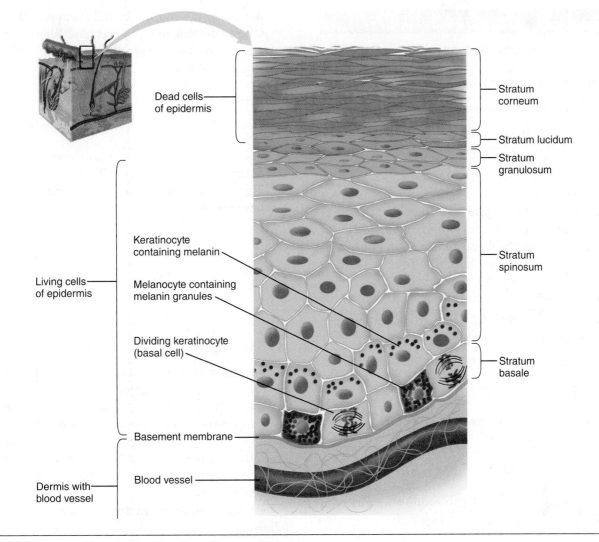

FIGURE 7.2 **The epidermis, showing five epidermal strata.**

Labels in figure:
- Dead cells of epidermis
- Stratum corneum
- Stratum lucidum
- Stratum granulosum
- Keratinocyte containing melanin
- Living cells of epidermis
- Melanocyte containing melanin granules
- Stratum spinosum
- Dividing keratinocyte (basal cell)
- Stratum basale
- Basement membrane
- Dermis with blood vessel
- Blood vessel

2. Obtain a skin model of scalp, body skin, and palm or sole. Thick skin is exemplified by the palm or sole, whereas thin skin is the major type found on the rest of the human body. Compare the epidermis of the two skin types in the model with the slides of human skin in Figure 7.3. Comment on the ease or difficulty of seeing details of the epidermis, between a human skin slide and a three-dimensional model of human skin.

   _____

   _____

   _____

3. Now that you have examined the epidermis in models and in photomicrographs, let's look at some real, living skin—your own skin, of course. Obtain a magnifying glass or hand lens and look at the skin on your arm (thin skin) and

palm (thick skin). Compare the look and feel the texture of the two skin areas. Use terms like *smooth, rough, soft, hard, fine,* and *coarse* to describe the textures.

Arm: _____

Palm: _____

4. Use the magnifying glass or hand lens to examine and compare your skin (thick and thin) to your lab partner's skin. What are the texture differences? Are the reasons for texture differences related to age, gender, or other factors?

   _____

   _____

   _____

   _____

ACTIVITY 3

Observing Slides of Human Skin

Materials for This Activity

Prepared slides of human skin (both with and without hair follicles)

Compound light microscope

1. Under the light microscope, scan a human skin slide without hair follicles, using the scanning-power lens. Look for the surface of the skin. You may see what appears to be thin peeling flakes. After you have focused on the epidermis, change to the low- and then high-power lens. The thin flakes may now appear to be a bundle of thin lines. Remember to look toward the surface of the skin. This is the epidermis, the most superficial portion of the skin.

   What do the peeling flakes look like?

   _____

   _____

   Describe any structures you observe.

   _____

   _____

   How many layers of the epidermis can you observe? Which ones do you see?

   _____

   _____

   _____

   _____

2. Using the scanning-power lens of a light microscope, focus on a prepared human skin slide with hair follicles. Find the epidermis, and focus below it into the dermis. Change to the high-power lens.

   Compare your tissue slide to Figure 7.3a–d. Identify as many structures shown in the photomicrographs as you can.

   What features can you clearly observe in addition to the hair follicles?

   _____

   _____

Are you able to distinguish between the dermis, epidermis, and hypodermis? What features make each tissue layer distinct?

_____

_____

_____

Recall your use of stain in Exercise 2 when preparing a wet mount slide. If you're now examining a particularly poor slide, how would staining improve it?

_____

_____

_____

Structure of the Dermis

The primary feature of the dermis is the fibers: **collagen, elastic,** and **reticular fibers** embedded in a ground substance. The fibers allow the skin to stretch when we move, and they give it strength to resist abrasion and tearing. As we age, our skin becomes less flexible and more wrinkled because the number of fibers in the dermis decreases.

Other structures in the dermis include the following:

1. **Hair follicle cells:** Our hair is formed at the root of hair cells, and a shaft pushes upward toward the skin's surface. The longer the hair, the older and less maintained it is, so the ends of long hair may show up as "split ends."
2. **Smooth muscle:** This is also called the arrector pili muscle. It contracts when you are frightened or cold, causing your hair to become more erect. Because our body hair tends to be fine, the result is "goose bumps."
3. **Sebaceous glands:** These are also called oil glands. The oily fluid serves to moisten and soften the hair and skin. If the oil is trapped and accumulates, it causes "pimples."
4. **Sweat glands:** These were formerly called the odoriferous glands. The salty, watery fluid contains dissolved ions and metabolic wastes.

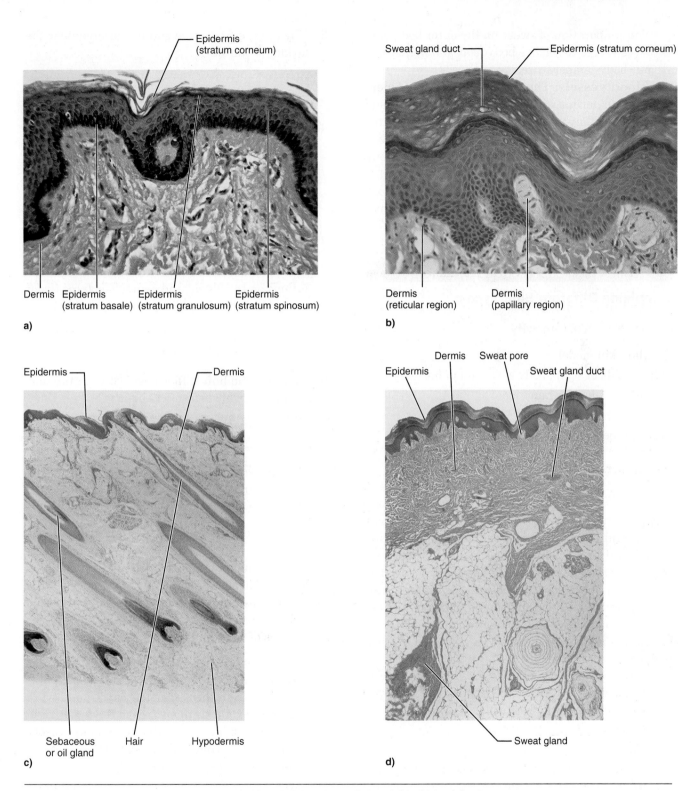

FIGURE 7.3 **Photomicrographs of human skin. a)** Thin skin. **b)** Palm (thick skin). **c)** Scalp.
d) Sweat gland.

The evaporation of sweat on the outer body surfaces serves to remove body heat and help lower our body temperature.

5. **Blood vessels:** These supply both the epidermis and dermis with nutrients and remove their wastes. They also regulate body temperature through dilation when we are too hot and constriction when we need to conserve heat for our internal organs.

6. **Sensory nerve endings:** Specialized receptors detect heat, cold, pain, light touch, deep pressure, and vibration.

ACTIVITY 4

Examining Different Skin Types

Materials for This Activity

Regular skin model
Skin models of scalp, body skin, and palm/sole
Magnifying glass or hand lens
Ice cubes
Blunt probe

Obtain a regular skin model, and compare it to Figure 7.1 on page 72. Look at the three skin layers—epidermis, dermis, and hypodermis—and then magnify the dermis. Identify the skin derivatives found in the dermis: hair cells, arrector pili muscle, oil and sweat glands, blood vessels, and sensory nerve endings. Notice the relationship between form, function, and location. For example, the sweat gland ducts have an opening in the skin in order to release sweat. The evaporation of sweat cools the skin surface, which leads to a reduction in body temperature.

1. What is the relationship between the arrector pili muscle and the hair cells?

   _____

   _____

2. What is the relationship between the oil glands and the hair cells?

   _____

   _____

3. Why are the free nerve endings located near the surface of the skin?

   _____

   _____

4. Obtain a skin model of scalp, body skin, and palm/sole. Review the differences between the dermis of thick and thin skin.

   _____

   _____

5. Use the magnifying glass to examine and compare some real, living skin—your own skin and your lab partner's skin. Compare the hair distributions on the dorsal and ventral surfaces (front and back) of the hand.

   _____

   _____

6. What are the hair differences between the dorsal and ventral surfaces of the arm?

   _____

   _____

7. *Make a map of the cold and touch sensations of the hand* and obtain a container of ice cubes and a blunt probe.

   Again, find a lab partner and one student will close his or her eyes, while the other student will systematically touch parts of the dorsal and ventral surfaces of the hand with the corner of an ice cube and a blunt probe. Use "c" for cold and "t" for touch to mark the dorsal and ventral outlines of the same hand in Box 7.1.

   (*Hint:* Think before you start so that the whole hand will be systematically covered. One idea is to start from the upper-left corner and work from left to right until you reach the lower-right corner. Test a bit so that the ice and blunt probe press is perceptible and consistent but not uncomfortable.)

8. What are the differences between the cold maps of the two surfaces of the hand?

   _____

   _____

9. What are the differences between the touch maps of the two surfaces of the hand?

_____

_____

Sensory adaptation is a decline in sensation due to continuous stimulation or no change in stimulus strength. For example, we may smell an odor upon entering a room and later may not notice it at all, even though the odor is still present in the room. Our sensory perception gives priority to information of immediate importance, especially new information. Once the information is evaluated, understood, and particularly if it is determined not to be dangerous, we "adapt" by decreasing our response to that specific stimuli. The notable exception to this is pain sensation.

10. Propose a reason why it is beneficial for pain sensation not to "adapt," whereas most other skin sensations do?

_____

_____

11. In the cold and touch mapping experiment, what did you notice about sensory adaptation? How long did it take before you were no longer aware of these features? Seconds? Minutes?

_____

_____

12. What were the differences between your map and your lab partner's map?

_____

_____

Box 7.1 Dorsal and Ventral Outline of a Hand

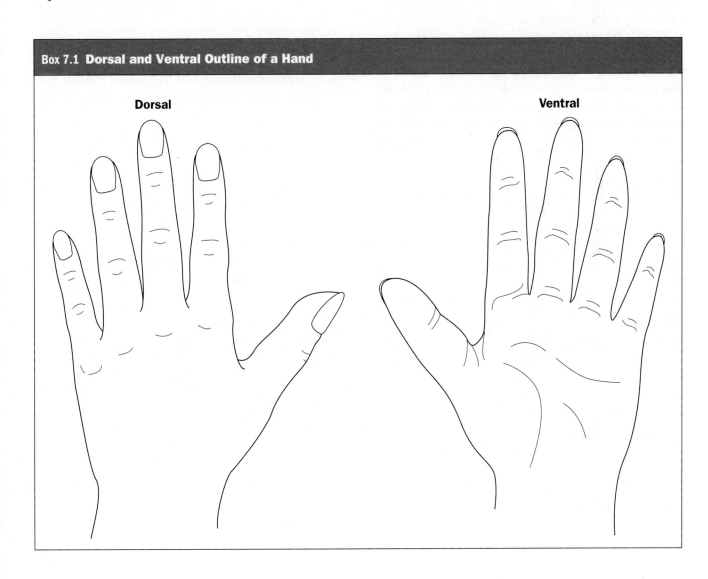

Dorsal

Ventral

Nails

Nails form at the fingers and toes, where they protect and strengthen the tips, especially when you use your hands and feet (Figure 7.4). Nails are clear and colorless, but they appear pink because of the blood supply in the underlying skin, except for a white crescent-shaped area called the **lunula** (Figure 7.4a).

Structure

The nails consist of a **nail body,** which is primarily composed of dead keratinized cells. It covers the **nail bed,** which is the skin underneath the nails (Figure 7.4b). There is a **free edge** at the most distal portion of the nail body and a **nail root** at the most proximal portion, which is the site of nail production. An epithelial skin fold **(eponychium or cuticle)** projects onto the nail body.

The **nail matrix,** which is a portion of the skin underneath the nail, produces the modified tissue we call nails (Figure 7.4b). As the nail cells grow out and extend from the base of the nails, they become keratinized (producing a harder variety of keratin) and die. Thus, like hairs, nails are mostly nonliving material.

ACTIVITY 2

Examining Nail Structures

1. Look at your fingernails. Compare them with those of your lab partners. Identify the nail structures described previously.
2. What are some general differences between your fingernails and toenails? Compare the size, shape, thickness, and texture of the thumb.

_____

_____

_____

3. What are the differences (size, shape, thickness, and texture) between each of your individual fingernails?

_____

_____

_____

4. Compare the thumb and last finger. Why do you think the thumbnail is thicker and larger?

_____

_____ ∎

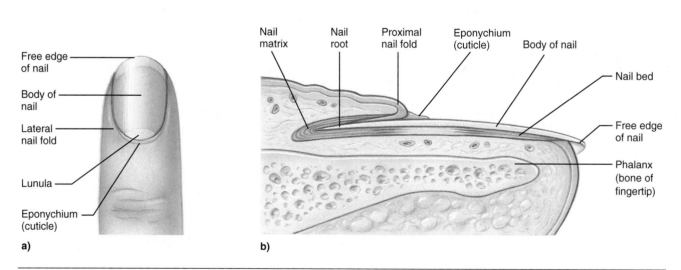

FIGURE 7.4 **Structure of a nail. a)** Frontal view of nail. **b)** Sagittal view of nail.

THE INTEGUMENTARY SYSTEM

Critical Thinking and Review Questions

1. Label the following structures of the skin and underlying hypodermis.

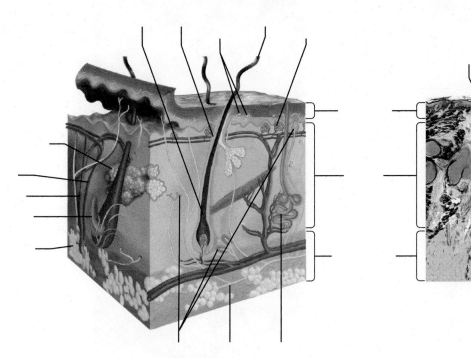

2. Label the five strata of the epidermis.

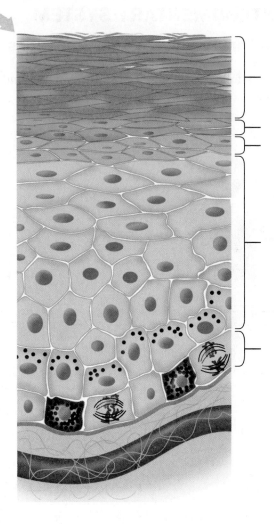

3. Label the indicated structures of a nail.

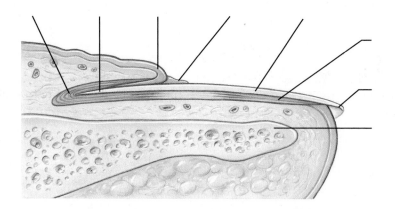

4. Compare the functions of the stratum corneum of a human and the bark of a tree.

   _____

   _____

5. Compare and contrast the features of the stratum corneum in the thin skin and thick skin.

   _____

   _____

6. You fall off your skateboard and scrape your knee. After checking your injury, you note that you are not bleeding. What layers of the integument are the most damaged? Which layers are not?

   _____

   _____

7. Speculate on the reasons for differences in skin texture and hair between men and women.

   _____

   _____

8. Speculate on the reasons for the absence of hair on the ventral surface of the hand (palm).

   _____

   _____

9. Think about all the times you have put on gloves, a hat, a scarf, or sat down on a chair. How long does it take for you to stop noticing that those items are in contact with your skin? Describe the principle that is in operation when you no longer sense that contact.

   _____

   _____

10. Again, consider all the times you have put on gloves, a hat, a scarf, or sat down on a chair. Does the sensation and your awareness of contact persist longer if the chair or clothing is hard, rough, or otherwise uncomfortable? Propose a reason for this. (*Hint:* Are all sensations equal with regard to receptor adaptation?)

    _____

    _____

    _____

11. Hair, nails, and the epidermis of the skin are all mostly made of keratin, but are different with regard to durability. What does this suggest about the nature of the keratin produced in different areas of the body?

    _____

    _____

EXERCISE

8

The Skeletal System

Objectives

After completing this exercise, you should be able to

1. Describe the five functions of the skeletal system.
2. Analyze the features of a long bone.
3. Describe key features of the axial skeleton: skull, vertebral column, and rib cage.
4. Describe key features of the appendicular skeleton: two girdles, arms, and legs.
5. Compare and contrast the three types of joints.
6. Demonstrate synovial joint movements.

Materials for Lab Preparation

Equipment and Supplies

○ Loose (disarticulated) human bones, stored in individual boxes

○ Fresh (if available) or clean, dry, long bones (cow, pig, or sheep, with most of the meat, ligaments, and tendons removed), both whole and cut longitudinally

○ Dissection equipment

○ Disposable gloves

○ Paper towels

○ Disinfectant in spray bottle

○ Large dissection tray or shallow pan

Models

○ Whole (articulated) human skeletons

○ Leg or knee muscles

Introduction

If you consider the wide scope of activities that you may perform in a single day—crawling out of bed, climbing stairs, the work you do, and so on—that will suggest the remarkable maneuverability and complexity of the human skeletal system.

However, the skeletal system has several important functions in addition to providing the levers and support for **locomotion.** Think, for a moment, about the bones in your body and about the role they might play in all of the systems of the whole body.

Support: Bones form a structure for attachment of skeletal muscles and support of soft organs (e.g., the hip bones help support the pelvic organs).

Movement: Bones provide leverage for muscles to change the amount and direction of muscle force during body movements.

Protection: Bones are hard, and can protect soft organs (e.g., the skull protects the brain).

Formation of blood cells: The spongy bone tissue in our bones makes red blood cells, white blood cells, and platelets that assist in blood clotting.

Mineral and fat storage: Bones store minerals, like calcium and phosphate, which regulate body metabolism. Some bones store fat in the form of yellow marrow.

Bone Structure

The human skeleton is composed of two connective tissues: cartilage and bone. In this exercise, we will focus primarily on bone.

Bones are the hard elements of the human skeleton. Muscles attach to bones to facilitate movement. Bones also provide support and protection for internal organs. Bones consist of living bone cells, nerves, and blood vessels.

There are four categories of bone based on shape: **long, short, flat,** and **irregular.** Long bones include the bones of the limbs and the fingers; short bones, which are short and wide, are the bones of the wrist; flat bones are thin and flattened, such as the cranial bones and the sternum; and irregular bones, such as the coxal (or hip) bones and the vertebrae, have shapes that do not fit into the other categories. Though the bones of the human skeleton vary in size, shape, and composition, we can learn something about the structure of all of them by examining the long bone.

Long Bone Anatomy

Long bones, for example, the femur, consist of a long, slender shaft called the **diaphysis** (Figure 8.1).

Most long bones are a little larger in diameter at the ends, or **epiphyses** (singular, epiphysis). The following activity will examine a long bone in detail.

ACTIVITY 1

Examining Long Bone Anatomy

Materials for This Activity

Fresh or prepared long bones from a large animal (e.g., pig, cow, or sheep), whole bones, and bones cut longitudinally

Large dissection tray or shallow pan

Disposable gloves

Dissection tool kit

Paper towels

Disinfectant in spray bottle

Note: If preserved bones are used, consider steps to minimize contact of preservative with skin, clothing, and surfaces.

1. Obtain a dissection tray containing a long bone, and bring it to your work area; your instructor will indicate what the specific bone is (usually the femur, also known as the thighbone) and from which animal it comes (e.g., pig, cow, or sheep). Visualize the proportion of the femur to the whole body. In which mammal—human, pig, sheep, or cow—would you expect that proportion to be the greatest?

   COW

   Looking first at the whole long bone, notice that if the bone is very fresh, the bits of tendon and muscle will still be pink or reddish and soft, the bone will be slightly sticky to the touch, and it will feel heavier and wetter. There may also still be marrow in the cavities and spongy bone. If the bone is older, the nonbone material will tend to be brownish and hardened, somewhat like beef jerky; the bone will have a rough, nonsticky touch, and it will feel lighter and drier.

2. Check the ends of the whole long bone for a thin translucent line, made mostly of hyaline cartilage, called the **epiphyseal (growth) plate.** During puberty, the osteoblasts in the epiphyseal plate become active and form new bone material, which contributes to the length of long bones. If the bone came from an older animal, this line would be barely discernible as a thin line of bone.

3. With forceps or tweezers, carefully pull away the fibrous, transparent membrane, called the

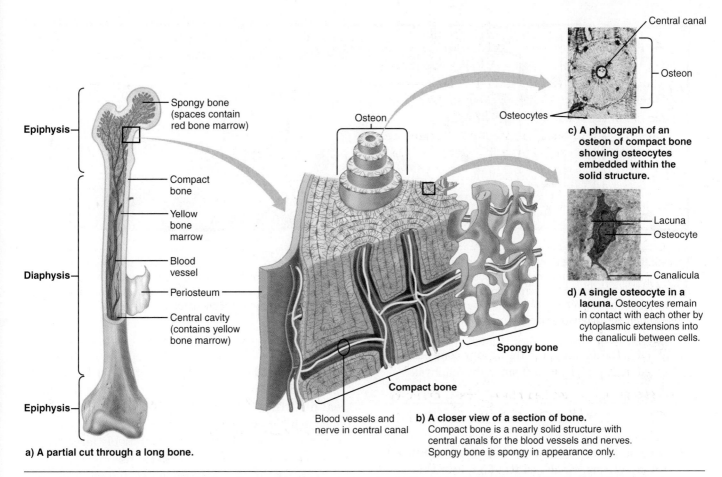

FIGURE 8.1 **A long bone (humerus). a)** Partial cut to show the interior. **b)** Epiphysis—enlarged view of spongy and compact bone. **c)** Diaphysis—enlarged view of the endosteum and periosteum layers. **d)** Enlarged view of a single osteocyte cell.

periosteum, which covers the outside of the bone. This layer contains active bone cells (**osteoblasts**) that contribute to the diameter or girth of long bones. Have you ever wondered why dogs lick and chew bones? They are scraping the periosteum off the bone and trying to crunch through to get at the marrow (Figure 8.2).

As bones grow longer and thicker, they become not only stronger but also heavier. How would heavier bones affect your movements?

Heavier bones add weight and overall lower movement efficiency

4. Using a long bone cut lengthwise, use a scalpel or knife to tease or scrape off the transparent membrane, called the **endosteum,** which lines the cavity in the shaft of the bone. This **central cavity** is used to store **yellow marrow,** a fat deposit that serves as an energy reserve. During puberty, the **osteoclasts,** located in the endosteum, erode the inner surface of the bone to counteract the additional weight caused by the bone formation on the outer surface of the bone.

The effect is to increase the size of the central cavity, while keeping the wall of the diaphysis an ideal weight-supporting thickness.

As the size of the central cavity increases, the weight of the bones decreases. How does this affect bone strength and body movement?

This lowers bone strength but allows movement to improve

5. Examine the cut long bone, and notice the pinkish **spongy bone** in the epiphysis. The color is derived from **red marrow,** which produces the blood cells for the circulatory system. Blood cell production is so important for gas transport and defense that spongy bone is found throughout the body, including the thicker portions of skull bones, where it exists sandwiched between two layers of compact bone.

Run your fingers over the cut portion of the spongy bone. Notice the rough, needlelike texture. Spongy bone consists of a lattice of interconnected calcified rods and plates. The open areas inside the lattice provide space for the red marrow.

FIGURE 8.2   Good thing they do love us so much! Although we usually remove or discard bones in our food, the fat (yellow marrow), minerals, and other nutrients found in bone tissue and bone marrow don't go to waste in the canine diet. Of course, their teeth are a little better at crunching through bone than ours. This does help us understand why grandma's old soup recipe that called for adding a large ham bone wasn't so strange after all. We use the cooking process instead of our teeth to get at those nutrients! (Hillary B. Price, Rhymes with Orange, June 25, 2009)

What is the advantage of the spongy bone being located in the epiphysis instead of the diaphysis?

Allows marrow to move and walk easily

6. Look at the white or creamy **compact bone** in the diaphysis. Knock on it, and notice that it is rigid and strong. Why are these characteristics important in terms of bone function?

Bones are known for strength and structure. Strong bones allow to resistance to injury

Run your fingers over the surface of the shaft of the bone. Notice that the texture of compact bone is considerably smoother than spongy bone. The smooth texture reduces the friction as bone surfaces rub against the surrounding muscles, tendons, and ligaments.

7. If you have been examining fresh long bones during this activity, now notice their weight. To what would you attribute this? How do you think their weight would compare to that of dry bones?

Fresh bones still contain blood and fresh marrow increasing weight

8. Return the bones to the supply area. Clean your counter with disinfectant and paper towels. Dispose of your gloves and wash your hands. ∎

Bone Tissue

As already introduced in the previous section, we find spongy and compact bone tissue in long bones and most other bones. Spongy bone looks like a magnified, dried sponge, with lots of spaces between the threadlike pieces of bone. However, note that in "real life," those spaces would likely be filled with red bone marrow. Compact bone looks more "solid"—so solid that the cells must form relatively large "cooperative units" of cells called **osteons** (Figure 8.1).

Each osteon has a central canal that contains a blood vessel and rings of bone cells radiating around the central canal. The cells link together through tiny canals called canaliculi so that they can share nutrients and return wastes to the blood vessel.

ACTIVITY 2

Examining Bone Tissue

Materials for This Activity

Microscope

Prepared slide of compact bone

Place a microscope slide of compact bone on your microscope and look for the osteons, which should appear as small target-like or dartboard-like structures. You need not magnify this slide beyond your 10× objective to see sufficient detail. Make a sketch of one or two osteons below, and label the central canal, osteocytes, and canaliculi.

Why do you suppose the central canal is the most conspicuous feature of an osteon?

The canal is supposed to hold marrow

Describe the appearance of the canaliculi and state their purpose.

These are circular holes that extend throughout the bone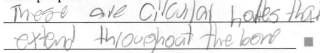

Organization of the Skeletal System

The adult human skeleton is formed from 206 bones and various other types of connective tissue that hold them together. The skeleton provides general structural support, as well as support for soft organs. It also protects certain organs from injury, such as the brain, which is enclosed within the bones of the skull. Another vital function of the human skeleton is that it allows for flexible movement at the joints. Finally, the human skeleton is organized into the **axial skeleton** and the **appendicular skeleton** (Figure 8.3).

Axial Skeleton

The axial skeleton consists of 80 bones, including the skull, vertebral column, ribs, and sternum (or breastbone). The **skull,** which consists of cranial and facial bones is the most superior feature, and it rests securely on top of the vertebrae. The **vertebral column**

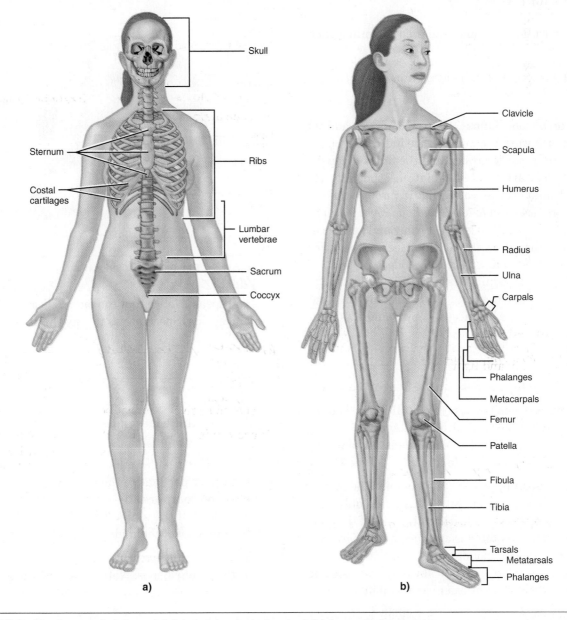

a)   b)

FIGURE 8.3   The human skeleton. a) Axial skeleton (anterior view). **b)** Appendicular skeleton.

provides a central axis that supports the head, protects the spinal cord, and anchors the body trunk.

The **rib cage** is formed by the ribs and sternum, and attaches posteriorly to the thoracic vertebrae. It surrounds and shields the chest organs, such as the heart and lungs. Speculate on the reason for the cartilage connecting the ribs to the sternum. Would elongation of the rib bones to replace the cartilage be a better design?

No, the rib cage needs to be strong for protection, but flexible for organ expansion

ACTIVITY 3

Examining the Skull

Materials for This Activity

Whole (articulated) human skeleton on rolling rack

or

Loose (disarticulated) individual human bones in boxes

1. Obtain a whole human skeleton on a rack and roll it to your work area, or use the individual bones. Be careful not to jar or catch the limbs if you are using the articulated skeleton, especially the hands and legs, as you navigate through the lab. *We encourage you to touch and handle the individual bones, but do so gently.* Firmly and gently run your fingers along the surfaces, feeling the different textures, bumps, holes, and ridges.

2. Examine the skull first (see Figure 8.4). The cranial portion of the skull is sometimes referred to as the braincase. Why would this be an appropriate description?

   This holds the brain and is encased

3. The **facial bones** make up the front of the skull.

   The **maxilla** is the largest facial bone. The upper portion forms part of each eye socket, and the bottom portion forms the upper jaw and contains the sockets for the upper row of teeth.

   The **mandible,** or lower jaw, contains the sockets for the lower row of teeth. This is the only bone that is not tightly joined with the rest of the skull bones. The mandible is attached by the jaw joint to the temporal bone, which is discussed later. This separation from the skull allows us a full range of motions, for example, the ability to chew

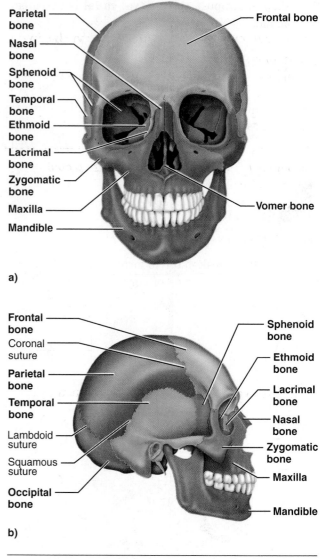

a)

b)

FIGURE 8.4   **The skull. a)** Anterior view. **b)** Right lateral view.

and speak. Notice that the mandible is quite thick and strong. Speculate as to why this is so.

The jaw needs to chew and break hard foods without breaking bones

4. The **cranial** bones make up the top, sides, and back of the skull.

   The **frontal** bone makes up the forehead and the upper portions of the eye sockets. The **parietal** and **temporal bones** are located on the left and right of the skull. Together, they make up the majority of the middle and back of the skull.

   Examine the texture, ridges, and depressions of these four bones. Notice the smooth surfaces, interior and exterior, as opposed to the rough material inside the skull bones. The rough material is called spongy bone and serves to produce blood.

5. A small portion of the **sphenoid** is visible in the temple area. It also forms some of the back of the eye socket. What other bones make up the eye socket?

Temporal, ethmoid, lacrimal, sphenoid

6. The back and base of the skull is formed by the **occipital bone,** which contains a large opening called the **foramen magnum.** This opening is where the vertebral column connects to the skull and the spinal cord enters the skull to communicate with the brain.

7. The two **zygomatic bones** form the cheekbones and the lateral portion of the eye socket. They are relatively thick and protect the sides of our face.

8. The **nasal bones** form the bony bridge of the nose; the rest of the nose is made of cartilage and other connective tissues. Notice these bones are not that thick and can easily be broken. If your nose is broken, speculate as to why it may not heal properly without reconstructive surgery.

The nasal area isn't the priority and the area is mainly cartilage with low blood flow and isn't very well fixed

ACTIVITY 4

Examining the Vertebral Column

Materials for This Activity

Whole (articulated) human skeleton on rolling rack

1. Looking at the whole articulated skeleton, you can see that the **vertebral column** is composed of 33 vertebrae—24 single bones and 9 fused bones.
2. The vertebral column has a curving, or sinusoid, pattern. Speculate on the advantages of a curved, versus a straight, backbone.

A curved back bone allows for a good center of gravity and movement

Vertebral Regions and Spinal Curves

1. Compare the skeleton to Figure 8.5, and identify each region starting at the top. Notice the **cervical curve** in the neck is formed from the

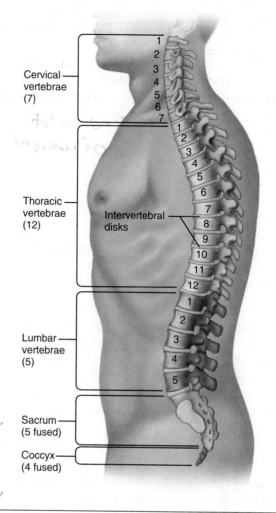

FIGURE 8.5 **The vertebral column.**

first seven bones, the **cervical vertebrae.** Why are the cervical vertebrae the smallest and most delicate of all vertebrae in the human skeleton?

The cervical vertebrae holds the head which isn't a large amount of weight and needs minimum support

The thoracic curve in the upper back is formed from the next 12 bones, the **thoracic vertebrae.** The **lumbar curve** in the lower back follows, and it is formed from the next five bones, the **lumbar vertebrae.** Notice that the lumbar vertebrae appear larger and stronger than the vertebrae above them. Speculate as to why this is so.

The lumbar needs to hold a large amount of weight and requires large, strong bones

The **sacral curve** is last, and it is formed from the **sacrum,** which is a single bone composed of five fused bones. The **coccyx,** or tailbone, continues the sacral curve. Speculate as to why it is ok for sacral vertebral to be fused, but not those above the sacrum.

The sacrum doesn't need to move so can be curved and fused and alot of weight

Spinal Curvatures

Primary curvatures. The thoracic and sacral curvatures are well developed at birth and resemble the curves of a four-legged animal. If you got down on your knees, you would see that these two curves "bump out," or as we say in anatomy, they are "convex posteriorly."

Secondary curvatures are "concave anteriorly," and begin to develop after birth. As the baby starts to lift its head, the cervical curvature also begins to develop, thus aligning the weight of the head over the vertebral column. If you have ever tried to balance a basketball on your fingertip, you will appreciate the difficulty of this task. A few months later, as the baby begins to walk, the lumbar curvature begins to develop. Its job is to position the weight of the trunk over the body's center of gravity, thus providing the proper balance when standing, walking, and later on, running.

What happens when these curvatures do not develop properly? During childhood, when there is uneven growth of the bones and muscles, vertebral problems, like scoliosis or lordosis, may appear. For example, lordosis (strutting posture, like a lord) may appear during preschool years, then disappear as the abdominal muscles become stronger, and the pelvis tilts forward.

Examine Figure 8.6. Compare the abnormal spinal curvatures with the normal position of the vertebral column, which has been outlined in dots. Briefly describe the posture of the individual with each vertebral problem.

Sco, standing lopsided
Kyph, hunched
Lord, lower back curved in

2. Notice that the individual vertebrae are stacked, but they are cushioned from each other by the **intervertebral disks,** which are made of a strong connective tissue called fibrocartilage. Speculate

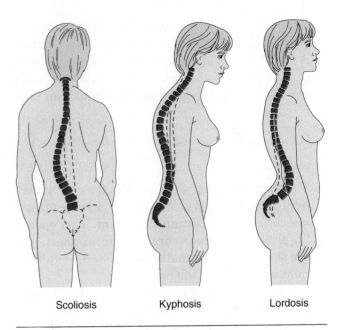

Scoliosis   Kyphosis   Lordosis

FIGURE 8.6   Abnormal spinal curvatures.

on the function of these disks (think about your various physical activities during the day).

These disks undergo constant movement

Sudden impact or movement can compress an intervertebral disk, forcing the softer center to balloon outward and cause intense back pain. This condition is called a "herniated disc." What would the effects be, if any, on mobility if the damaged disk was surgically removed?

The mobility would be reduced since the disks arent a val.

Protection of the spinal cord is a crucial function of the vertebral column. When injured on the playing field or during an automobile accident, injured parties are always required to lie absolutely still until they have been examined by a medical professional. ■

ACTIVITY 5

Examining the Rib Cage

Materials for This Activity

Whole (articulated) human skeleton

1. Humans have 12 pairs of **ribs,** which are anchored between the vertebral column and the sternum (Figure 8.7). The **sternum** is more commonly known as the breastbone, and is a

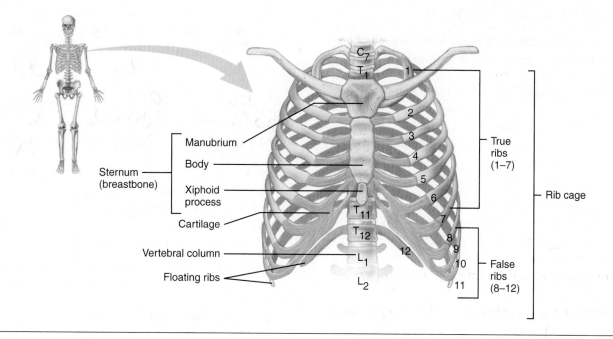

FIGURE 8.7   The rib cage. Notice the blue-colored costal cartilage that attaches the rib cage to the sternum.

flat blade-shaped bone, consisting of three separate bones that have fused. The **manubrium** is at the top of the sternum, followed by the **body,** which is the bulk of the sternum, and then the **xiphoid process.** This process serves as an attachment for abdominal muscles.

2. Now look at the ribs. The topmost seven ribs are called **true ribs** because their terminal cartilage pieces attach *directly* to the sternum. The second set of five ribs are called **false ribs** because they do not attach directly to the sternum. The first three false ribs attach *indirectly* to the sternum, in conjunction with the cartilage of rib 7. The last set of two ribs, which are also grouped with the false ribs, are called **floating ribs** because they do not attach at all to the sternum.

Which ribs are the longest? What might be the reason for this?

The upper true ribs, protects heart.

Why would it not make sense for the rib cage to be formed of two solid sheets of bones instead of individual rib bones?

The bones can move for organs instead of being fixed

Notice that the rib cage does not extend into the abdominal area. Speculate on the reasons for this.

The abdomin doesn't need protected, but needs to be flexible and open for the digestive system

Appendicular Skeleton

The appendicular skeleton consists of those parts of the body that attach, or are *appended* to, the axial skeleton, which forms the longitudinal axis of the body. The pectoral and pelvic girdles and the bones of the upper and lower limbs of the body make up the appendicular skeleton. The **pectoral** (shoulder) **girdle** attaches the arms to the axial skeleton and allows for a lot of mobility of the arms. In contrast, the **pelvic** (hip) **girdle,** which attaches the legs to the axial skeleton, must support the weight of the upper body against gravity, and allows a lot less mobility of the legs.

Try swinging your arms and legs in as big an arc as possible. Which one swings more easily? Try to explain the reason for differences in mobility between arms and legs.

The arms move easily to grab items, but the legs help with movement

Compare the flexibility and stability of the pectoral and pelvic girdles.

The pelvic is much more stable to hold up the body, but pectoral needs to be flexible for movement

ACTIVITY 6

Examining the Pectoral and Pelvic Girdle

Materials for This Activity

Whole (articulated) human skeleton

Pectoral Girdle, Arm, and Hand

1. The pectoral girdle (Figure 8.8) consists of the **clavicle** or collarbone, which extends across the top of the chest and connects to the sternum or breastbone. The clavicle anchors many muscles and serves to brace the shoulders away from the chest. When the clavicle is broken, the shoulder region will collapse toward the center.

   Look at the slender clavicle. Does it look very strong? If you were to fall on your upper body, what might happen to your clavicles? If they don't heal, what might happen to your shoulder?

   They would easily break,
   but the shoulder would basically lose all function

2. The second part of the pectoral girdle is the **scapula** or shoulder blade. It connects to the clavicles and provides a socket for the arms. Notice the scapula is somewhat triangularly shaped.

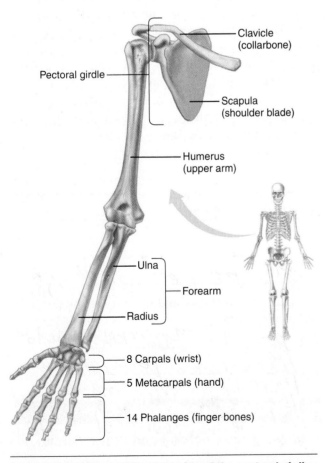

- Clavicle (collarbone)
- Pectoral girdle
- Scapula (shoulder blade)
- Humerus (upper arm)
- Ulna
- Forearm
- Radius
- 8 Carpals (wrist)
- 5 Metacarpals (hand)
- 14 Phalanges (finger bones)

FIGURE 8.8 **Bones of the right side of the pectoral girdle and the right arm and hand.**

In a dislocated shoulder, the arm is no longer anchored to the scapula. How does this affect the effective use of the arm?

The bone is no longer effective since it isn't anchored

3. Each arm consists of the **humerus, ulna,** and **radius.** What joint connects these three bones?

   _____

4. Each hand is made of two regions: 8 **carpals** and 5 **metacarpals** make up the wrist and palm, and 14 **phalanges** make up the thumb and fingers.

   What joint connects the hand to the arm? _____ How many bones make up the two hands? _____ Which finger has only two bones? _____

5. Examine the arrangement of the "opposable thumb." What are the benefits of this hand arrangement? (*Hint:* Pick up an object with just the fingers.)

   _____
   _____
   _____

Pelvic Girdle, Legs, and Feet

1. Examine the **pelvic girdle** and compare it to the representation in Figure 8.9. The pelvic girdle supports the weight of the upper body, protects the pelvic organs, and attaches the leg to the vertebral column.

2. The left and right **coxal bones,** or hip bones, make up the pelvic girdle. During childhood, each coxal bone was composed of three separate bones: the ilium, the ischium, and the pubis. By adulthood, the three bones are firmly fused, and their boundaries are indistinguishable—an example of **bone fusion.** Notice, these bones connect the lower extremities to the axial skeleton via the sacrum. In adult women, this girdle is broader and shallower than in men, and the opening is wider. Speculate on the reasons.

   _____
   _____

3. Examine the thigh, which is formed by the **femur,** the longest and strongest bone in the body. Below the knee, the leg is formed by the

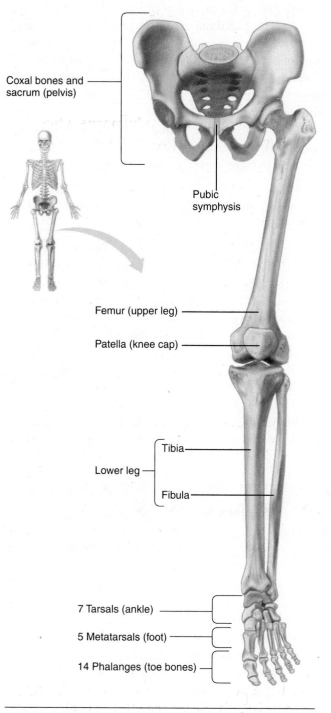

Coxal bones and
sacrum (pelvis)

Pubic
symphysis

Femur (upper leg)

Patella (knee cap)

Tibia

Lower leg

Fibula

7 Tarsals (ankle)

5 Metatarsals (foot)

14 Phalanges (toe bones)

FIGURE 8.9 **Bones of the pelvic girdle and the left leg and foot.** Notice the blue-colored pubic symphysis that connects the left and right coxal bones.

tibia and **fibula.** At its proximal end, the tibia connects with the femur and **patella,** or knee-cap, to form what structure?

_____

At their distal ends, the tibia and fibula connect with the tarsals to form what structure?

_____

4. Each foot is formed by three regions: 7 **tarsals** make up the ankle and heel, 5 **metatarsals** make up the sole of the foot, and 14 **phalanges** form the toes. Like the thumb, the big toe only has two bones, yet it is thicker than the other phalanges. Speculate on the reason.

_____

_____ ∎

Whole Body Growth

During childhood and adolescence, we "grow up." In other words, we get bigger and taller; but notice that there are also changes in our body proportions, as illustrated in Figure 8.10. At birth, the head and trunk are about ⅔ of the length of the body, while the legs are about ⅓. The legs grow faster than the head and trunk, so by about age 10, the head and trunk are about 50–50 with the legs. Later on, the proportions become about 40–60, depending on genetic disposition.

After adulthood, as we approach old age, we begin to "grow down." As the vertebral discs become thinner, the vertebral column can indeed get shorter. Osteoporosis, or porous bones, is another cause of loss of height. Diet and exercise may help prevent or reduce development of osteoporosis.

All bones lose mass with age, not just the vertebral column. For example, what are the effects of shrinking the facial bones, smaller jaws, losing your teeth, and having porous bones?

_____

_____

_____

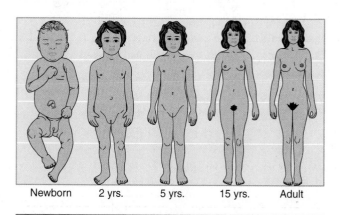

Newborn   2 yrs.   5 yrs.   15 yrs.   Adult

FIGURE 8.10 **Whole body growth.** As we grow, during adolescence, the legs grow faster than the head and torso.

Joints

Movements make up our human world—cutting a piece of paper, writing, dancing, opening an envelope, or chopping food. Movements are possible because of joints, the way the 206 bones in the human body are connected. Joints, also called articulations, are formed wherever one bone comes in contact with another bone.

Although joints are essential in permitting movement of the skeleton, not all joints are movable. Some, like the sutures of the skull, are simply for strong connection of bones.

ACTIVITY 7

Examining Joints

Materials for This Activity

Whole (articulated) human skeleton

There are only three types of joints, yet they determine not only type of movement but also range of motion. **Fibrous joints** are the simplest. Like ropes, the strong fibers in the fibrous connective tissue firmly connect and stabilize the adjacent bones, and they allow no movement. **Cartilaginous** joints are just a bit more complex. Like rubber or plastic, the cartilage connects bones, but also provides a cushion for compression. They allow limited movements. **Synovial joints** are the most complex. If you name the first joint you think of, you'll probably come up with a synovial joint, such as the hip, shoulder, or elbow. The structure of synovial joints includes a joint cavity filled with fluids, which not only allows bones to move easily past one another, but also to have a greater range of motion.

As a cautionary note, recall that not all joints are freely movable. So move the parts of the skeleton with care.

1. Let's look at how the bones are connected on the human skeleton. Remember that the living joints are connected with actual tissues, but the model will not have the connecting tissues and may simply be connected by a steel pin. Nevertheless, it is possible to use a model to observe the relationship between bones and deduce the joint type and the range of motion.
2. Refer to Box 8.1, Joint Movements. Find the first joint under "Bones Involved." Before touching the model, try to move that joint in your own body and observe the range of motion. Do this

Box 8.1 **Joint Movements**		
Bones Involved	**Range of Motion** (none, limited, free)	**Joint Type** (fibrous, cartilaginous, synovial)
1. Between skull bones		
2. Distal end of radius and ulna		
3. Distal end of tibia and fibula		
4. Between vertebral discs		
5. Sternum and ribs		
6. Between coxal bones		
7. Hip: femur and coxa		
8. Shoulder: humerus and scapula		
9. Elbow: humerus, radius, and ulna		

for each of the bones involved listed. After you have determined which joints will move easily, and observed the proper movement in your own body, try the same on the model, keeping in mind the caution notes above. Observing the joint directly should give you a better understanding of joint movement.

3. Complete Box 8.1. From your observations of the possible range of motion for each joint, fill in the second column, "Range of Motion," and deduce the joint type (fibrous, cartilaginous, synovial).

What did you learn about joint differences from observing fibrous, cartilaginous, and synovial joints? Why should you not try to move fibrous and cartilaginous joints on the model, as cautioned above?

_____

_____

_____ ■

ACTIVITY 8

Exploring Synovial Joint Movements

1. Let's perform some simple joint movements. Step to a clear area of the lab, and take turns demonstrating the movements depicted in Figure 8.11 with your lab partner. Notice that joint movements usually occur in pairs. A movement away is complemented with a movement that returns the body part to the original point.

2. **Abduct and Adduct:** Perform these joint movements using the following joints: wrist, hip, and shoulder. To *abduct* means to move away from midline. Think of abduct to mean "take away," and adduct to "add" to the midline of the body.

3. **Rotate and Circumduct:** Perform these movements from the following joints: shoulder and hip. Notice that rotation is a tighter, smaller pinpoint move than circumduction.

4. **Flex, Extend, and Hyperextend:** Use the neck, elbow, and knee joints to perform these movements. Hyperextension is not a natural anatomical position, so perform it carefully.

5. **Supinate and Pronate:** This palm up or palm down motion is mostly due to rotation of the radius on the ulna. A good memory trick is to visualize that your hand must be **supinated** to hold a bowl of **soup**.

Perform the activities in Box 8.2 and record the joint movements used to perform each activity. ■

Box 8.2 **Synovial Joint Movements**		
Activity	**Joint Movement**	**Bones Involved**
Lift a cup		
Put down a cup		
Twist your arm (turn a doorknob)		
Turn your palm upward		
Bend your head way back		
Move a leg to the side		
Move a leg to the center		
Turn your palm downward		

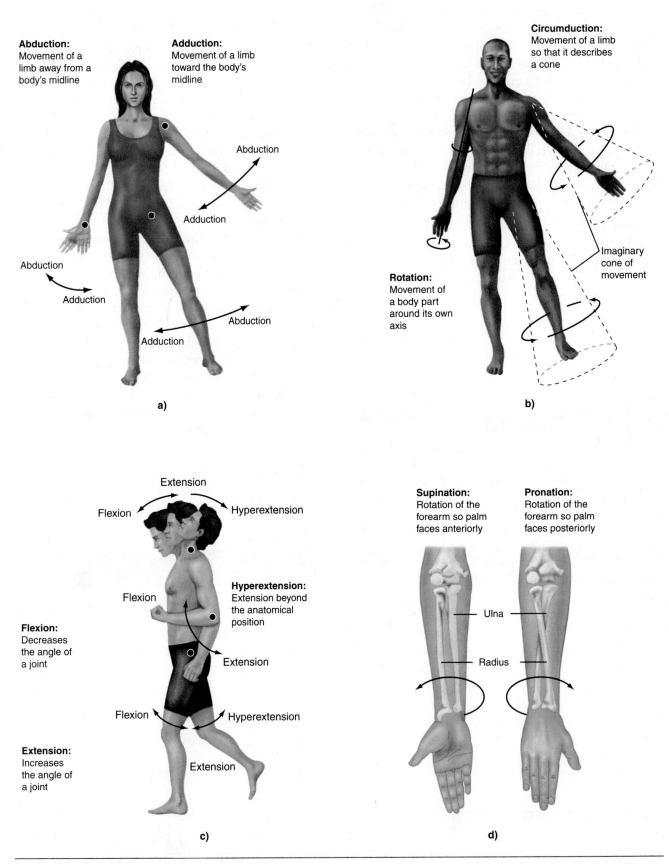

FIGURE 8.11   Synovial joint movements. a) Abduction and adduction. **b)** Rotation and circumduction. **c)** Flexion, extension, and hyperextension. **d)** Supination and pronation.

THE SKELETAL SYSTEM

Critical Thinking and Review Questions

not done!

1. Label the parts of the human skeleton.

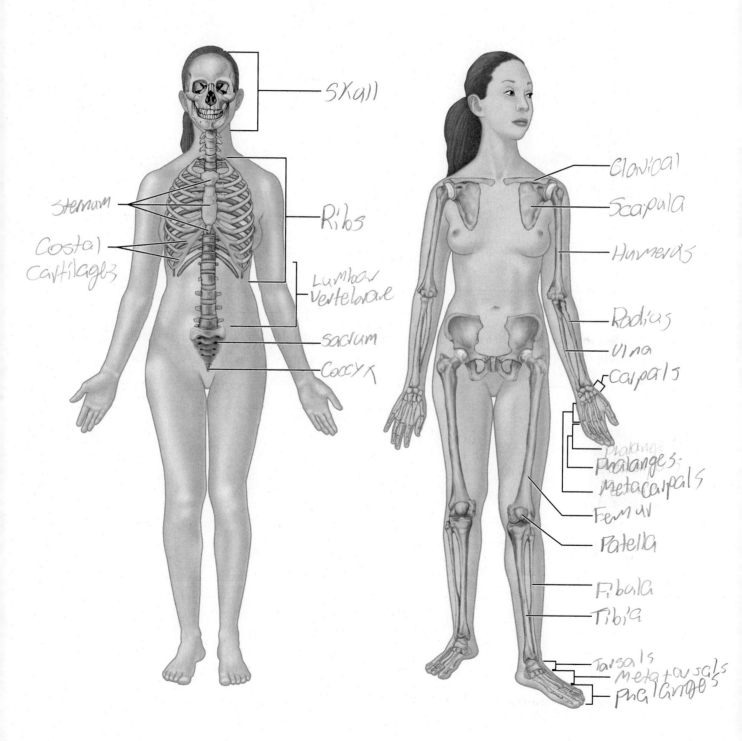

- Skull
- Sternum
- Costal Cartilages
- Ribs
- Lumbar Vertebrae
- Sacrum
- Coccyx
- Clavical
- Scapula
- Humerus
- Radius
- Ulna
- Carpals
- Phalanges
- Metacarpals
- Femur
- Patella
- Fibula
- Tibia
- Tarsals
- Metatarsals
- Phalanges

2. During adolescence, our bones become longer and heavier. What is the role of the osteoclasts in counteracting this additional body weight?

_____

_____

_____

3. Imagine a line drawn from the middle of the lower jaw to the back of the head as shown below.

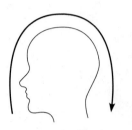

   a. Identify four facial bones on this midline.

   _____

   _____

   b. Identify four cranial bones on this midline.

   _____

   _____

4. What are the bones that form the jaw? Open your mouth. Did you elevate the maxilla or depress the mandible?

_____

_____

5. Label the parts of the vertebral column.

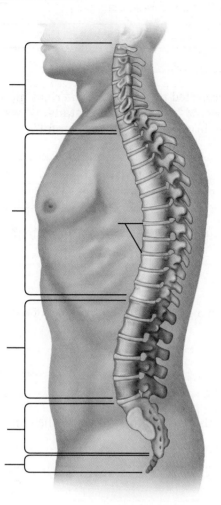

6. Why does the human vertebral column have four curves? Speculate on the effects of having a perfectly straight backbone.

_____

_____

_____

_____

7. Refer to Figure 8.6, Abnormal spinal curvatures. Which condition would affect walking the most? Why?

_____

_____

_____

8. Identify the three types of ribs. Which ones break most easily? Why?

_____

_____

_____

9. Why is the pectoral girdle more free than the pelvic girdle?

    _____

    _____

    _____

10. A hip replacement is not performed properly, and the right hip is not aligned to the left hip by just 1 cm. What would be the effects on standing, walking, and climbing stairs? Do you think this condition might be quite painful?

    _____

    _____

    _____

11. Why is falling more dangerous for the elderly than the young?

    _____

    _____

12. Name the opposite joint movement and give an example of each.

    a. Supinate _____

    b. Adduct _____

    c. Flex _____

13. For each joint, name a bone that articulates with the following bone:

    shoulder: humerus with   _____

    elbow: radius with   _____

    wrist: radius with   _____

    hip: coxa with   _____

    knee: femur with   _____

    ankle: tibia with   _____

The Muscular System

Objectives

After completing this exercise, you should be able to

1. Explain the role played by muscle tissue in the human body.
2. Describe the microstructure of skeletal muscle cells.
3. Describe how skeletal muscle is stimulated.
4. Identify the major skeletal muscles, and describe their functions.
5. Explain the basic mechanism of muscle contraction.
6. Explain the basic physiology involved in muscle contraction.

Materials for Lab Preparation

Equipment and Supplies

○ Compound light microscope
○ Gripper or other dynamometer capable of measuring the force of contraction
○ Assorted dumbbells of various weights
○ Cloth tape measure

Prepared Slides

○ Skeletal muscle (teased)
○ Neuromuscular junction
○ Muscle (cross section)

Models

○ Muscle cell showing parts of a myofibril
○ Muscle cell and motor neuron showing the neuromuscular junction
○ A whole muscle and/or skinned chicken leg with muscle attached to bone
○ The major human muscles

Introduction

As you learned in Exercise 5, Tissues, muscle tissue is unique in its ability to shorten or contract. Contraction of muscle cells is used by the human body to do work in a variety of ways. **Smooth muscle** is used to push things through many internal organs, and it is also used to regulate the pressure and flow of blood throughout the body. **Cardiac muscle** is formed into a hollow structure that is used to pump blood. The heart is a major topic, which is covered later in this lab manual. Finally, **skeletal muscle** is the voluntary muscle tissue that is typically attached to the skeleton for movement. This exercise focuses on skeletal muscle tissue and the organs it forms. However, it may be useful for you to review the muscle tissue section of Exercise 5, Tissues, to help put this exercise in proper perspective.

Skeletal Muscle Cell Microstructure

In Exercise 5, you learned that skeletal muscle cells are **striated,** that is, they have a cross-striped appearance (Figure 9.1a). The reason for these alternating bands of light and dark areas is a precisely arranged pattern of thin and thick contractile proteins, or **myofilaments.** The thin myofilaments are mostly made of a protein called **actin,** and the thick myofilaments are mostly made of a protein called **myosin.** These myofilaments are bundled together within the muscle cell into structures called **myofibrils** (Figure 9.1b).

Figure 9.2 depicts how these proteins are arranged in a myofibril. Note that the area where the thick proteins are located has a darker appearance, while the

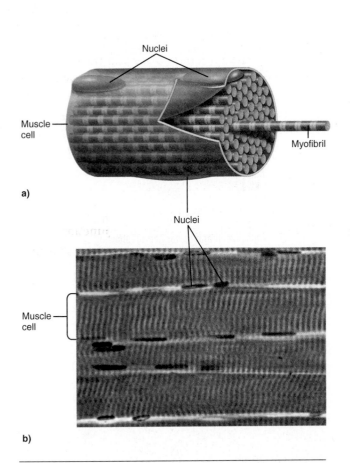

a)

b)

FIGURE 9.1 **Muscle cells. a)** A single skeletal muscle cell contains many individual myofibrils and has more than one nucleus. **b)** A photograph (×2,000) of portions of several skeletal muscle cells.

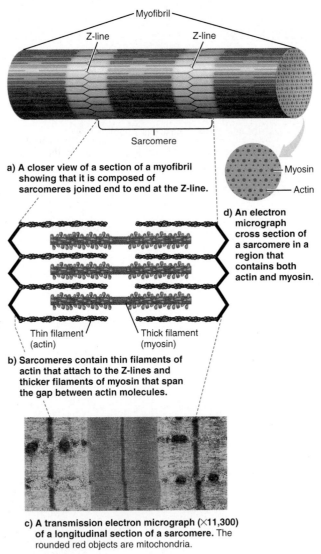

a) A closer view of a section of a myofibril showing that it is composed of sarcomeres joined end to end at the Z-line.

d) An electron micrograph cross section of a sarcomere in a region that contains both actin and myosin.

b) Sarcomeres contain thin filaments of actin that attach to the Z-lines and thicker filaments of myosin that span the gap between actin molecules.

c) A transmission electron micrograph (×11,300) of a longitudinal section of a sarcomere. The rounded red objects are mitochondria.

FIGURE 9.2 **a)** Structure of a myofibril. **b)** Structure of a sarcomere. **c)** Photomicrograph of a sarcomere. **d)** Cross section of a sarcomere showing actin and myosin.

area containing the thin proteins has a lighter appearance. Also note the jagged actin backbone, which is called the **Z-line.** The area between any two Z-lines is the basic contractile unit of skeletal muscle, called the **sarcomere.**

ACTIVITY 1

Learning Skeletal Muscle Microstructure

Materials for This Activity

Compound light microscope

Prepared slide of skeletal muscle (teased)

Model of muscle cell showing parts of a myofibril

View a slide of skeletal muscle, and carefully focus and adjust the light so that you can see the striations, or alternating light and dark areas. The light areas are where we mainly have the thin actin proteins. The darker areas are where we mainly have the thick myosin proteins. Now view a model of a muscle cell showing the details of a myofibril. Identify the thick and thin myofilaments, Z-lines, and sarcomeres. Sketch a sarcomere in the space provided below, labeling the previously mentioned structures. ■

Stimulation of Skeletal Muscle Cells

Nearly all skeletal muscle is under the direct control of our conscious brain, which is why it is also referred to as *voluntary* muscle. Although the diaphragm and some of the muscles in the neck are sometimes exceptions, we usually consciously direct movement of skeletal muscles. The **motor neuron** is the nerve cell that carries the message from the brain to the muscle cells we want to stimulate. The specialized area where the motor neuron interacts with a muscle cell is called the **neuromuscular junction.**

Figure 9.3 illustrates a neuromuscular junction. The neuron's long, slender process, the **axon,** may branch and lead to more than one muscle cell. A motor neuron and all of the muscle cells that it will stimulate is called a **motor unit.** Also see Figure 9.4, which shows this branching of the motor neuron.

Also note that the motor neuron and muscle cell do not quite touch, leaving a space in between called the **synaptic cleft.** Unlike electrical wires and the devices they supply, the stimulus from a motor neuron is brought to the muscle cell by using chemicals called **neurotransmitters.** In the case of most muscle cells, the neurotransmitter is acetylcholine.

ACTIVITY 2

Learning About the Neuromuscular Junction

Materials for This Activity

Compound light microscope

Prepared slide of a neuromuscular junction

Model of a muscle cell and motor neuron showing the neuromuscular junction

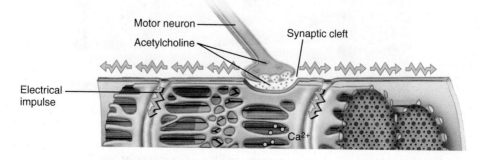

FIGURE 9.3 **The neuromuscular junction.**

View a slide and/or a model of the neuromuscular junction. Note the branching of the motor neuron to several muscle cells (Figure 9.4a–b) and the small gap between the neuron and the plasma membrane of the muscle cell. Make a sketch of the neuromuscular junction in the space below. ■

Whole Skeletal Muscle Organization

Motor neurons may stimulate only a small percentage of skeletal muscle cells when lifting or moving something that is lightweight. Nonetheless, the muscle must contract as a whole. To accomplish this, the muscle cells must be fastened together with connective tissue.

Figure 9.5 illustrates how a whole muscle is organized. Note the individual muscle cells, each one surrounded by a thin sheath of connective tissue (called the endomysium). Muscle cells are bundled into **fascicles** and are also surrounded by connective tissue (called the perimysium). Finally, all of the fascicles are wrapped together by connective tissue

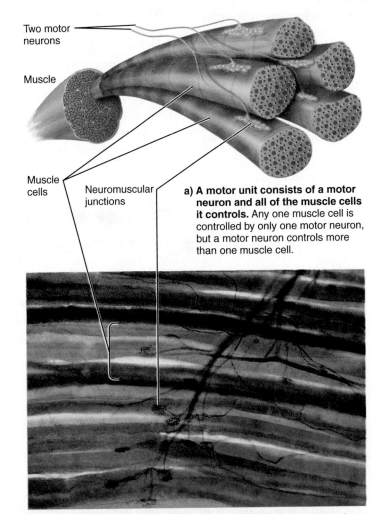

Two motor neurons

Muscle

Muscle cells

Neuromuscular junctions

a) **A motor unit consists of a motor neuron and all of the muscle cells it controls.** Any one muscle cell is controlled by only one motor neuron, but a motor neuron controls more than one muscle cell.

b) **Photograph of the muscle cells in a motor unit, showing branches of the motor neuron and neuromuscular junctions.**

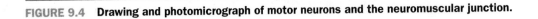

FIGURE 9.4 **Drawing and photomicrograph of motor neurons and the neuromuscular junction.**

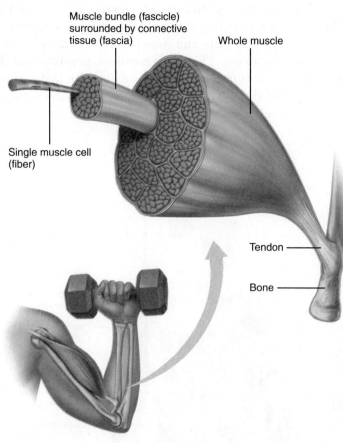

Muscle bundle (fascicle)
surrounded by connective
tissue (fascia)

Whole muscle

Single muscle cell
(fiber)

Tendon

Bone

FIGURE 9.5   Organization of whole muscle.

(called the epimysium). These connective tissues typically merge to form a **tendon,** which attaches the muscle to a bone or some other structure.

ACTIVITY 3

Observing Whole Muscle Organization

Materials for This Activity

Compound light microscope

Model of a whole muscle and/or skinned chicken leg with muscle attached to bone

Prepared slide of muscle (cross section)

1. Observe the muscle model and/or a skinned chicken leg. Observe the connective tissue, which surrounds the muscle and tapers down to the tendon. If using a chicken leg for examination, feel the muscle, fascia, and tendons. *Note:* Wash your hands after handling the chicken leg to prevent the spread of bacteria.

2. View a prepared slide of a muscle cross section. Note that the muscle cells are grouped into bundles (fascicles) and are surrounded by connective tissue. Sketch the muscle cross section in the space provided below, labeling the fascicles, perimysium, and epimysium. ■

Gross Anatomy of Skeletal Muscles

Muscles are most commonly studied by learning their **origin** (usually the stationary or anchoring point of attachment), **insertion** (usually the more movable point of attachment), and **action** (what movement they cause when they contract). This movement is described using the terminology covered in the joints section of Exercise 8. Review Figure 8.11 to ensure that you are familiar with the terms that describe movement at a joint.

ACTIVITY 4

Observing the Gross Anatomy of Skeletal Muscles

Materials for This Activity

Models displaying the gross anatomy of the major human muscles

1. Observe the models of human muscles available for study in your laboratory, and consult Figure 9.6 to aid you in identifying the selected muscles.
2. Complete the final column (Example of Use) in Table 9.1 using the knowledge you have obtained so far in this lab. Figure 9.6 is also a good reference. Begin by visualizing the location of the muscles listed in the first column of the table on your own body. Contract that muscle and note the type of movement it produces. Think of a common, everyday use for this muscle. Write this use in the last column of the table. The first muscle, pectoralis major, serves as an example. ■

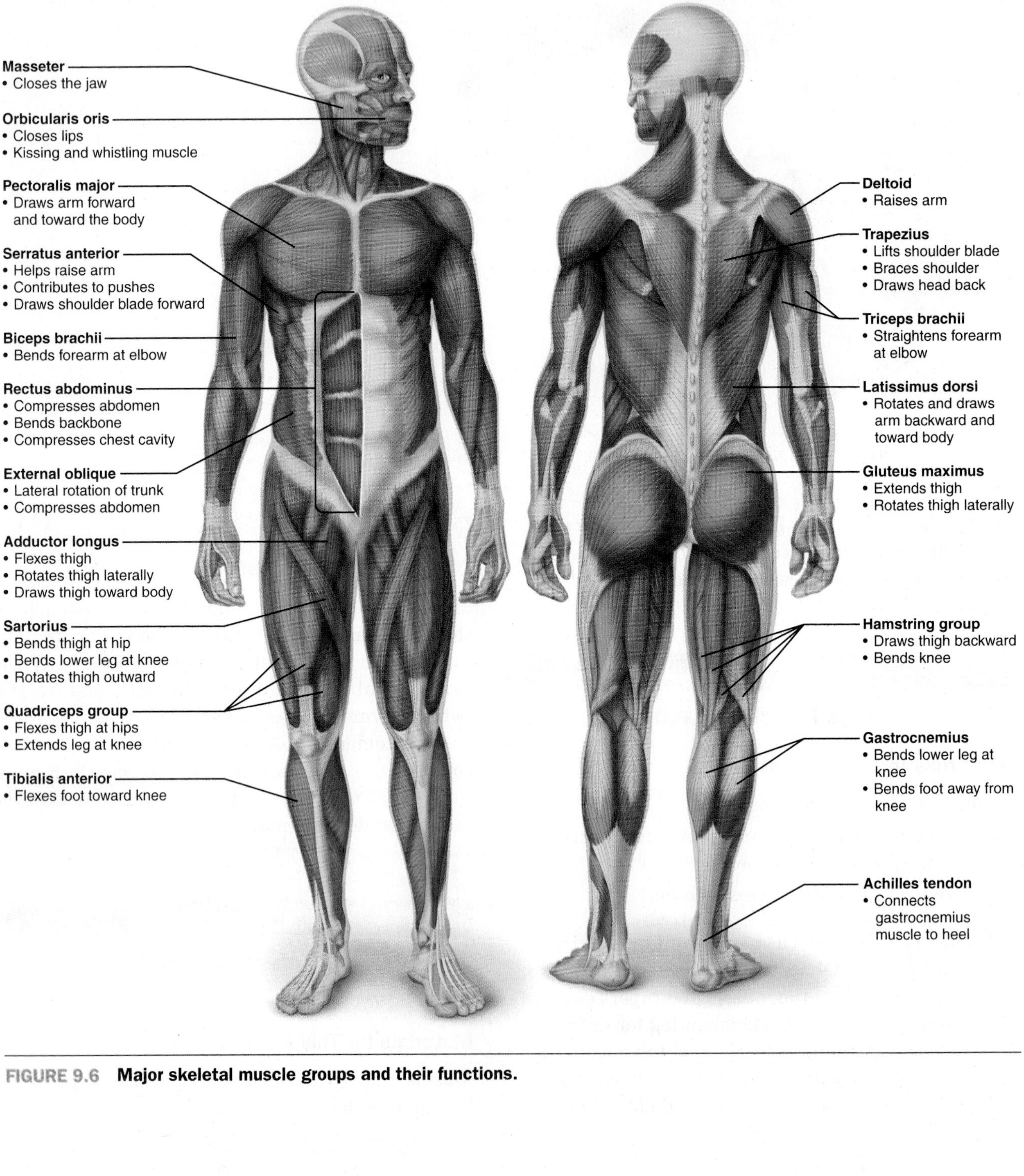

Masseter
• Closes the jaw

Orbicularis oris
• Closes lips
• Kissing and whistling muscle

Pectoralis major
• Draws arm forward and toward the body

Serratus anterior
• Helps raise arm
• Contributes to pushes
• Draws shoulder blade forward

Biceps brachii
• Bends forearm at elbow

Rectus abdominus
• Compresses abdomen
• Bends backbone
• Compresses chest cavity

External oblique
• Lateral rotation of trunk
• Compresses abdomen

Adductor longus
• Flexes thigh
• Rotates thigh laterally
• Draws thigh toward body

Sartorius
• Bends thigh at hip
• Bends lower leg at knee
• Rotates thigh outward

Quadriceps group
• Flexes thigh at hips
• Extends leg at knee

Tibialis anterior
• Flexes foot toward knee

Deltoid
• Raises arm

Trapezius
• Lifts shoulder blade
• Braces shoulder
• Draws head back

Triceps brachii
• Straightens forearm at elbow

Latissimus dorsi
• Rotates and draws arm backward and toward body

Gluteus maximus
• Extends thigh
• Rotates thigh laterally

Hamstring group
• Draws thigh backward
• Bends knee

Gastrocnemius
• Bends lower leg at knee
• Bends foot away from knee

Achilles tendon
• Connects gastrocnemius muscle to heel

FIGURE 9.6 **Major skeletal muscle groups and their functions.**

Table 9.1 **Selected Skeletal Muscles**

Muscle	Location	Function	Example of Use
Pectoralis major	Mostly under the breast area; spans from sternum to humerus	Pulls the arm forward and down	Hugging someone around the waist
Serratus anterior	On the side of the body under the armpit (axillary) area	Stabilizes the shoulder and contributes to arm movements such as pushing	
Biceps brachii	On the anterior side of the upper arm (humerus)	Bends (flexes) the arm at the elbow joint	
Rectus abdominis	Runs from the ribs to the pelvis on the anterior side of the body	Bends (flexes) the backbone; compresses the abdomen	
External oblique	Lateral sides of the abdomen; angles down toward the center of the abdomen	Acts the same as the rectus abdominis when both are contracted together; rotates the trunk laterally	
Adductor longus	Inner thigh near the groin	Flexes the thigh; rotates thigh laterally; draws the thigh toward the body	
Sartorius	Long, belt-shaped muscle that runs diagonally across the thigh from hip to inside of the knee area	Bends the thigh at the hip; bends the lower leg at the knee; rotates the thigh outward	
Quadriceps (femoris) group	Group of four large muscles on the anterior side of the thigh	Raises (flexes) the thigh; extends the lower leg at the knee	
Tibialis anterior	In the front and slightly to the side of the tibia	Raises (flexes) the foot	
Deltoid	Covers most of the shoulder	Raises (abducts) the arm	
Trapezius	Large, diamond-shaped muscle of the upper back	Pulls the head back (extension); moves and braces the shoulder	
Triceps brachii	On the posterior side of the upper arm (humerus)	Extends the arm at the elbow joint	
Latissimus dorsi	Large muscle of the middle back	Pulls the arm back and down (adducts the arm)	
Gluteus maximus	Covers most of the buttocks area	Kicks and rotates the leg backward (extends the leg)	
Hamstring group	Group of three large muscles on the posterior side of the thigh	Flexes the leg (bends the knee)	
Gastrocnemius	Posterior calf muscle	Bends the foot downward (extension)	

Contraction of Skeletal Muscles

As explained earlier, skeletal muscles contract when stimulated by neurons. But what exactly makes the cells shorten or contract? The answer lies in our previous explanation of the microstructure of skeletal muscle cells. The thick myosin proteins have small projections called **myosin heads** that grab the actin fibers and pull them toward the center of the sarcomere (Figure 9.7). Thus, the myofilaments (actin and myosin) appear to slide along one another. The myosin heads have enzymes that split ATP molecules. It is the energy from the splitting and release of ATP that allows the myosin heads to cock back and then pull the actin fibers. Consult your textbook for a more thorough description of the **sliding filament mechanism** of muscle contraction.

The ATP-splitting enzymes seem to work more efficiently at elevated temperatures, which is one of the several reasons why a muscle that has been properly warmed up by light exercise performs better and is less likely to suffer injury.

ACTIVITY 5

Measuring the Force of Skeletal Muscle Contraction

Materials for This Activity

Gripper or other dynamometer capable of measuring the force of contraction

Work in pairs. Obtain a gripper or other dynamometer, and read the directions for use.

1. Squeeze the gripper five times, and record the average strength of contraction.

   _____

2. Rest for one minute, squeeze the gripper five times, and record the average strength of contraction.

   _____

3. Rest for one minute, squeeze the gripper five times, and record the average strength of contraction.

   _____

   Which average was the lowest?

   _____

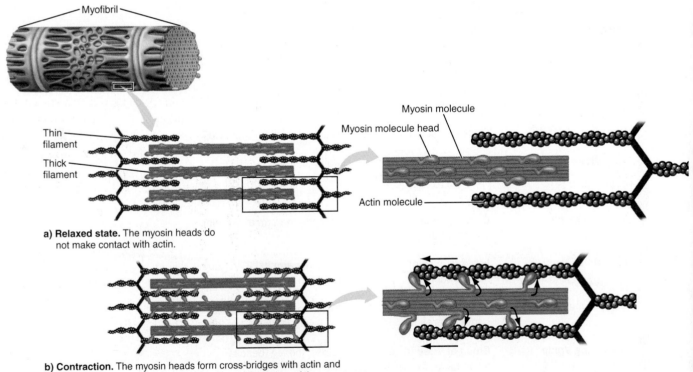

a) **Relaxed state.** The myosin heads do not make contact with actin.

b) **Contraction.** The myosin heads form cross-bridges with actin and then bend, pulling the actin filaments toward the center of the sarcomere.

FIGURE 9.7 Sliding filament mechanism of contraction. a) In the relaxed state, the myosin heads do not make contact with actin. **b)** During contraction, the myosin heads form cross-bridges with actin and bend, pulling the actin filaments toward the center of the sarcomere.

Propose a reason for your results.

_____

_____

_____

_____ ∎

Energy Use by Skeletal Muscles

As explained earlier, myosin heads require a molecule of ATP in order to cock back and then move the actin fibers. Considering that there are many millions of myosin heads in the smallest of muscle cells, contraction clearly requires an enormous expenditure of energy. In addition, there are finite limits to the amount of ATP that a cell can store. These two facts taken together suggest that muscle cells will cease to operate if they cannot rapidly replace the ATP they use. Creatine phosphate and an inefficient mechanism of sugar metabolism called anaerobic respiration provide the fastest ways to regenerate ATP. But these energy sources are also quickly depleted, and anaerobic respiration produces **lactic acid.**

This is not all bad. By triggering pain receptors, lactic acid tells us that we should slow down, if possible, and allow the stressed muscles to recover. Lactic acid also dilates blood vessels that supply skeletal muscles, bringing in the additional nutrients and oxygen needed by the overtaxed muscle cells. However, once ATP levels fall too low in a muscle cell, its strength of contraction may weaken to the point where it cannot perform the task it is being called upon to do. This is referred to as **muscle fatigue.**

ACTIVITY 6

Measuring the Energy Use of Contracting Skeletal Muscles

Materials for This Activity

Assorted dumbbells of various weights

Cloth tape measure

Part 1

Work in pairs. Obtain a tape measure and dumbbell of sufficient weight that it will be challenging for the designated lifter.

1. Measure the arm of the designated lifter around the thickest portion of the upper arm. Do this during both relaxation (extended) and forced contraction (flexion) of the biceps brachii. Record your measurements.

   Relaxed/extended   _____

   Contracted/flexed   _____

2. Perform 10 curls (flexing the arm by contracting the biceps brachii) with the arm that was measured in the previous step. If you cannot do 10 repetitions (reps), switch to a lighter weight; if 10 reps did not noticeably tire your arm, switch to a heavier weight until 10 reps leave your arm feeling at least slightly tired. Measure the arm as you did in the previous step, and record your measurements.

   Relaxed/extended   _____

   Contracted/flexed   _____

Explain any differences you note in the measurements made in steps 1 and 2. *Note:* If you did not obtain any difference, repeat step 2 until you do.

   _____

   _____

Part 2

Continue to work in pairs, and use the same dumbbell used in the previous part.

1. Using the same arm as in the previous experiment, perform as many curl repetitions as you can until discomfort or a mild burning sensation is felt in the biceps brachii muscle. Have your partner record the number of reps, and rest for no more than 10 seconds. *Note:* If you can do more than 20 reps, you should switch to a more challenging weight.

   Number of reps   _____

2. After no more than 10 seconds of rest, perform as many curl repetitions as you can until the same level of discomfort or mild burning sensation is felt in the biceps brachii muscle. Have your partner record the number of reps.

   Number of reps   _____

   Which number was lower?   _____

   Propose a reason for your results.

   _____

   _____ ∎

EXERCISE

9

THE MUSCULAR SYSTEM

Critical Thinking and Review Questions

1. Match the item to its basic description:

   _____ Z-line   a.   the bundle of contractile proteins in a muscle cell

   _____ Actin   b.   the thick myofilaments

   _____ Myosin   c.   the jagged "actin backbone"

   _____ Sarcomere   d.   the thin myofilaments

   _____ Myofibril   e.   the contractile unit between Z-lines

2. Define the following terms:

   Motor unit _____

   _____

   Synaptic cleft _____

   _____

   Neurotransmitter _____

   _____

   Fascicle _____

   _____

3. Match the muscle to its function:

   _____ Biceps brachii   a.   pulls the arm back and down (adducts the arm)

   _____ Gluteus maximus   b.   raises (abducts) the arm

   _____ Deltoid   c.   pulls the head back (extension); moves and braces the shoulder

   _____ Latissimus dorsi   d.   bends (flexes) the arm at the elbow joint

   _____ Trapezius   e.   kicks and rotates the leg backward (extends the leg)

4. Why does a muscle perform better when it is warmed up?

   _____

   Summarize the experiment you performed to demonstrate this principle.

   _____

   _____

   _____

5. Briefly explain the connection between lactic acid and "pumping up" a muscle.

   Lactic acid dilates vessels entering muscles to
   increase blood flow and pumps more blood in a
   muscle

6. What is muscle fatigue?

When a muscle cell has used all, or almost all ATP and can't contract

Summarize the experiment you performed to demonstrate this principle.

We used Activity, Part 2 to exercise a muscle until it got tired and used all or almost all atp.

7. Label the following diagram.

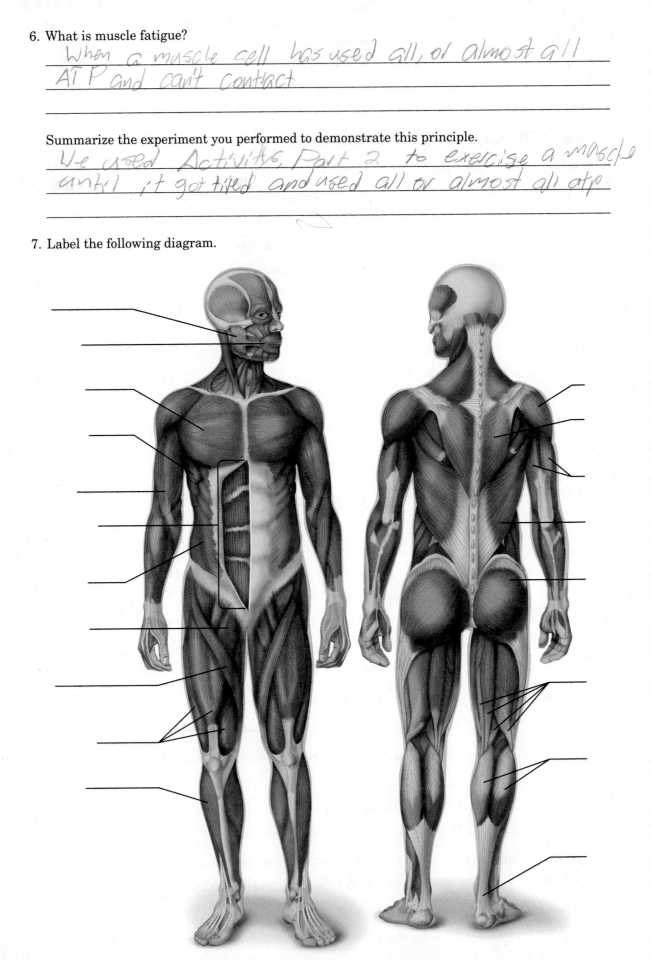

10

The Nervous System I: Organization, Neurons, Nervous Tissue, and Spinal Reflexes

Objectives

After completing this exercise, you should be able to

1. Explain the organization of the nervous system.
2. Describe the supporting cells, and explain basically what they do.
3. Describe the different classes of neurons, and explain basically what they do.
4. Explain how neurons work together to transmit information.
5. Explain how a simple spinal reflex works.

Materials for Lab Preparation

Equipment and Supplies
○ Compound light microscope
○ Reflex hammer

Prepared Slides
○ Giant multipolar neurons
○ Peripheral nerve (cross section)

Models
○ Vertebral column with nerves and plexuses
○ Sensory and motor neurons

Introduction

As you learned in Exercise 5, **neurons** are used to rapidly transmit information throughout the body. However, there is more to nervous tissue than that. First, not all cells found in nervous tissue transmit impulses. There are also supporting cells called the **neuroglia.** Second, there are different kinds of neurons, each modified for the job it must do. Finally, neurons are only useful when connected to other neurons. So, we must first understand how the nervous system is organized and how its neurons work together.

Organization of the Nervous System

The nervous system is divided into two major segments: the **central nervous system (CNS)** and the **peripheral nervous system (PNS).** The CNS contains the **brain** and **spinal cord,** which are also referred to as **integration centers** because they process incoming information and effect a response. The incoming information and response must be carried to and from the appropriate body parts by the PNS. **Nerves** are the typically long structures that carry the extensions of neurons to and from often-distant body parts.

The PNS is divided into **sensory** (input) and **motor** (output) divisions (Figure 10.1). The sensory division brings impulses, or **action potentials,** from receptors to the CNS. The motor division brings action potentials from the CNS to **effectors** such as muscles and glands. This motor division is further divided into **somatic** and **autonomic** divisions. The somatic division's nerves primarily bring information to skeletal muscles, so it is often described as the *voluntary* motor division. As the name implies, the autonomic division mainly stimulates *involuntary* structures such as smooth muscle and glands. The autonomic division will be covered in detail in Exercise 11.

ACTIVITY 1

Identifying the Organization of the Nervous System

Consult Figure 10.1, and match the following items to their places in the nervous system:

_b_ Spinal cord   a. sensory division
_d_ Nerves to the stomach   b. CNS
_c_ Nerves to the deltoid   c. somatic motor division
_a_ Nerves from the skin   d. autonomic motor division
_b_ Brain

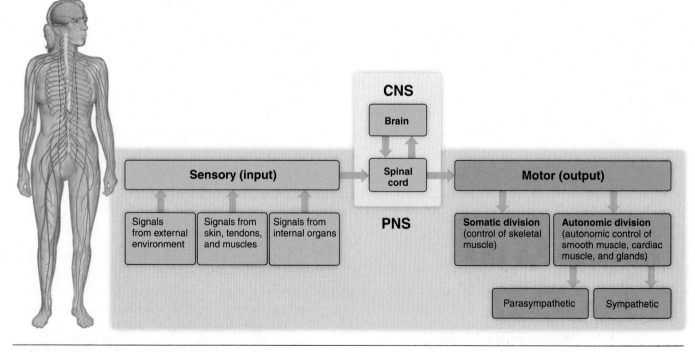

FIGURE 10.1 Organization of the nervous system.

Nervous Tissue

The **neuroglia,** or supporting cells of the nervous system, perform a variety of functions. Some, such as **astrocytes** and **microglial cells,** attach to neurons to support and connect them to each other and to other structures. Others, such as **oligodendrocytes** and **Schwann cells,** wrap an insulation-like material called **myelin** around neurons. They fulfill important accessory roles, but they do not conduct impulses. Myelin protects neurons and helps to speed impulse conduction by causing the impulse to skip, rather than slowly flow, down the axon. Consult your textbook for a more detailed description of the role played by myelin in the nervous system.

Neurons come in several varieties, but each has the same basic components (Figure 10.2). They have a large **cell body** that holds the nucleus, several short **dendrites** that pick up stimuli, and a long, slender **axon,** which can transmit action potentials over great distances.

Sensory neurons typically have a cell body attached to an axon with a single, short process (Figure 10.2). For this reason, sensory neurons that look like this are called *unipolar* neurons; that is, they have *one* process attached to the cell body. Some sensory neurons are attached to a short dendrite at one end and a short axon at the other end of the cell body. However, these *bipolar* neurons are rare and found only in certain specialized nervous tissues such as the retina of the eyes. These neurons carry information toward the CNS.

Motor neurons have several branched dendrites attached directly to the cell body and one axon running from the tapered end of the cell body (Figure 10.2). For this reason, motor neurons are also called *multipolar* neurons; that is, they have *many* processes attached to the cell body. These neurons carry information from the CNS to muscles and glands.

Interneurons resemble very short motor neurons and would also be considered structurally to be multipolar (Figure 10.2). These short neurons connect other neurons together in the CNS.

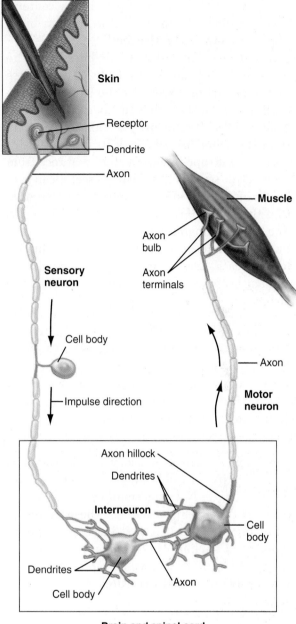

Brain and spinal cord

FIGURE 10.2 **Types and components of neurons.** Note the single process attached to the sensory neuron *(unipolar)* and numerous processes attached to the motor neuron and interneuron *(multipolar).*

ACTIVITY 2

Observing Neurons and Neuroglia

Materials for This Activity

Compound light microscope

Prepared slide of giant multipolar neurons

Models of sensory and motor neurons

1. View a slide of giant multipolar neurons, which is sometimes labeled "ox spinal cord smear." This slide may look familiar as one of the slides selected for the "Observing Cellular Diversity" activity in Exercise 3, or from Figure 5.13.

2. Look more carefully at the large blue-stained cells. Although you may be able to see them with

your low-power or scanning-power lens, center a clearly visible cell, and view it using your high-power (high-dry) objective lens.

3. Note the nucleus in the cell body and the branches coming off the cell body. The branches are all dendrites, except for the longest, thickest branch. This branch is the axon. However, it is often difficult to identify the axon separately from the other branches, as its relative thinness, once it tapers down a short distance from the nucleus, makes it difficult to see; it can also break off during slide preparation. Also note that there are many smaller cells in the smear. These are neuroglial cells.

4. Sketch a small portion of this slide in the following space, labeling the cell bodies, dendrites, and axons where you can discern them from the dendrites.

5. Now view models of sensory and motor neurons. Identify the cell body, axon, dendrites, and myelin sheath. Note the small gaps or "cracks" between the myelinated areas. These **nodes of Ranvier** allow for the fast impulse skipping mentioned earlier. Make a general sketch of these models in the following space, labeling the previously mentioned structures. ▪

The PNS and Nerves

Nerves are simply bundles of axons wrapped by connective tissue into structures that resemble strings and function a little like telephone wires. The layout

of a nerve is remarkably similar to that of a muscle. Individual axons are wrapped by a thin layer of connective tissue, a group of axons are wrapped into **fascicles,** and the fascicles are wrapped into the whole nerve.

There are two categories of nerves. They are the **cranial nerves,** which are attached to the brain, and the **spinal nerves,** which are attached to the spinal cord. More details about these nerves and the way they are connected to the brain and spinal cord will follow in the next exercise. For now, we will focus on the spinal nerves specifically where they have exited the spinal cord (Figure 10.3).

Note that the nerves are initially named for the region of the vertebral column from which they exit. That is, they are named **cervical nerves** in the neck, **thoracic nerves** in the chest region, **lumbar**

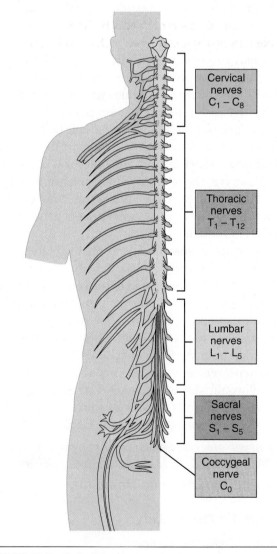

Cervical nerves $C_1 - C_8$

Thoracic nerves $T_1 - T_{12}$

Lumbar nerves $L_1 - L_5$

Sacral nerves $S_1 - S_5$

Coccygeal nerve C_0

FIGURE 10.3   Spinal nerves.

nerves in the lower back, **sacral nerves** in the area of the sacrum, and a pair of coccygeal nerves.

ACTIVITY 3

Learning About the Spinal Nerves

Materials for This Activity

Compound light microscope

Model of the vertebral column with nerves and plexuses

Prepared slide of a peripheral nerve (cross section)

1. Observe the model of the vertebral column with nerves. Find the cervical, thoracic, lumbar, and sacral nerves.
2. Look carefully at Figure 10.4, and then view the peripheral nerve cross section slide.
3. Use the lower-power magnification, and note how the nerve is arranged in fascicles similar to muscle as described in Exercise 9. Then center the lens area on one fascicle so that you may switch to your high-power (high-dry) objective lens. Now observe these individual axons. The axons will appear as dark dots surrounded by a small area of "empty" space. This space is not empty at all but is where the myelin sheath wraps around the axon. Because myelin does not easily pick up the stain used to color the slide, the area containing the myelin *appears* empty.
4. In the space provided, sketch the low-power view of the nerve and the high-power view of the axons. Compare your observations with Figure 10.4. ∎

Low-power view of nerve.

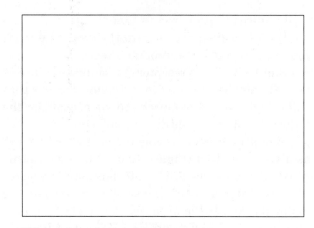

High-dry view of several axons.

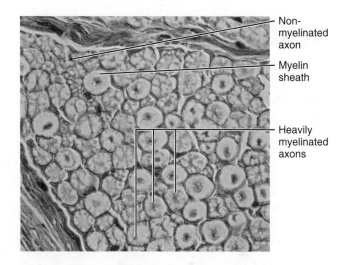

FIGURE 10.4   **A photomicrograph of a cross section of a peripheral nerve (560×).**

Spinal Reflexes

Although the spinal cord is not covered until Exercise 11, it seems important and appropriate to finish this exercise with an example of a way that neurons are connected and used in the human body. A spinal reflex is the simplest example of this use.

We usually think of the CNS as the place where sensory information is processed so that a conscious decision can be made and action taken in response to an event. That is not always the case, however. There are situations where no thinking or decision making is necessary, only a rapid response. When we contact something harmful (e.g., touching a sharp object or a hot burner on a stove) or when we need to balance when walking, the extra second or more that the brain would take to process a response may be far too long. In these cases, the spinal cord is already prewired to connect a sensory impulse from a pain receptor to the motor neurons that will execute the appropriate response.

Similarly, stress receptors in joints and muscles may already be prewired to the motor neurons that will rapidly stimulate muscles to compensate for the failure of balance. This direct connection of sensory and motor neurons in the spinal cord is referred to as a **spinal reflex** (Figure 10.5). As shown in this illustration, the stimulus information is immediately transferred from the sensory neuron to the motor neuron. In some cases, an interneuron may also be involved in the pathway. However, the result is the same—rapid communication of a response without waiting for a decision from the brain.

ACTIVITY 4

Testing Spinal Reflexes

Materials for This Activity

Reflex hammer

Part 1: Knee Jerk Reflex

1. Work in pairs, and obtain a rubber reflex hammer. Have one person sit on a chair or table so that the lower legs dangle freely, with a few inches of space below and behind the foot. First, find the softer area directly below the patella (kneecap). Then, tap this softer area with the reflex hammer. This may take several attempts. Describe your observations.

   <u>Slight jerk on knee</u>

2. Consider what was happening to the nervous system of the subject during this reflex, and compare it to the reflex in Figure 10.5. How is this reflex similar?

   <u>Stimuli under the knee</u>
   <u>which tells brain to move</u>

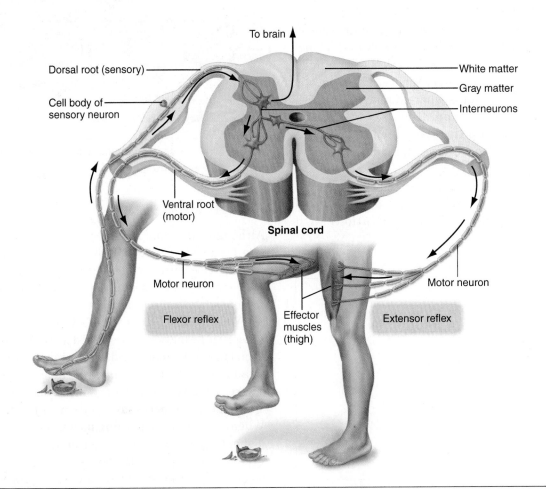

FIGURE 10.5 A spinal reflex: pain withdrawal reflex.

To brain

Dorsal root (sensory)

Cell body of
sensory neuron

White matter

Gray matter

Interneurons

Ventral root
(motor)

Spinal cord

Motor neuron

Motor neuron

Flexor reflex

Effector
muscles
(thigh)

Extensor reflex

How is this reflex different?

It does not flex

3. Repeat this procedure while the subject concentrates on some mental activity (such as counting backward from 100). How does the reflex response compare with the first trial?

There was no response

Part 2: Ankle Jerk Reflex

1. Similar to Part 1, the foot should be freely suspended. Use the wider edge of the reflex hammer to strike the Achilles tendon about one inch above the heel bone. Repeat several times and describe your observations below.

The foot moves each time

2. Compare this reflex to the one in Part 1 and the reflex illustrated in Figure 10.5. How is this reflex similar to each of the above-mentioned reflexes?

It was more of a slight movement, not extension

From which *one* is it different? How so?

Jerk seemed more painful than the knee jerk

Why is it important for both pain withdrawal and balance reflexes to have pathways using heavily myelinated neurons? *Note:* It would be unethical for us to ask you to actually perform a pain withdrawal reflex, so base your answer upon your memories of experience with pain stimuli.

Pain always gets a pain reception, but balance

THE NERVOUS SYSTEM I
Critical Thinking and Review Questions

1. List the organizational parts of the CNS and PNS, and define what each part does.

   CNS: brain, spinal, nerves

   PNS: effector, somatic, autonomic

2. Define the following terms:

   Neuroglia   supporting cells

   Schwann cell   fulfill important accessory roles

   Sensory neurons   short process

   Motor neurons   control most of the body, long branching

   Myelin   insulation like material

3. Match the following structures to their functions or descriptions:

   d   Cell body   a. transmits impulses over long distances

   a   Axon   b. are short connecting neurons in the CNS

   e   Dendrite   c. wrap myelin around an axon

   c   Schwann cells   d. contains the nucleus

   b   Interneurons   e. is the receptive region of the neuron

4. List the four main categories of spinal nerves.

   Cervical, lumbar, thoracic, sacral

5. a. Rearrange the following terms so that they reflect the proper order of a spinal reflex: *pain receptor, effector muscle, motor neuron, sensory neuron, spinal cord.*

   Sensory neuron, motor neuron, effector muscle, spinal
   cord, pain receptor

   b. Which of the above terms would be different if the reflex in question were a balance reflex? _Pain receptor_ What term would you substitute? _Interneurons_

6. Using the following figure, label the parts of the spinal reflex.

   1. _Nerve_
   2. _Dorsal Root_
   3. _Spinal Cord_
   4. _Ventral root_
   5. _motor neuron_

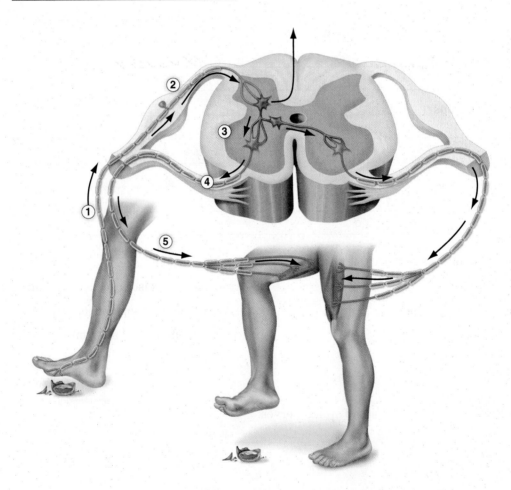

EXERCISE
11

The Nervous System II: The Spinal Cord, Brain, and Autonomic Nervous System

Objectives

After completing this exercise, you should be able to

1. Describe the meninges and cerebrospinal fluid.
2. Describe the structure of the spinal cord, and explain its basic functions.
3. Identify the selected parts of the brain.
4. Describe the major parts of the brain, and explain the basic function of each.
5. Explain the basic functions of the two parts of the autonomic nervous system.

Materials for Lab Preparation

Equipment and Supplies

○ Compound light microscope
○ Sheep brain with and without meninges (if possible)
○ Large, nonserrated knife
○ Dissection microscope

Prepared Slide

○ Spinal cord (cross section)

Models

○ Spinal cord (cross section)
○ Human brain

Introduction

In the previous exercise, you learned that the nervous system is made of many neurons connected together to communicate information. Integration centers are the complex areas of the nervous system where neurons interact to process and store information. The brain and spinal cord are these integration centers in humans. Before we specify how the brain and spinal cord work, we must first study how these delicate organs are protected.

The Meninges and Cerebrospinal Fluid

Three layers of connective tissue surround and protect the brain and spinal cord. These layers of connective tissue are called the **meninges** (Figure 11.1). The **dura mater** is the tough outer layer of connective tissue, the loose **arachnoid** (also called the **arachnoid mater**) is in the middle, and the soft, thin **pia mater** lies directly on the brain and spinal cord. These are the tissues that become infected with viral or bacterial meningitis, which can result in dangerous inflammation/swelling of the brain.

The meninges alone would not be completely effective in protecting delicate organs such as the brain and spinal cord; thus, these organs are also surrounded by a thin fluid called **cerebrospinal fluid (CSF).** The CSF is produced by specialized capillaries in the brain, and it resembles the fluid found around most of the tissues in the body (Figure 11.2). Consult your textbook for a more detailed description of CSF production.

The CSF fills hollow spaces called **ventricles** in the brain, and it also surrounds the brain and spinal cord in the space between the arachnoid and pia mater (Figure 11.2), where it acts as a shock absorber.

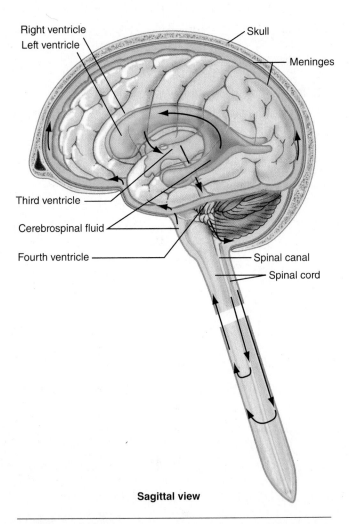

Sagittal view

FIGURE 11.2 **A spinal cord cross section showing dorsal and ventral roots.**

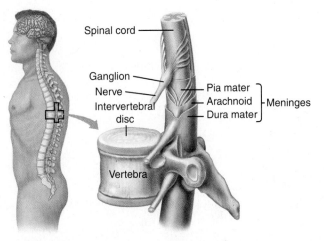

FIGURE 11.1 **The spinal cord and meninges.**

ACTIVITY 1

Observing the Meninges and CSF

Materials for This Activity

Sheep brain with meninges

Spinal cord model

1. Observe a model of the spinal cord. You will examine this model in more detail later, but for now, notice the relationship of the bone to the dura mater. Observe the arachnoid and pia mater if they are present on the model.

2. Observe a sheep brain with the dura mater still intact. Notice its toughness. Is the dura mater easy or difficult to tear?

   easy

   What would be the advantage of having this tissue between the brain and the bones of the skull?

   helps keep brain out

3. Use a probe to lift a portion of the dura mater to observe the loose arachnoid under it. Describe the arachnoid below.

   different color

From the information already provided in this exercise, what substance would you find in this loose tissue?

Cerebral spinal fluid

Does "form follow function" (i.e., does it make sense that the arachnoid is a loose tissue rather than a denser one)? Explain your answer:

Yes, since its directly on the brain

The Spinal Cord

The spinal cord is, in part, comparable to a huge nerve that carries information to and from the brain. But as you learned in Exercise 10, it also makes connections and handles reflexes. This suggests that we may find two very different functional components in the spinal cord, and we do. The **white matter** is packed with myelinated axons that rapidly transmit information to the brain, in ascending tracts, and down from the brain, in descending tracts. The **gray matter** is where we find the cell bodies of motor neurons, as well as interneurons, which make connections between neurons for effects such as reflexes (Figure 11.3).

The spinal cord also serves as the "junction box" for the spinal nerves, which split into dorsal and ventral roots just before attaching to the spinal cord (Figure 11.3). The **dorsal root** veers toward the back of the spinal cord and contains the sensory neurons. Most sensory neurons do not have their cell

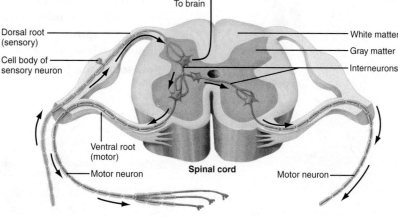

FIGURE 11.3   A spinal cord cross section showing dorsal and ventral roots.

bodies positioned near one end of the neuron as do motor neurons. Because of this, the cell bodies of sensory neurons are gathered together just outside the spinal cord in a bulging area of the dorsal root called the **dorsal root ganglion.** The **ventral root** veers toward the front of the spinal cord and contains the axons of motor neurons. Cell bodies of motor neurons are positioned near its end, and these cell bodies are found in the gray matter of the spinal cord (Figure 11.3). Thus, there is no ventral root ganglion.

Because it partly functions as a junction device for spinal nerves, the spinal cord is not uniform in thickness throughout its length. Instead, it is thicker where more neurons enter and exit and thinner where there is less traffic. These thicker areas, which must handle the extra neuron traffic from the arms and legs, are respectively called the **cervical enlargement** and **lumbar enlargement** (Figure 11.4).

Another interesting feature of the spinal cord is that it ends in the lumbar region of the vertebral column instead of extending all the way down into the sacral region. The spinal nerves that must exit from the sacrum and lower lumbar region are found projecting down in this area (Figure 11.4). This cluster of spinal nerves is called the **cauda equina** due to its resemblance to a "horse's tail."

Suggest a reason why the spinal cord appears to "come up short" in the vertebral column.

_____

_____

ACTIVITY 2

Examining the Spinal Cord

Materials for This Activity

Compound light microscope
Dissection microscope
Prepared slide of the spinal cord (cross section)
Model of a spinal cord cross section

1. Observe the model of a spinal cord cross section. Note the "butterfly-shaped" area of gray matter surrounded by white matter. Also note the dorsal and ventral roots of the spinal nerve and the dorsal root ganglion.
2. Using the dissection microscope, view a prepared slide of a spinal cord cross section. You should again be able to see the "butterfly-shaped" area of gray matter surrounded by white matter. Compare your slide to Figure 11.5.
3. Use your compound light microscope to view the same prepared slide. You may not be able to see the entire spinal cord in your field of view, even under the low power, so scan around the slide to become familiar with this view. Then, center

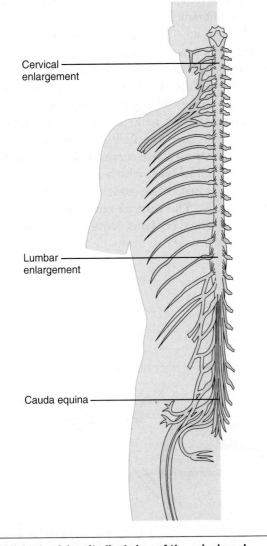

FIGURE 11.4 A longitudinal view of the spinal cord, showing the cervical and lumbar enlargements.

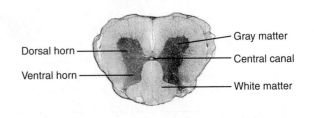

FIGURE 11.5 A photomicrograph of a spinal cord cross section.

the slide on the thickest area of gray matter and switch to a higher magnification. You should now be able to see the large cell bodies of motor neurons in this area of the gray matter. Are these neurons multipolar, bipolar, or unipolar?

The Brain

The brain is the complex integration center that processes, stores, and makes decisions about the information we obtain from the millions of sensory neurons in the body. There are several specialized regions of the brain, each representing a level of brain evolution and/or embryonic development. Although valid schools of thought differ on how to divide the regions of the brain for study, we will use the functional divisions of the **hindbrain, midbrain,** and **forebrain** found in your textbook.

The Hindbrain

The hindbrain, which makes up most of what is referred to by some as the **brain stem,** consists of the **medulla oblongata,** the **pons,** and the **cerebellum** (Figure 11.6). The medulla oblongata controls various autonomic life-support functions such as breathing (respiratory center), blood pressure (vasomotor center), and heart rate (cardiac center). The pons is a bridge connecting neurons from the spinal cord with higher brain centers.

The cerebellum coordinates basic and complex movements.

The Midbrain

Sometimes referred to as part of the brain stem, along with the pons and medulla oblongata, the midbrain relays information (mainly vision and hearing) to higher brain centers (Figure 11.6). It is also the control center for some movements and for regulating our degree of alertness or wakefulness. The midbrain and hindbrain are considered to be the most primitive parts of the brain.

The Forebrain

The forebrain contains the parts of the brain that are more highly evolved. The **thalamus, hypothalamus, cerebrum,** and various diffuse systems coordinate a huge array of sensory information, and they produce our complex responses (Figure 11.6).

Together, the thalamus and hypothalamus mostly constitute an area often referred to as the **diencephalon.** The thalamus receives a great deal of sensory information, and it is responsible for the initial processing of the information before relaying it on to the cerebrum for more precise interpretation. The hypothalamus is the center of highly sophisticated life-support systems. Control over body temperature, hunger, thirst, and other features critical to homeostasis is regulated by the hypothalamus.

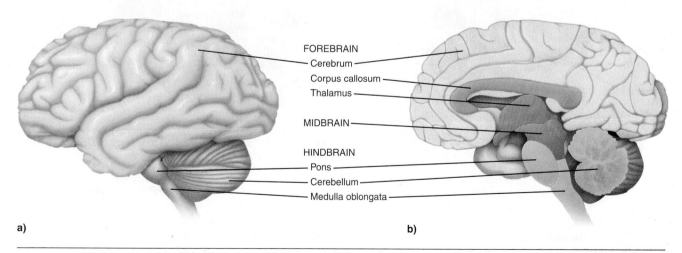

a)   b)

FOREBRAIN
Cerebrum
Corpus callosum
Thalamus

MIDBRAIN

HINDBRAIN
Pons
Cerebellum
Medulla oblongata

FIGURE 11.6   **The human brain. a)** A surface view. **b)** A midline sagittal section (front to back on a vertical plane).

The highly folded cerebrum is where sensory information is processed, stored, and sometimes forwarded for decision making (Figure 11.7a). The folding is a neat trick of Mother Nature to increase surface area on the outside (cortex) where the gray matter is found. This way, the amount of gray matter, which is used for storing and processing information, could be increased without expanding the overall size of the brain. Some of these folds separate the cerebrum into several **lobes,** each with a specialized function. These lobes are named for the bone that covers them.

The **occipital lobe** receives and processes visual information. The **temporal lobe** is the center of our hearing interpretation. Because spoken language came about long before written language, the temporal lobe is also the area where language interpretation is centered. The **parietal lobe** is where information from the skin, primarily the sense of "touch," is located. The largest of the lobes, the **frontal lobe,** is involved in nonsensory functions such as initiating movement, speaking, and thinking (Figure 11.7a).

A deep fold separates the right and left halves, or hemispheres, of the cerebrum (Figure 11.7b). Communication between the right and left cerebral hemispheres occurs via a band of white matter called the **corpus callosum** (Figure 11.6).

ACTIVITY 3

Observing the Brain

Materials for This Activity

Model of the human brain
Sheep brain
Large, nonserrated knife

Part 1: Human Brain Model

1. Observe the model of the human brain, and use Figures 11.6 and 11.7 to assist you in locating the cerebellum and the four lobes of the cerebrum discussed previously.
2. Separate the halves of the brain model, and locate the medulla oblongata, which appears to be a simple enlargement of the spinal cord. Find the somewhat rounded pons, directly above the medulla oblongata. Directly above the pons, find the midbrain.

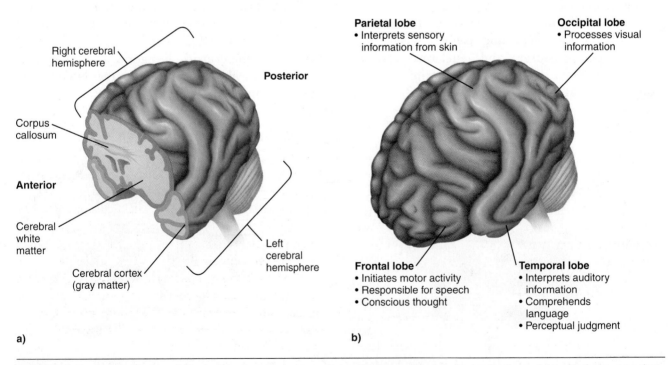

a)

b)

FIGURE 11.7 **The cerebrum. a)** The cerebral cortex (gray matter; outer layer, shown in four colors) consists of interneurons that integrate and process information. White matter (inner core) consists of ascending and descending nerve tracts. The two separate hemispheres are joined by the corpus callosum. **b)** The functions of the four lobes of the cerebral cortex are location-specific.

3. Next, locate the cerebellum directly behind the pons, medulla oblongata, and midbrain. Notice that the cerebellum has a highly branched core of white matter.

After reviewing the function of the cerebellum, propose a reason why the branched white matter is found there.

The pons connects the midbrain and brains

Continue your examination of the midsagittal view of the brain model. Note the somewhat rounded thalamus above the midbrain. Below the thalamus, angling down toward the marble-like pituitary gland, you will find the hypothalamus. Above the thalamus, locate the curved strip of white matter called the corpus callosum. Finally, observe the cerebral hemisphere and its highly folded appearance. Does the corpus callosum seem as if it would be connected to both cerebral hemispheres?

yes

What would happen if the corpus callosum were cut in a living human being?

have no connection to the brain

Part 2: Sheep Brain

1. Obtain a preserved sheep brain (for this activity, a specimen *without* meninges may be preferable). If the brain still has the meninges intact, carefully remove the dura mater, noting its toughness. Observe the whole brain from the side. You will note that it looks more straight or linear than the human brain. Suggest a reason for this.

they don't have as many functions

2. View the whole sheep brain from the top (the dorsal view), and note the deep fissure that separates the left and right cerebral hemispheres

(Figure 11.8c). Although the separation of the lobes is less distinct than it appeared in the human brain model, approximate the location of each lobe. Observe the cerebellum.

3. View the sheep brain from the bottom (the ventral view), and identify the medulla oblongata, the rounded pons, and the midbrain (Figure 11.8a–b). In the area of the midbrain, you should see two thick, tough trunklike structures crossing over each other. These are the optic nerves, pointed out here as an example of the **cranial nerves,** which are attached directly to the brain rather than the spinal cord. Some of the cranial nerves are too tiny to be easily seen or may have become detached when the meninges were removed. Observe any other cranial nerves that you can find and the flaplike olfactory bulbs near the anterior end.

4. Next, use a nonserrated knife (a serrated knife is more likely to tear the brain tissue) to slowly and carefully cut the sheep brain along the midsagittal line. From the inside of this midsagittal view, identify the medulla oblongata, pons, midbrain, cerebellum, thalamus, hypothalamus, corpus callosum, and cerebral hemisphere (Figure 11.9 on page 133). Which of these structures are considered part of the more primitive hindbrain?

42

The Autonomic Nervous System

The autonomic branch of the nervous system is composed of **sympathetic** and **parasympathetic** divisions, and it controls a wide variety of involuntary actions such as heart rate and digestion (Figure 11.10, on page 134). Many autonomic functions are controlled by cranial nerves, particularly in the parasympathetic division.

Although the sympathetic and parasympathetic divisions seem at odds, these two systems have a lot in common. They work together to maintain homeostasis, and they share an unusual two-neuron pathway scheme (Figure 11.10).

The sympathetic division is commonly referred to as controlling our "fight or flight" response; that is, it automatically prepares the body for some

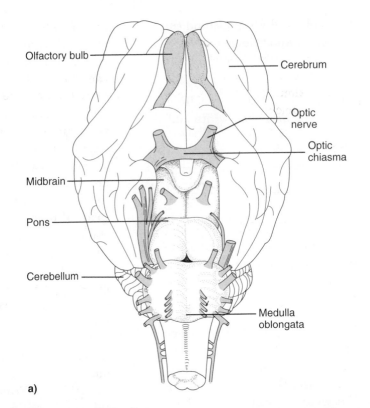

a)

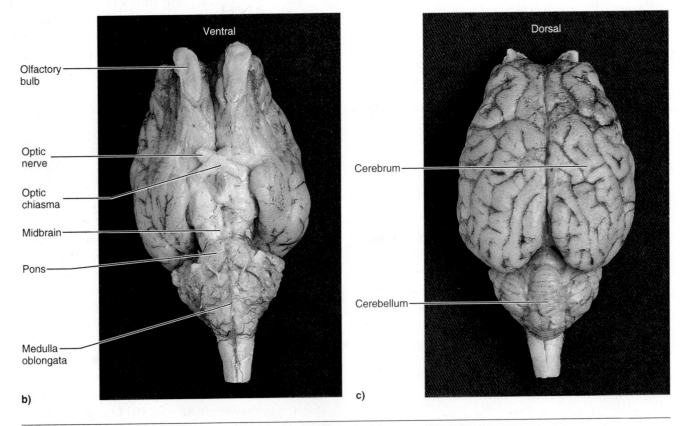

b)

c)

FIGURE 11.8   **Intact sheep brain. a)** Ventral view. **b)** Photograph of ventral view. **c)** Photograph of dorsal view.

kind of emergency response. Speeding heart rate, increasing blood pressure, and activating sweat glands are just a few activities stimulated by the sympathetic nerves within seconds of perceiving a potentially dangerous situation. Most sympathetic neurons originate in the thoracic and upper lumbar region of the spinal cord.

The parasympathetic division is the opposite of the sympathetic. It controls maintenance functions such as stimulating the digestive system and slowing the heart rate. Most parasympathetic neurons originate from the cranial nerves, with a small percentage coming from the sacral region of the spinal cord (Figure 11.10).

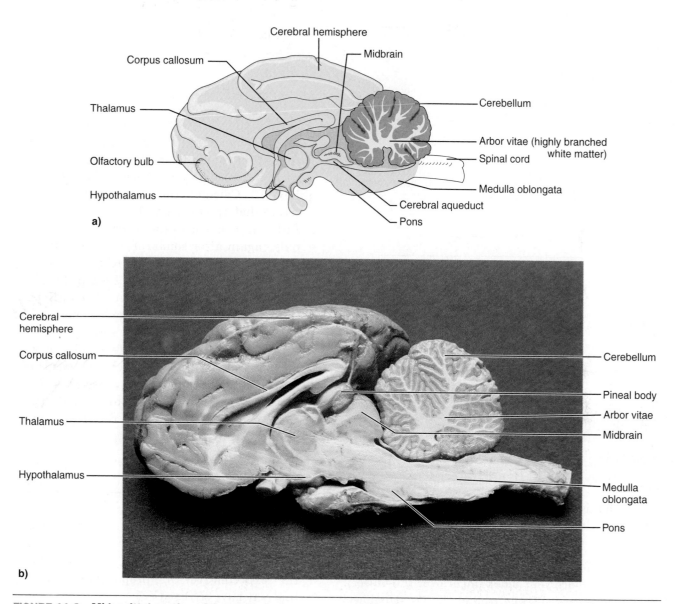

a)

b)

FIGURE 11.9 **Midsagittal section of the sheep brain.**

Autonomic Nervous System

Sympathetic
"emergency response"

A short first neuron, a synapse through a ganglion, a second long neuron, then stimulation via norepinephrine.

Parasympathetic
"maintenance"

A long first neuron, then a synapse near or in the target organ, a second short neuron, then stimulation via acetylcholine.

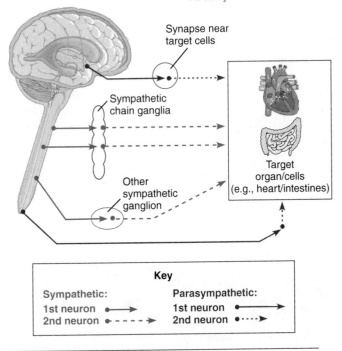

FIGURE 11.10 **A schematic drawing of the autonomic nervous system.**

ACTIVITY 4

Measuring the Effect of the Autonomic Nervous System on Heart Rate

Although it is possible to perform this activity in the laboratory, it may be better to do it at home for the best results.

1. Sit comfortably with your right or left arm on a table. Place your index finger over the pulse, which is the pulsing area of the wrist below the thumb. Count the number of pulses over a 15-second period, and then multiply by 4. For example, if you count 20 pulses during a 15-second

Box 11.1	Pulse Rate Data			
	Trial 1	Trial 2	Trial 3	Average
Baseline	88			
Parasympathetic				
Sympathetic				

period, then $20 \times 4 = 80$. The heart rate is 80 beats per minute. Take your pulse three times and calculate the beats per minute for each. Record it as your "baseline" pulse rate in Box 11.1. Calculate the average pulse rate, and record that number as well.

2. Either at home or in the laboratory, take your pulse again after some relaxing activity such as eating a meal, listening to *relaxing* music, or breathing slowly and deeply. Record this pulse rate in the " parasympathetic" row. Repeat these activities twice more, and then calculate and record your average pulse rate.

3. After some stressful activity (e.g., *after* driving home in heavy traffic or watching a suspenseful movie), take your pulse and record it in the "sympathetic" row. Repeat this activity twice more, and then calculate and record your average pulse rate.

Which average was the highest?

_____

Which average was the lowest?

_____

Explain your results below. If you did not obtain the "expected" results, propose a reason why this may have occurred.

_____

_____

THE NERVOUS SYSTEM II
Critical Thinking and Review Questions

1. What are the three meninges, and why are they important?

   Dura: tough outer layer
   Arachnoid: loose middle
   Pia: soft, thin, directly on brain and spinal

2. Where would we find cerebrospinal fluid, and what role does it play in the CNS?

   It is in most of the body, and it protects
   organs for function

3. Contrast the roles of the gray matter and the white matter of the spinal cord.

   White sends info, gray reacts to info

4. What is a dorsal root ganglion? Why are there no ventral root ganglia?

   Cell bodies in a group together. Because the
   bodies are found at the end

5. Match the following parts of the brain to their functions:

   _b_ Hypothalamus
         a. controls basic life-support functions such as control of breathing

   _a_ Medulla oblongata
         b. maintains higher life-support functions such as body temperature regulation

   _e_ Temporal lobe
         c. serves as movement coordination center

   _f_ Occipital lobe
         d. initiates movement and thought

   _d_ Frontal lobe
         e. interprets hearing

   _c_ Cerebellum
         f. interprets vision

6. Describe the two branches of the autonomic nervous system. Summarize their differences and similarities.

   Dorsal roots have sensory neurons, and ventral
   roots contains axons of motor neurons

7. Label the following diagrams with the name of the structure and its major functions.

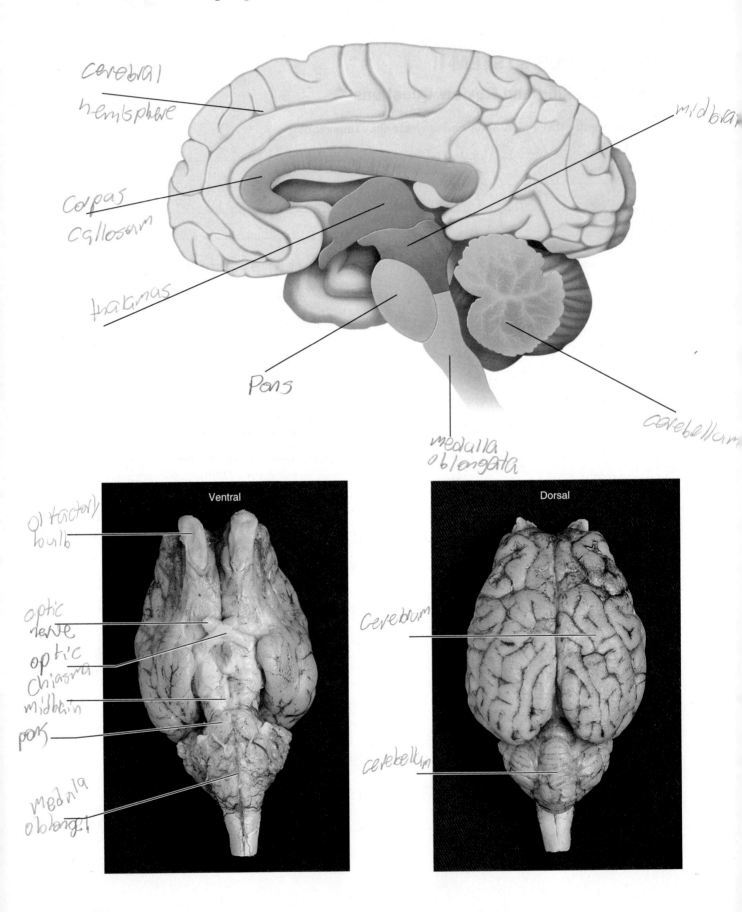

cerebral
hemisphere

midbrain

Corpus
callosum

thalamus

Pons

cerebellum

medulla
oblongata

Ventral

Dorsal

olfactory
bulb

optic
nerve

optic
chiasma

midbrain

pons

medulla
oblongata

Cerebrum

cerebellum

12

The Senses

Objectives

After completing this exercise, you should be able to

1. List and describe the receptors involved in general sensation.
2. List the main structures of the eye, and describe their role in vision.
3. Describe selected vision tests.
4. List the main structures of the ear, and describe their role in hearing and balance.
5. Describe selected hearing and balance tests.

Materials for Lab Preparation

Equipment and Supplies

- ○ Cotton swabs
- ○ Powdered or granulated sugar (sucrose)
- ○ Lemon juice
- ○ Seven flat-head dissecting pins
- ○ Three corks or Styrofoam cubes (about 1-inch cubes)
- ○ Metric ruler
- ○ Fine-point marker
- ○ Paper cups (8-ounce size)
- ○ Tap water
- ○ Beaker or flask
- ○ Sheep or cow eye
- ○ Dissection equipment
- ○ Dissection tray
- ○ Snellen eye chart
- ○ Astigmatism eye chart
- ○ Medium-size test tube
- ○ Pencil
- ○ Tuning fork
- ○ Chair with wheels
- ○ Stopwatch or clock
- ○ Thermometer

Models

- ○ Human ear
- ○ Human eye

Introduction

In Exercise 11, you learned that the brain interprets sensory information. But where does this sensory information come from? Specialized portions of sensory neurons called **receptors** are capable of responding to environmental stimuli and eventually converting the stimuli into action potentials. These receptors are usually grouped into two major categories for study: the **special senses** and **general sensation.** Receptors for the special senses include the eye and ear, organs that will be covered as separate topics in this exercise. Taste and smell are also special senses based on chemoreceptors. General sensation includes the category of **somatic sensation,** which is detected by the sensory receptors in skin and muscle that are generally spread out over the entire body (Figure 12.1).

Receptors

Receptors are classified according to the stimuli to which they are sensitive, and their names are often self-explanatory. **Photoreceptors,** which we will study during examination of the eye, are sensitive to light. **Chemoreceptors** are sensitive to chemicals.

Thermoreceptors respond to temperature, and **pain receptors** are sensitive to pain. Technically, the mechanism for thermoreceptors and pain receptors may also qualify them as chemoreceptors, as it seems to be the chemicals released from damaged or stressed cells and tissues that trigger these receptors.

The largest class of receptors is the **mechanoreceptors,** which include all receptors that respond to mechanical energy. This category includes the following:

- **Touch** or **tactile** receptors, which are usually found in the skin and detect contact and/or vibrations (Figure 12.1).
- **Pressure** receptors, which detect pressure on the skin or joints and even subtle changes in blood pressure (Figure 12.1).
- **Stretch** receptors, which are found in muscles, tendons, joints, and the lungs and detect stretching or stress (Figure 12.2).

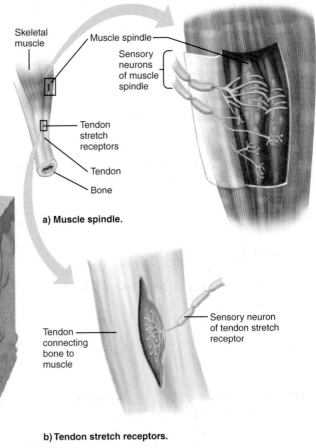

a) **Muscle spindle.**

b) **Tendon stretch receptors.**

FIGURE 12.2 Muscle spindles and tendon receptors that monitor muscle length. Reflexes initiated by these receptors help us maintain balance.

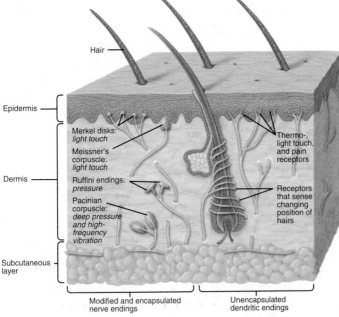

FIGURE 12.1 Sensory receptors in skin for sensing touch, temperature, pain, pressure, and vibration.

- **Hearing** and **balance** receptors, which are found in the inner ear and will be covered later in this exercise.

With the notable exception of pain receptors and some receptors found in joints and muscles, most receptors are more sensitive to *changes* in stimuli and will eventually stop responding to the same level of stimulus. This phenomenon is referred to as **receptor adaptation.** You probably are not aware of the clothes you are wearing (unless they are uncomfortable) because the receptors in your skin *adapted* to the stimuli induced by the clothing. Touch receptors, as well as the chemoreceptors in the nose, adapt rapidly. If you have ever encountered someone who has put on far too much cologne, or someone who smokes tobacco regularly, you know they do not seem to notice these odors.

ACTIVITY 1

Exploring Touch Receptors

Materials for This Activity

Seven flat-head dissecting pins

Three corks or Styrofoam cubes (about 1-inch cubes)

Metric ruler

Fine-point marker

It would be wasteful and impractical for touch receptors to be placed all over the body in the same density that we find them in our fingertips. Instead, we find a much lower density of touch receptors in areas of the skin where feedback from touch receptors is less essential to our needs. An easy way to determine the relative density of touch receptors is a **two-point discrimination test.** Follow the instructions below to perform a simple two-point discrimination test.

1. Using your marker and ruler, place two dots 5 mm apart near the center of the cork. Insert the pins into these marks, making sure that the pins are straight and level. Then, repeat this procedure for the remaining two corks, placing your marks 10 mm and 25 mm apart. Insert the pins into these marks as well. You will have one pin left over. Place this single pin aside to use as your control.

2. Have your partner close his or her eyes, and place one hand on the table palm up. In random order, touch the heads of the pin sets and single pin to your partner's index finger. After each application of the pin head(s) to your partner's finger, ask if he or she feels one point or two. Do not tell your partner the correct answer. Repeat this experiment until you obtain a consistent trend.

3. Was your partner able to accurately and consistently judge the number of points?

   2 wrong out of 8

   Which pin set or sets (if any) was the subject most consistently *incorrect* in identifying as two points?

   2

Note: It is important that the subject answer carefully and honestly. The single pin serves as a control so that the subject is not tempted to answer "two" every time.

4. Repeat step 2, this time using your partner's palm. Was your partner able to accurately and consistently judge the number of points?

   Very wrong

   Which pin set or sets (if any) was the subject most consistently *incorrect* in identifying as two points?

   1 ~ 8

5. Repeat step 2, this time using your partner's forearm. Was your partner able to accurately and consistently judge the number of points?

   all wrong but one

   Which pin set or sets (if any) was the subject most consistently *incorrect* in identifying as two points?

   1 ~ 8

A higher density of receptors makes it easier to discriminate the two different points of contact, as more receptors increase the likelihood that each point will stimulate entirely different receptors. Based upon your results, rank the three tested areas in order of their density of receptors (starting with the highest density).

   Finger, forearm, palm

Speculate as to why this test would be useful or practical.

to test different nerves in the body

ACTIVITY 2

Exploring Proprioceptor Function

Materials for This Activity

Three paper cups (8-ounce size)

Tap water

Beaker or flask

Mechanoreceptors in joints, tendons, and muscles that help to convey information about the position of our body parts and the stress on joints are categorized as **proprioceptors.** Proprioceptors contribute to our sense of balance by helping the brain rapidly monitor the effect of movement or stress on joints.

To test the function of these receptors, work in pairs, and obtain three paper cups and a beaker or flask.

1. Add water to two of the paper cups so that one is about ¼ full and the second is about ⅔ full. Fill a third cup about ⅓ full, and put it aside for later use.

2. Your partner must close his or her eyes while you carefully place one cup in each hand. Now ask your partner which cup is heavier or has more water. Take the cups away, and repeat the process (perhaps in a different order of right or left) twice more to reduce the possibility of lucky guessing. Can your partner determine which cup is heavier at least two out of three times?

no times, both incorrect

3. Pour water from the heavier cup into the lighter cup until both are approximately ⅓ full. The exact amount of water is not critical as long as both cups are as close to equal in weight as possible. Once again, after your partner closes his or her eyes, carefully place one cup in each of his or her hands. Pour the contents of the third cup, which you had previously put aside, into one of the cups, and ask your partner to identify which cup received the water. Repeat this procedure twice more to reduce

the possibility of lucky guessing. Can your partner determine which cup you added water to at least two out of three times?

no

Explain your results.

the mind feels it is being tricked

ACTIVITY 3

Exploring Receptor Adaptation

Materials for This Activity

Thermometer

Two paper cups (8-ounce size)

Tap water

Stopwatch or clock

1. Prepare two cups of water so that one is approximately room temperature and one is very warm but not uncomfortably hot—approximately 43 to 44°C. The cups should be about ¾ full.

2. Have your partner place the index finger of one hand in the room temperature water. Ask your partner to describe the temperature sensation of the immersed finger, and record the comments.

Comfortable and chill

3. Next, have your partner remove and immediately place the same finger in the warm water. Ask your partner to describe the temperature sensation of the immersed finger.

tingle/hot

After one minute, ask your partner to describe the sensation again, now that the finger has been immersed for a little while.

like it isnt in water

4. Finally, have your partner remove and immediately place the same finger in the room temperature water again. Ask your partner to describe the temperature sensation of the immersed finger.

feels the same

Does this description exactly match the comments you recorded in step 2? Explain.

no, get used to the water

ACTIVITY 4

Exploring Oral Chemoreceptors

Materials for This Activity

Cotton swabs

Powdered or granulated sugar (sucrose)

Lemon juice

1. Wet a cotton swab, and coat it with sugar. Using a mirror, or the assistance of a lab partner, touch the sugar-coated swab on the surface of the tongue near the tip. *Note:* The tongue should be moist, not dry, for this experiment to work. Do you immediately detect a significant sweet sensation?

   yes

2. Wet the cotton swab with lemon juice until it is saturated. Place the lemon juice-saturated cotton swab on the tongue near the sides. Do you immediately detect a significant sour sensation?

   yes

3. Considering that molecules and chemoreceptors must make a lock-and-key fit, speculate as to whether the same or different chemoreceptors were used during steps 1 and 2 above.

   different

   See Figure 12.3 illustrating the tongue, and note that taste buds are located within grooves on the tongue. How is this feature related to the need for the tongue to be moist?

   _____

   _____

The Eye and Vision

The eye is very much like a video camera, with the photoreceptors of the retina acting as the instantly reusable videotape. The three tissue layers of the

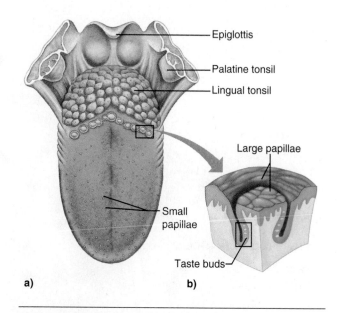

FIGURE 12.3   Structure of taste buds on the tongue.
a) Taste buds are associated with peglike projections (large and small papillae). **b)** An illustration of a section of large papillae with the taste buds in its lateral wall.

eye from outside to inside are the **sclera, choroid,** and **retina.** Most of the structures of the sclera and the choroid are simply there to ensure that a focused image is projected on the retina (Figure 12.4).

The Sclera

The sclera is the so-called white of the eye. It is the tough outer layer of connective tissue that is modified into the transparent **cornea** in the front (Figure 12.4). Thus, the delicate structures inside the eye are partly protected by the sclera, but the cornea still allows light to enter the eye.

The Choroid

The choroid is the dark, pigmented, middle layer of the eye (Figure 12.4). Similar to the dark inside of a camera, this dark pigmentation absorbs light and prevents the bouncing or scattering of light that would distort the image. Most of the blood supply needed to nourish the retina is contained in the choroid.

Several special structures are modified from the choroid coat (Figure 12.4). The **lens** is a transparent, somewhat oval, marble-sized piece of tissue that bends light that passes through it. The thickness of the lens is modified by the **ciliary muscle,** which causes the flexible lens to become thicker or thinner to assist in focusing on near or far objects. Another

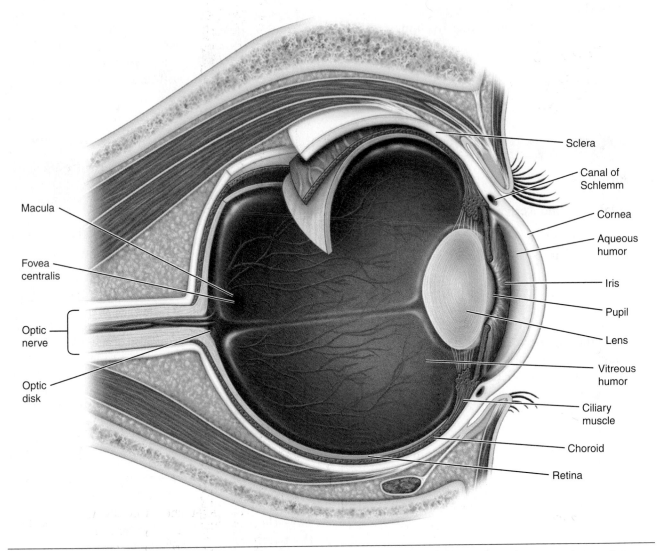

FIGURE 12.4 **Structure of the eye.** Light enters through the transparent cornea and passes through aqueous humor, the pupil, the lens, and vitreous humor before striking the retina at the back of the eye. The fovea is the area of the retina with the highest visual acuity. The nerves and blood vessels exit the eye through the optic disk.

muscular structure, the **iris,** adjusts the diameter of the **pupil** so that more light enters the eye when it is darker and less light is permitted to enter the eye under brighter conditions.

A thin liquid, called **aqueous humor,** is found between the cornea and the lens. The **vitreous humor** is a thicker fluid found in the large chamber behind the lens. These fluids are important for transporting nutrients and maintaining the shape of the eye (Figure 12.4).

The Retina

The retina, as previously mentioned, contains the photoreceptors of the eye (Figure 12.5). The **cones** (for color) are active in brighter light conditions, and the **rods** (for black and white) are active in dimmer light conditions. A high concentration of cones called the **macula lutea** is found in a small area in the back of the eye, called the **fovea centralis** *(center of focus).* When we squint to try to make out the details of a small object, we attempt to center the focus of the object on this area.

There are three layers of specialized neurons in the retina: the rod and cone cells (commonly known as photoreceptors), the **bipolar cells,** and the **ganglion cells.** Interestingly, the three layers are arranged backward from a design viewpoint; that is, the rod and cone cell layer points toward the back of the eye into the choroid, with the other two neuron layers sticking out toward the front. Fortunately, the retina is fairly transparent, and the other cells do not really

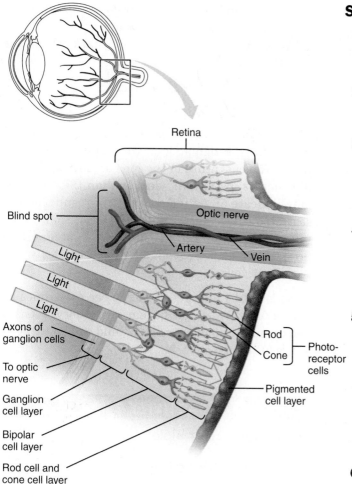

Retina

Blind spot

Optic nerve

Artery

Vein

Light

Light

Light

Axons of
ganglion cells

Rod

Cone

Photo-
receptor
cells

To optic
nerve

Ganglion
cell layer

Pigmented
cell layer

Bipolar
cell layer

Rod cell and
cone cell layer

FIGURE 12.5   Structure of the retina.

block the light on its way to the photoreceptors. However, this structure creates a problem in that the axons of a million ganglion cells are now in front of the photoreceptor layer, and they must somehow exit the eye in order to connect to the brain. All of these axons are bundled together and exit the eye in an area called the **optic disk,** or blind spot, because the exiting axons displace the photoreceptors (Figure 12.4).

ACTIVITY 5

Examining the Structure of the Eye

Materials for This Activity

Model of the human eye

Sheep or cow eye

Dissection equipment

Dissection tray

Sheep or Cow Eye

1. Obtain a sheep eye or cow eye, and place it cornea side up on a dissection tray. Cut away any muscles that may still be attached to the eye.
2. Using a sharp scalpel, carefully make a small incision in the sclera about ¼ inch to the outside of the transparent cornea.
3. Insert the point of a scissors blade into this incision, and cut a circle around the cornea, leaving a ¼-inch border of gray-white sclera. Once you have cut all the way around it, remove the front section of the eye.
4. First, examine the inside of the smaller front section. You will see a marble-shaped structure that is relatively hard. This is the lens, which would have been more flexible when the specimen was living.
5. Use forceps to remove the lens. Now you will see the pupil and some of the cornea from the inside. Forming the pupil is the thin, dark iris. To the outside of the iris is the dark ciliary body, which is made largely of the ciliary muscle previously described. The ciliary body is easily discernable from the iris because of its radiating grooves, while the iris is smooth in appearance.
6. Now examine the back section of the eye. The cream-colored or light yellow retina may have already become detached from the dark choroid. If not, peel the retina away from the choroid. You will notice that there is one place where the retina will not detach. This is the optic disk, where the retina attaches to the optic nerve. Note the position of the optic disk, flip the eye over, and find the optic nerve to the outside of this point. You may have to cut away some muscle or fat tissue in order to find the optic nerve on the outside. Next, carefully separate the choroid from the sclera in one small area so that you can observe the three tissue layers of the eye and note their relative thickness to each other.

You may notice that there is a lighter blue area in the back of the choroid coat. This is called the **tapetum lucidum,** a somewhat more reflective area found in the eyes of some animals, which improves night vision. This reflective area is a large part of the reason why a cat's eyes seem to "glow" at night. The eyes do not actually glow; they simply reflect more light than the cat's surroundings.

Eye Models

Using Figure 12.4 as a reference, locate the following structures on the eye models provided for you in the laboratory.

- Cornea
- Lens
- Pupil
- Iris
- Ciliary body/muscle
- Retina
- Choroid
- Sclera
- Optic nerve
- Optic disk

ACTIVITY 6

Testing Visual Acuity

Materials for This Activity

Snellen eye chart

Astigmatism eye chart

1. Stand 20 feet away from the Snellen eye chart and gently cover one eye. Begin reading the letters in each row, starting with the top row, while your lab partner stands next to the chart. Your partner will stop you when you make a mistake, and he or she will read the numbers to the left of the last row that you read correctly. Repeat this procedure for the other eye. Record the numbers for each eye.

Right _Level 5 F_

Left _Level 5 P_

A pair of numbers are found to the left of each line on the Snellen chart; the first represents the distance you are standing from the chart (20 feet), and the second is the distance a person with "normal" vision would stand to read the same line as his or her last correct line. Thus, someone who can read the large "E" at the top of the chart as his or her only correct line would be reading the line that a person with normal vision would see when standing much farther away. This person's vision would be rated as 20/200; that is, when standing at 20 feet (the first number), this person sees what a person with normal vision would be able to see at 200 feet (the second number).

Vision rated at 20/20 suggests normal vision—that the subject's eye is correctly reading a line at 20 feet that a person with normal vision would also be reading at that same distance. Second numbers larger than 20 may suggest nearsightedness (myopia), while second numbers smaller than 20 (e.g., 20/15) may either suggest farsightedness (hyperopia) or simply better-than-normal vision.

2. Next, stand 10 feet away from the astigmatism eye chart, and gently cover one eye. Note if any of the lines look *substantially* darker than the others, and record your results.

Right _distance 10 P_

Left _distance 10 2_

The lines in the astigmatism eye chart are all equal in thickness. Astigmatism is the result of irregularities in the curvature of the lens or cornea. If some lines look substantially darker than the others, it may indicate an astigmatism. As a result, light is scattered somewhat irregularly, rather than focusing evenly on the retina. Like nearsightedness or farsightedness, an astigmatism can be easily corrected with eyeglasses.

Note: Don't be alarmed if you do see some lines as darker. Very few people have *perfectly* curved lenses and corneas, and many cases of mild astigmatism do not require correction.

ACTIVITY 7

Testing Binocular Vision

Materials for This Activity

Medium-size test tube

Pencil

The brain is capable of judging depth perception partly because of learned behavior and partly because of the fact that we receive slightly different visual information from each eye. For example, we learned long ago that if a person looks larger than a tree, he or she is probably standing quite a bit closer to us compared to the tree. But for items where size and point of reference are less helpful, we rely on comparing data from both eyes. This is the concept of binocular vision.

1. Work in pairs, and have one partner hold a small test tube as steady as possible. The other partner should stand a full arm's length away from the test tube and hold a pencil with the eraser end *down*. Using a steady and gentle arc-like motion, the second partner should attempt to place the eraser end of the pencil in the test tube. The movement should not be excessively slow, considering this experiment is testing your ability to quickly evaluate the three-dimensional positions of the pencil and test tube. This action is supposed to be somewhat difficult to do, and it would be unrealistic

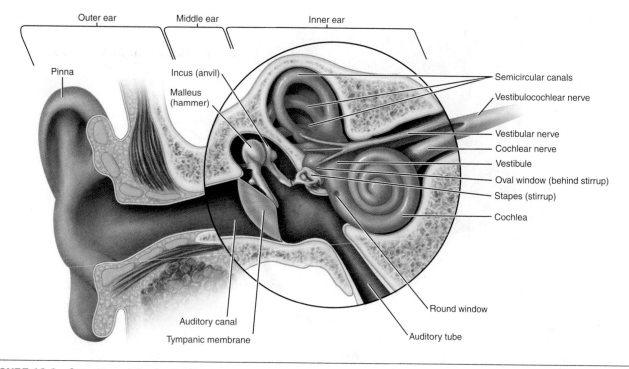

FIGURE 12.6 **Structure of the human ear.**

if you are able to get the pencil in the test tube every time. Record the number of times out of 10 that you successfully inserted the pencil into the tube.

2. Repeat step 1, but this time, cover one eye. Record the number of times out of 10 that you successfully inserted the pencil into the tube.

4 yes, 6 no

Propose a reason for your results.

because her balance was off since her eye was closed

The Ear and Hearing

The ear is an organ of both hearing and balance. The structures of the ear are divided into three sections: the **outer ear, middle ear,** and **inner ear** (Figure 12.6). We will cover the balance responsibilities of the inner ear in a separate section in this exercise. Most of the structures associated with the ear are involved in our hearing mechanism.

The visible portion of the outer ear, or **pinna,** collects and channels sound waves into the **auditory canal.** These sound waves strike the **ear drum** or **tympanic membrane** (Figure 12.6).

The middle ear begins on the inside of the eardrum and is an air-filled cavity (Figure 12.7 on page 146). If this cavity were not connected to the outside in some way, changes in air pressure would create painful stress on the eardrum. The **Eustachian tube,** or auditory tube, connects the middle ear cavity with the throat—which explains why swallowing may cause our ears to "pop" and relieve the pressure we feel in our ears when going up or down in elevation (Figures 12.6 and 12.7). In the middle ear, we find three little bones attached to the eardrum. These little bones, the **malleus** (hammer), **incus** (anvil), and **stapes** (stirrup), amplify the vibrations of the eardrum by concentrating the larger eardrum's vibrations down to a single, smaller point. This smaller area at the end of the stapes is called the **oval window,** and it brings us to the inner ear (Figure 12.6).

The inner ear consists of three main parts (Figure 12.6). The **semicircular canals,** as the name implies, are three small tubes bent into a curved, semicircular shape. The **vestibule** is the thick, middle portion of the inner ear. These two structures are mainly involved with the balance function of the inner ear and will be explained in more detail later. The coiled **cochlea** is the hearing portion of the ear, where vibrations are eventually converted into nerve impulses that the brain interprets as sound (Figures 12.6 and 12.7).

Fluid within the long, coiled ducts of the cochlea vibrates when the stapes vibrates against the oval

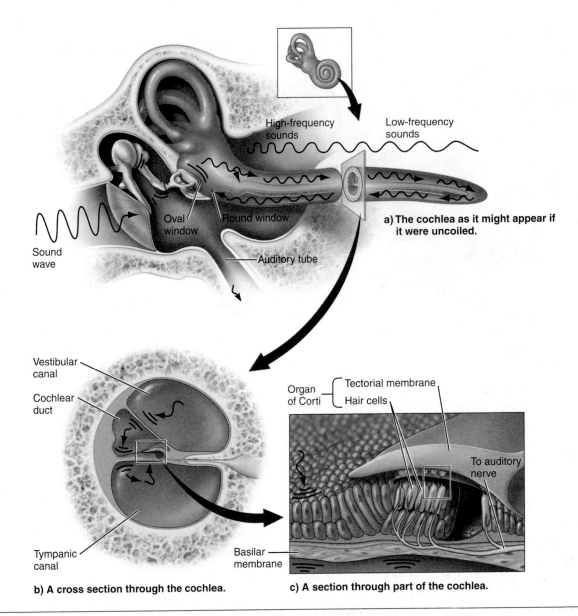

a) The cochlea as it might appear if it were uncoiled.

b) A cross section through the cochlea.

c) A section through part of the cochlea.

FIGURE 12.7   The middle and inner ear: converting vibrations into nervous impulses.

window. Mechanoreceptors, called **hair cells,** form a long line, running the length of one of these ducts (Figure 12.7c). Bending of the hairlike projections of the hair cells triggers an action potential that travels to the brain for interpretation. Sounds of different frequencies vibrate the cochlea in different areas, thus allowing the brain to interpret the frequency of the sound based upon which part of the cochlea is receiving impulses (Figure 12.7a).

Localizing Sounds

A sound that arrives at your right ear from 100 feet away does not lose much of its volume as it travels the few inches to the other side of your head. Yet,

with a fairly high degree of reliability, we can determine the direction of incoming sound. It does seem that the sound entering one's ear from the side closer to the origin of the sound is louder. This differentiation is yet another interesting feature of the human nervous system. Although there is only a small fraction-of-a-second delay in the sound that is farther from the source of the sound reaching the ear, the delay is enough to cause the nervous system to inhibit the impulses from the "second" ear. Thus, even though the volume of the sound reaching both ears is nearly identical in most cases, our *perception* is of a louder sound from the closer ear. We learn to interpret this difference as a way of judging the direction from which a sound originated.

ACTIVITY 8

Examining the Structure of the Ear

Materials for This Activity

Model of a human ear
Using Figure 12.6 on page 145 as a reference, locate the following structures on the ear models provided for you in the laboratory.

- Pinna
- Auditory canal
- Tympanic membrane
- Eustachian tube
- Malleus
- Incus
- Stapes
- Oval window
- Semicircular canals
- Vestibule
- Cochlea

ACTIVITY 9

Testing Hearing Conduction Pathways

Materials for This Activity

Tuning fork

1. Hold the tuning fork by its handle and strike one tine from the side on the thick portion of the palm of your hand (near the pinky side of your palm).
2. Place the base of the tuning fork handle on your temporal bone about 1 inch behind your earlobe. If you strike the tuning fork hard enough, you should be able to hear the sound because you are directly vibrating the fluid in your cochlea, which is found within the temporal bone. If you do not hear a sound, repeat the procedure of striking the tuning fork and placing it on your temporal bone again until you do. When the sound disappears, move the tuning fork so that the ends of the tines are in front of your pinna. Do you hear a sound after moving the tines to the front of your pinna?

   yes

3. Repeat step 2 in reverse order; that is, strike the tuning fork, place the tines in front of your pinna, and when the sound disappears, place the base of the tuning fork on your temporal bone.

Do you hear a sound after moving the base of the tuning fork onto the temporal bone?

yes

What does this response suggest about the value of your conduction pathway (the eardrum, malleus, incus, and stapes)?

they can pick up low frequencies well

ACTIVITY 10

Testing for Hearing Localization

Materials for This Activity

Tuning fork

1. Work in pairs, with one partner closing his or her eyes. Strike the tuning fork. In random sequence, place the tuning fork in the following positions, asking your partner each time to identify the position of the tuning fork: near the right ear, near the left ear, near the front of the nose, near the top of the head, and a few inches behind the head. Prompt your partner by saying "now" when you have the fork in position. If he or she hears nothing, strike the tuning fork harder and try again. Do not tell your partner if he or she is right.
2. Repeat this exercise until a pattern becomes clear.

At which location(s) was your partner able to most accurately and consistently judge the location of the tuning fork?

all

At which location(s) was the subject most consistently *incorrect* in identifying the location of the tuning fork?

none

Explain your results after considering the background information provided earlier regarding localization of sound.

Sound waves are sensitive

The Ear and Balance

With slight modification, the very same kind of hair cell receptors that we identified in the cochlea are also used by the inner ear to heavily contribute to our sense of balance, or equilibrium. There are two main categories for our sense of equilibrium: **static equilibrium,** or our sense of equilibrium when stationary, and **dynamic equilibrium,** or our sense of equilibrium when in motion.

The **vestibule** is most responsible for our sense of static equilibrium (Figure 12.8a–c). Gravity pulls downward on hair cells embedded in a weighted, gelatinous material. Whether we are moving or standing still, we always know which way is down based upon the pull of gravity on these hair cells.

The **semicircular canals** provide our sense of dynamic equilibrium (Figure 12.9). When we move in any direction, the fluid in at least one of our semicircular canals will slosh backward in the opposite direction, in the same way the water in a container would seem to move when you start moving. Actually, the fluid is trying to remain stationary (the principle of inertia), but the effect is the same. The apparent sloshing back of fluid through the semicircular canals bends hair cells, which alerts our brains to movement along that plane. It is no coincidence that there are three dimensions in normal space and that we have three semicircular canals at right

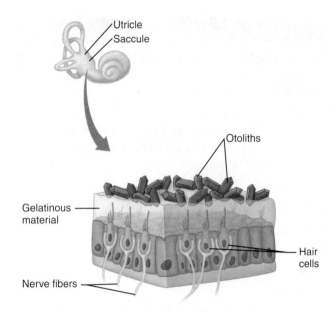

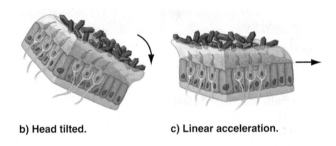

a) A cutaway view of the vestibule with the head at rest.

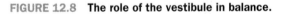

b) Head tilted.   c) Linear acceleration.

FIGURE 12.8   **The role of the vestibule in balance.**

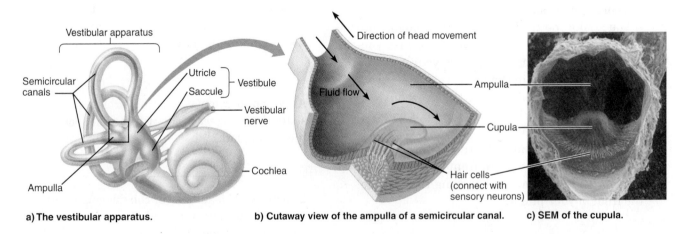

a) The vestibular apparatus.   b) Cutaway view of the ampulla of a semicircular canal.   c) SEM of the cupula.

FIGURE 12.9   **The role of the semicircular canals and vestibular apparatus in balance. a)** The vestibular apparatus (the organ of balance) is part of the inner ear, which also includes the cochlea. **b)** Cutaway view of the ampulla of a semicircular canal, showing the hair cells embedded in the base of the gel-like cupula. Rotational movement of the head causes fluid movement in the opposite direction in the semicircular canal. The fluid pushes against the cupula, bending it and activating the mechanoreceptor hairs of the hair cells. The hair cells release a neurotransmitter that stimulates sensory neurons of the vestibular nerve. **c)** SEM of the cupula.

angles to each other. It is important to note that our sense of balance is not restricted to input from the inner ear. You may recall that in our discussion of the cerebellum in Exercise 11, the inner ear's feedback is one of *several* pieces of sensory information used by the brain to help us maintain our balance.

ACTIVITY 11

Testing Equilibrium

Materials for This Activity

Chair with wheels

Stopwatch or clock

Part 1: Direction of Movement

1. Work in pairs. Have your partner sit in a chair with wheels in an area with some open space and close his or her eyes.
2. *Slowly and carefully,* move the chair a few feet randomly in the following directions, asking your partner each time to identify the direction of movement: right, left, forward, or backward. *Note:* For safety, remind your partner to hold on to the chair, and warn them right before you start to move the chair. Was your partner able to consistently identify the direction of movement?

   50/50

Explain your results after considering the background information provided earlier regarding dynamic equilibrium.

overly sensitive

Part 2: Maintenance of Equilibrium

Work in pairs. Have one person stand on one foot with his or her eyes open, and time how long he or she can maintain balance. Set a limit of two minutes. Repeat this activity with the subject's eyes closed, *standing by to "spot" your partner if he or she loses balance.* Record your results.

Eyes open   *1:05 mm left*

Eyes closed   *1:44 mm left*

Explain your results after considering the background information provided earlier regarding the role of the cerebellum and other factors involved in your sense of balance.

Sight is important for balance

THE SENSES

Critical Thinking and Review Questions

1. Match the receptor to its function or description:

   _c_ Tactile receptors   a. temperature receptors

   _e_ Stretch receptors   b. rods and cones

   _a_ Thermoreceptors   c. touch receptors found mostly in the skin

   _b_ Photoreceptors   d. taste receptors

   _d_ Chemoreceptors   e. receptors found in muscles, tendons, joints, and lungs

2. Describe at least two examples of receptor adaptation.

   = Ignoring clothing

   - Nose blind

3. Match the part of the eye to its function or description:

   _d_ Cornea   a. adjustable light-bending structure

   _a_ Lens   b. changes the tension on the lens

   _e_ Iris   c. blind spot

   _b_ Ciliary muscle d. transparent portion of the sclera

   _c_ Optic disk   e. modifies the size of the pupil

4. Explain what the term _20/30 vision_ means.

   means near sighted

5. Explain what the binocular vision experiment in Activity 7 taught you about depth perception.

   It taught me that depth perception is important

6. Summarize the roles of the outer ear, middle ear, and inner ear in contributing to our sense of hearing.

   Outer: transmit sound, middle: vibrates drum inner: sends impulse to nerves

7. Explain why it is difficult to localize a sound that originates from directly behind you.

No ear is closer to the sound yes!

8. Considering all that you have learned from this exercise, as well as the previous exercises on the nervous system, list and describe all of the structures that contribute to your sense of balance.

Every sense is important to keeping equilibrium. Nothing is more important

9. Label the following diagram.

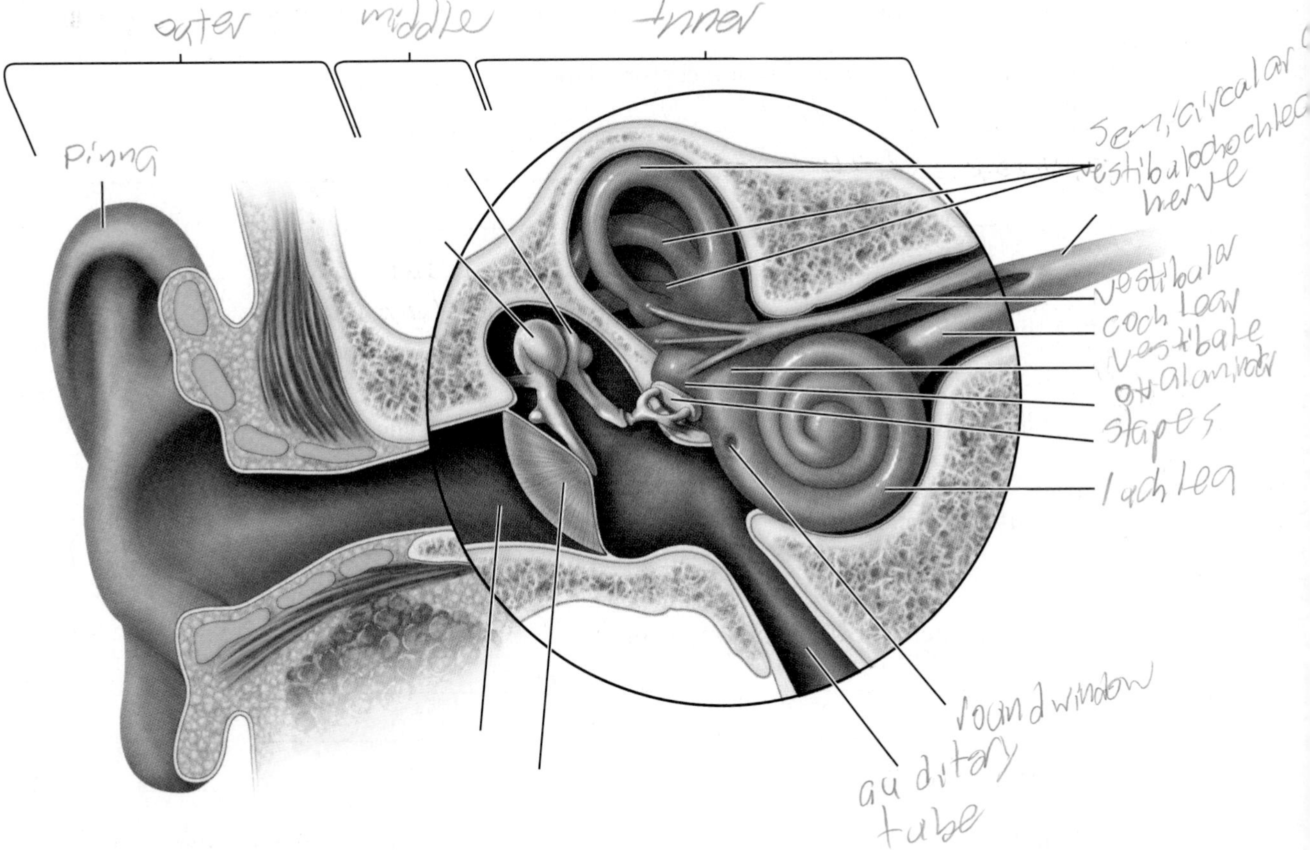

outer   *middle*   *Inner*

Pinna

Semicircular
vestibulocochlea
nerve

vestibular
cochlear
vestibule
oval window
stapes
cochlea

round window

auditory tube

EXERCISE

13

The Endocrine System

Objectives

After completing this exercise, you should be able to

1. List and describe the major endocrine glands and the major hormone(s) each releases.
2. List and describe the function of the major hormones of the endocrine system.
3. Identify the major endocrine glands in the fetal pig and/or human models.
4. Identify microscopic tissue sections of the pancreas and pituitary gland.

Materials for Lab Preparation

Equipment and Supplies

○ Sheep brain
○ Compound light microscope
○ Fetal pig
○ Dissection equipment
○ Dissection tray
○ Disposable gloves (optional)
○ String
○ Cleaning disinfectant or detergent solution in a spray bottle
○ Storage container or bag (to keep pig for later use)

Prepared Slides

○ Pancreas
○ Pituitary gland

Models

○ Human torso
○ Human brain
○ Sheep brain

Introduction

You have already learned that the nervous system is the body's primary communication system. However, another system using chemical messengers is also important in maintaining homeostasis. The endocrine system uses the blood to carry its chemical messengers, called **hormones,** to target cells. The target cells have receptors capable of receiving the hormones and translating their messages. Because hormones are powerful chemical messengers, endocrine glands need not be particularly large or produce large amounts of hormone in order to have powerful effects on the human body.

Although there are numerous hormones, including some secreted by individual cells and tissues rather than a distinct gland, the scope of this exercise mainly focuses on the major endocrine glands and their hormones.

The Hypothalamus and the Pituitary Gland

The close relationship between the nervous system and the endocrine system is demonstrated by the hypothalamus and the pituitary gland (Figure 13.1). The hypothalamus uses the *posterior pituitary* as its release site for two hormones, **antidiuretic hormone (ADH)** and **oxytocin.** These hormones are actually made in cell bodies of neurons in the hypothalamus and transported to the end of the axons in the posterior pituitary, where they can be released. ADH stimulates water reabsorption by certain kidney tubules, and it is released as part of the hypothalamus's thirst response. Oxytocin stimulates smooth muscle contraction in the uterus and stimulates contraction of cells in the mammary glands.

The *anterior pituitary* makes and releases several hormones, some of which control other endocrine glands (Figure 13.1b). These **tropic hormones** (hormones that control other endocrine glands) give the pituitary its somewhat undeserved reputation as the "master gland." The hypothalamus tightly controls the release of hormones from the anterior pituitary through hormones it releases into a short network of blood vessels connecting the hypothalamus to the pituitary.

The tropic hormones released from the anterior pituitary are **thyroid stimulating hormone (TSH),** which stimulates the thyroid to produce thyroid hormone; **adrenocorticotropic hormone (ACTH),** which stimulates the adrenal cortex; and **follicle-stimulating hormone (FSH)** and **luteinizing hormone (LH),** which stimulate the ovaries (and testes). The nontropic hormones are **growth hormone (GH),** which stimulates growth, particularly of bone and muscle tissue, and **prolactin (PRL),** which stimulates the mammary glands.

ACTIVITY 1

Observing the Hypothalamus and the Pituitary Gland

Materials for This Activity

Compound light microscope

Prepared slide of the pituitary gland

Sheep brain and/or human and sheep brain models

You should already be familiar with the hypothalamus from Exercise 11, which covers the brain.

1. Observe the pituitary gland, located directly under the hypothalamus, in the sheep brain and/or brain models provided by your instructor.

2. Next, observe a prepared slide of the pituitary gland using your microscope's 10× objective lens (100× total magnification). Your instructor may elect to set this up as a demonstration slide in order to ensure that both portions of the pituitary are in your field of view.

You should notice that the two portions of the pituitary have very different textures. The *anterior pituitary* mostly consists of glandular tissue, and you should see it spotted with the nuclei of the cuboidal epithelial cells that produce its hormones. The *posterior pituitary* is mostly composed of axons of neurons that originate in the hypothalamus. So, instead of a spotted, glandular appearance, you should see many "streaks" running through the posterior pituitary.

Which portion of the pituitary gland is to the left of your slide, the anterior or posterior?

_____

Which is to the right?

_____

Your instructor will verify your answer after you have had an opportunity to view this slide. ■

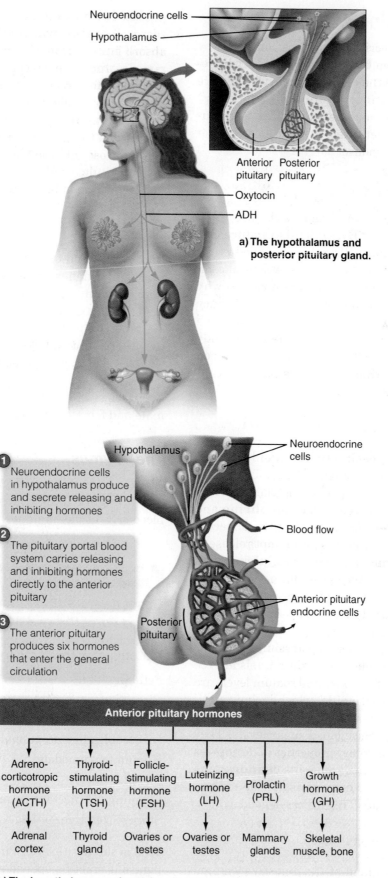

Neuroendocrine cells

Hypothalamus

Anterior pituitary   Posterior pituitary

Oxytocin

ADH

a) The hypothalamus and posterior pituitary gland.

Hypothalamus

Neuroendocrine cells

1 Neuroendocrine cells in hypothalamus produce and secrete releasing and inhibiting hormones

2 The pituitary portal blood system carries releasing and inhibiting hormones directly to the anterior pituitary

3 The anterior pituitary produces six hormones that enter the general circulation

Blood flow

Anterior pituitary endocrine cells

Posterior pituitary

Anterior pituitary hormones

Adreno-corticotropic hormone (ACTH)	Thyroid-stimulating hormone (TSH)	Follicle-stimulating hormone (FSH)	Luteinizing hormone (LH)	Prolactin (PRL)	Growth hormone (GH)
Adrenal cortex	Thyroid gland	Ovaries or testes	Ovaries or testes	Mammary glands	Skeletal muscle, bone

b) The hypothalamus and anterior pituitary gland.

FIGURE 13.1   **The hypothalamus in relation to the posterior and anterior pituitary glands.**

Other Endocrine Glands

The remaining endocrine glands are controlled by either hormones from the pituitary or simply by responding directly to the chemical or factor they are charged with regulating (Figure 13.2). The exception to this is the adrenal medulla, which is stimulated by the sympathetic nervous system as part of the "fight or flight response."

The Thyroid Gland

The thyroid produces two hormones (Figure 13.2). **Thyroid hormone (TH),** which is produced by the thyroid's follicular cells, stimulates body cells to increase metabolism. Iodine is an essential component of the TH molecule—the main reason why salt is iodized in order to prevent iodine deficiency and dysfunction of the thyroid. **Calcitonin,** which is produced by parafollicular cells, helps the body store calcium in bones. This hormone lowers blood calcium levels and, therefore, plays an important role in calcium homeostasis.

The Adrenal Glands

Like the pituitary, each adrenal gland has two very different portions (Figure 13.2). The **adrenal medulla** is the inner part of the gland derived from nervous tissue. It releases **epinephrine** and **norepinephrine.** The medulla is directly connected to the hypothalamus by nerves for prompt release of these hormones during the "fight or flight" response.

The **adrenal cortex** is the outer part of the adrenal gland, and it is made from glandular epithelial tissue (Figure 13.2). It releases three main hormones. **Aldosterone** stimulates kidney tubules to reabsorb sodium and excrete potassium. The body releases aldosterone when potassium levels are too high, blood pressure is too low, and sodium levels are too low. The **glucocorticoids,** which include cortisone and cortisol, increase glucose production from other molecules and raise blood sugar levels. They also have anti-inflammatory properties and suppress the immune system. Small amounts of **androgens** (male sex hormones) and **estrogens** (female sex hormones) are also produced, regardless of gender.

The Pancreas

Although mainly a gland of the digestive system, the pancreas contains small clusters of cells called pancreatic islet cells (or islets of Langerhans) that produce hormones (Figure 13.2). Some of these cells secrete **insulin,** which stimulates many cells to absorb glucose from the blood. In particular, liver, muscle, and adipose cells absorb large amounts of glucose for processing. This results in a drop in blood sugar levels. Other pancreatic islet cells secrete **glucagon,** the antagonist of insulin, which stimulates the liver to break up glycogen and release glucose into the blood. Production of new glucose molecules is also encouraged. These activities result in a rise in blood sugar levels.

The Gonads

Both testes and ovaries are stimulated by the pituitary hormones LH and FSH (Figure 13.1). LH mainly stimulates hormone production in the gonads, whereas FSH is primarily involved in gamete production. **Testosterone** from the testes and **estrogen** and **progesterone** from the ovaries (Figure 13.2) have a wide variety of effects on the body associated with masculine and feminine characteristics, including proper development of the male and female reproductive systems.

The Thymus

The thymus secretes **thymus hormone,** or thymosin (Figure 13.2), which activates certain white blood cells of the immune system (T cells).

ACTIVITY 2

Observing the Endocrine Glands in the Fetal Pig and Models

Materials for This Activity

Fetal pig

Dissection equipment

Dissection tray

Disposable gloves (optional)

Human torso models

String

Cleaning disinfectant or detergent solution

Storage container or bag

This is the first exercise in which you use the fetal pig, so it must be opened so that you may view the structures specified in this and some of the remaining exercises.

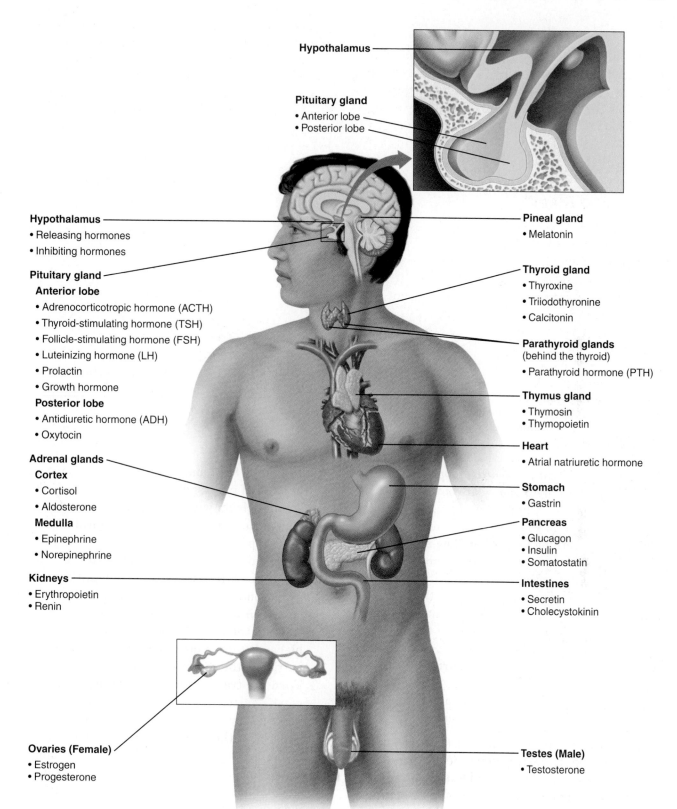

Hypothalamus

Pituitary gland
• Anterior lobe
• Posterior lobe

Hypothalamus
• Releasing hormones
• Inhibiting hormones

Pituitary gland
  Anterior lobe
  • Adrenocorticotropic hormone (ACTH)
  • Thyroid-stimulating hormone (TSH)
  • Follicle-stimulating hormone (FSH)
  • Luteinizing hormone (LH)
  • Prolactin
  • Growth hormone
  Posterior lobe
  • Antidiuretic hormone (ADH)
  • Oxytocin

Adrenal glands
  Cortex
  • Cortisol
  • Aldosterone
  Medulla
  • Epinephrine
  • Norepinephrine

Kidneys
• Erythropoietin
• Renin

Ovaries (Female)
• Estrogen
• Progesterone

Pineal gland
• Melatonin

Thyroid gland
• Thyroxine
• Triiodothyronine
• Calcitonin

Parathyroid glands
(behind the thyroid)
• Parathyroid hormone (PTH)

Thymus gland
• Thymosin
• Thymopoietin

Heart
• Atrial natriuretic hormone

Stomach
• Gastrin

Pancreas
• Glucagon
• Insulin
• Somatostatin

Intestines
• Secretin
• Cholecystokinin

Testes (Male)
• Testosterone

FIGURE 13.2   **Endocrine glands.**

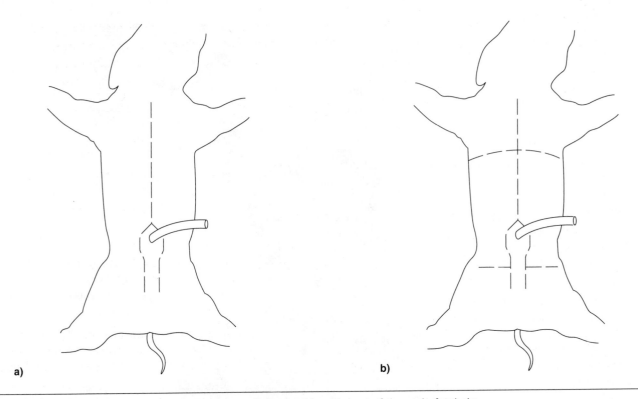

FIGURE 13.3 **Fetal pig dissection. a)** Diagram outlining a midsagittal cut of the male fetal pig.
b) Diagram outlining transverse cuts of the male fetal pig.

First, consider a few words about safety. Eye protection, especially if you wear contact lenses, is always a good idea when working with anything that could splash in your eyes. Your instructor will advise you concerning proper handling and disposal of the fluid that comes with your specimen, including the policy regarding gloves and other precautions. Your dissection equipment, particularly the scalpel, is extremely sharp. Always make certain that your fingers are clear of an area that you are cutting toward, and never use a scalpel to point to an area, particularly if you or a partner are holding that area open with your fingers. Read and follow the directions below slowly and carefully—impatience and sharp cutting tools do not mix.

1. Obtain a fetal pig, and place it faceup on a dissection tray. Use string to tie one of the forelegs, run the string around the underside of the tray, and tie the other foreleg. Repeat this procedure for the hind legs. After you cut the pig open, this tying procedure and additional wrapping of loose string around the legs will help to keep the inside of the pig open and easily accessible.
2. Before beginning the actual dissection, see Figure 13.3a–b for a diagram of the cuts you

will make for a male fetal pig. If you have a female fetal pig, read the note in step 3. Make a small incision with a scalpel in the abdominal wall about ½ inch above the umbilical cord. Put the scalpel aside, and use your scissors to continue this cut all the way up into the middle of the neck. It is important that you cut while gently pulling upward with the scissors to avoid cutting internal organs. It is necessary to cut through the ribs (use bone snips here if available) and sternum as you cut through the ventral portion of the thoracic cavity.
3. Turn the tray around to make the next series of cuts easier. Continuing to use your scissors, make a "keyhole-shaped" cut around the umbilical cord of the male pig, leaving about ½ inch of material around the cord. This incision is important to avoid cutting blood vessels under the umbilical cord that you may be instructed to view in a later exercise. The keyhole-shaped cut prevents cutting the penis of the male fetal pig, which is not yet developed and barely visible as a thin tube running from the pelvic area to the umbilical cord. *Note:* If you have a female fetal pig, the keyhole-shaped cut is unnecessary,

and you may simply circle around the umbilical cord, leaving a ½-inch border around it.

4. Next, make lateral cuts with scissors as shown in Figure 13.3b. These cuts should be just below the forelegs and just above the hind legs. So that you may more easily open up the body cavity, snip some of the diaphragm, which will be found inside the pig above the liver.

5. Observe Figure 13.4, and other fetal pig figures found in this manual (Figures 17.5, 18.3, and 19.7). Familiarize yourself with the major organs of the fetal pig. Although the focus of this exercise is the endocrine organs, you may find this step is helpful so that you will be able to follow the directions for locating the endocrine glands in the following steps.

6. Directly over and above the heart is a loose, tan-colored covering of glandular tissue. This is the **thymus,** which is proportionally much larger in the fetal pig than it would be in an adult pig. What is the function of the thymus?

_____

_____

7. Above the thymus in the lower neck area is the **thyroid,** which is often covered by the upper part of the thymus, muscle tissue in the neck, and possibly other tissues as well. Use your forceps to gently pull away the muscle tissue and other tissue covering the lower neck region until you see the small, brown, bean-shaped structure. What two hormones are produced by the thyroid?

_____

Do these hormones have similar or different functions? Explain.

_____

_____

_____

8. Using your fingers and/or a blunt probe as needed, lift up (do *not* cut and remove) the liver on the pig's left side (your right). Then, lift up the saclike stomach, and use forceps to break and remove the thin, plastic like material under the stomach. You should now be able to see the **pancreas,** which looks somewhat like a

miniature ear of corn or many tiny beads stuck together. Why do you suppose the pancreas is so close to the stomach and intestines?

_____

_____

_____

9. Lift, but again do *not* cut, the stomach and intestines on the *pig's* left side. Push them as far to the *pig's* right as you can without breaking them. You should be able to see the "kidney bean-shaped" kidney stuck into the back body wall. Use forceps to break and remove the thin, plastic like material covering the kidney. Near the top and middle section of the kidney, you should see a lighter strip of material that might have become partially disconnected from the kidney. This is the **adrenal gland,** which is named for its close relationship to the kidney (*renal* = kidney). Which hormone of the adrenal gland is most closely associated with the function of the kidney?

_____

10. Next, find the gonads. The **ovaries** are very small bean-shaped structures attached to the wavy uterus in the lower abdominopelvic cavity of the female fetal pig. To find the **testes,** use a scalpel to carefully cut the surface of the scrotum, which appears as two pockets of loose skin in the groin area of the fetal pig. It is only necessary to cut the skin and connective tissue directly underneath the skin. Palpate the area, and find the small, brown, bean-shaped testes covered by a membrane. Find a classmate with a fetal pig of the opposite sex, and compare the two gonad types.

11. When finished, slide the string out from the bottom of the tray (do *not* cut or untie the string). Carefully clean and dry your dissecting tools. Spray your work area with the disinfectant and wipe it down. Your instructor will tell you how to prepare your pig for storage for use in future lab exercises.

12. Using Figure 13.2 as a reference, locate the following endocrine glands in the human torso models.

- Adrenal glands
- Ovaries
- Pancreas
- Testes
- Thymus
- Thyroid ■

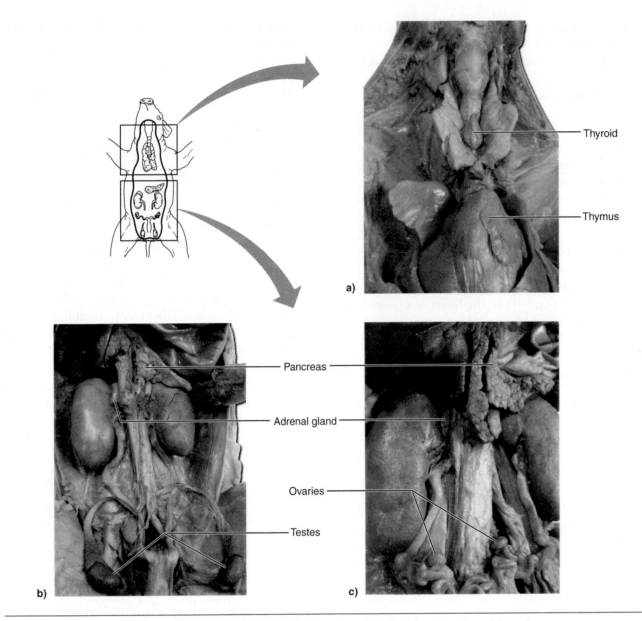

FIGURE 13.4 **Major endocrine glands of the fetal pig. a)** Thoracic cavity. **b)** Male abdominopelvic cavity. **c)** Female abdominopelvic cavity.

ACTIVITY 3

Observing Selected Endocrine Glands—The Pancreas and Thyroid Gland

Materials for This Activity

Compound light microscope

Prepared slide of the pancreas

Prepared slide of the thyroid gland

Observe a prepared slide of the pancreas, using your microscope's 10× objective lens (100× total magnification). Your instructor may elect to set this up as a demonstration slide in order to ensure that a pancreatic islet (islet of Langerhans) is in your field of view. Note that the islets are relatively small clusters of cells, and they make up a relatively small proportion of pancreatic cells.

What two hormones are produced by the pancreatic islets? Summarize the function of each hormone.

_____

_____

_____

_____

Considering the importance of these hormones, how is it that so few cells can do such a big job?

_____

_____

_____

Observe a prepared slide of the thyroid gland using your microscope's 10× objective lens. The large, somewhat round structures are not cells, but **thyroid follicles** that store the materials needed to make thyroid hormone. With higher magnification, you should also clearly see the **follicular cells,** which are the small cells with (usually) round nuclei that border the follicles. Clumps of cells not bordering the follicles are **parafollicular cells.**

What hormone is produced by the parafollicular cells?

_____

_____

What aspect of homeostasis is controlled by this hormone?

_____

_____

_____

Using your textbook or other source, identify what other gland and hormone plays a role in this aspect of homeostasis.

_____

_____

_____

THE ENDOCRINE SYSTEM
Critical Thinking and Review Questions

1. Why can an endocrine gland be a great distance from its target tissue or organ?

   _____

   _____

2. Why do some people refer to the pituitary as the "master gland," and what part of the body controls the pituitary?

   _____

   _____

   _____

3. Match the hormone to its endocrine gland.

   ____ Insulin   a. adrenal medulla

   ____ Calcitonin   b. pancreas

   ____ Epinephrine   c. pituitary

   ____ Glucocorticoids   d. thyroid

   ____ Growth hormone   e. adrenal cortex

4. Match the hormone to its function.

   ____ Thyroid hormone   a. stimulates smooth muscle contraction in the
                                              uterus

   ____ Aldosterone   b. stimulates kidney tubules to reabsorb sodium
                                              and excrete potassium

   ____ Antidiuretic hormone (ADH)   c. stimulates the mammary glands

   ____ Oxytocin   d. stimulates body cells to increase metabolism

   ____ Prolactin (PRL)   e. stimulates water reabsorption by certain
                                              kidney tubules

5. Why is it best to minimize the use of your scalpel during dissection activities?

   _____

   _____

   _____

   _____

6. Which dissecting tools should be used for pointing to structures, removing connective tissue, and separating structures, and why?

_____

_____

_____

_____

7. Which two endocrine glands are made of a mixture of nervous and epithelial tissues?

_____

8. Fill in the following chart, describing the location and hormones produced by each gland.

Gland	Location	Hormone(s)
Adrenal		
Ovary		
Pancreas		
Testis		
Thymus		
Thyroid		

The Cardiovascular System I: Blood

Objectives

After completing this exercise, you should be able to

1. Describe the functions and components of blood.
2. Perform the hematocrit test and calculate the hematocrit values.
3. Sketch the red blood cell and five white blood cells.
4. Explain the terms *antigen, antibody,* and *antigen-antibody complex.*
5. Describe the characteristics of the ABO blood type.
6. Describe the characteristics of the Rh blood type.
7. Perform the blood typing procedure.
8. Properly dispose of blood-contaminated materials.

Materials for Lab Preparation

Equipment
- Compound light microscope
- Calculator

Prepared Slides
- Normal blood
- Pathological blood

Supplies

- Heparinized capillary tubes
- Microhematocrit centrifuge
- Microhematocrit reader
- Seal-Ease or modeling clay
- Sterile lancets
- Alcohol prep pads
- Band-Aids
- Plastic centimeter ruler
- Disposable gloves
- "Sharps" container for sharp objects
- Autoclave bag
- Cleaning disinfectant or detergent solution in a spray bottle
- Paper towels
- Container of 10% bleach solution
- Freshly drawn animal blood (alternative choice for hematocrit test)
- Simulated human blood (alternative choice for blood typing)
- Anti-A, anti-B, and anti-D blood typing sera
- Disposable slides for blood typing
- Toothpicks
- Wax marker

Introduction

The cardiovascular system is composed of the heart, blood vessels, and blood. In this exercise, we will explore the different aspects of blood.

Blood serves several critical functions for the body. Our blood transports oxygen and nutrients in the form of blood glucose to all the cells of the body, and on the return trip, takes away the waste products of cells' metabolic processes. Hormones from the endocrine system, which serve to regulate growth and development, are also delivered by the blood. Other important functions of blood include the regulation of body temperature, fluid volume, and pH balance. Finally, blood defends us in two important ways: The white blood cells protect us against infections and diseases and the platelets initiate a clotting mechanism to prevent excessive blood loss.

Components of Blood

There are four components of blood: red blood cells (RBCs), white blood cells (WBCs), platelets, and plasma (which consists of water, ions, proteins, hormones, gases, nutrients, and wastes). Table 14.1

lists the components of blood and their specific functions.

When anemia is suspected, a **hematocrit test** is performed to determine the relative percentage of red blood cells in the blood. A blood sample is centrifuged, or spun, until the lightest components are at the top and the heaviest components are at the bottom. The hematocrit test, also called **packed cell volume (PCV)**, measures the volume of each blood component. As indicated in Table 14.1, the main function of red blood cells is to carry oxygen, so the hematocrit is often used as a measure of the oxygen-carrying capacity of an individual.

ACTIVITY 1

Measuring Blood Components: The Hematocrit Test

Part 1: Performing a Hematocrit Test

Note to Instructor: Following are two activities that demonstrate the hematocrit test. Part 1 is recommended for students using labs with the proper equipment and supplies. Part 2 is designed for students

---

Table 14.1 **Components of Blood**

Blood Component	Examples and Functions
Formed Elements (45%)	
Red blood cells	Transport oxygen and carbon dioxide to and from body tissues.
White blood cells	Defend body against invading microorganisms and abnormal cells.
Platelets	Initiate clotting to prevent blood loss.
Plasma (55%)	
Water	The primary component of blood plasma. Water serves as the major transport medium in plasma.
Electrolytes (ions)	Sodium, potassium, chloride, bicarbonate, calcium, hydrogen, magnesium, and others. Ions contribute to the control of cell function and volume, to the electrical charge across cell membranes, and to the function of excitable cells (nerve and muscle).
Proteins	Albumins maintain blood volume and transport electrolytes, hormones, and wastes. Globulins serve as antibodies and transport substances. Fibrinogens contribute to blood clotting.
Hormones	Insulin, growth hormone, testosterone, estrogen, and others. Hormones are messenger molecules that initiate metabolic processes.
Gases	Oxygen is needed for metabolism; carbon dioxide is a waste product of metabolism. Both are dissolved in plasma as well as carried by RBCs.
Nutrients and wastes	Glucose, urea, heat, and other raw materials. Nutrients and wastes (including heat) are transported by plasma throughout the body.

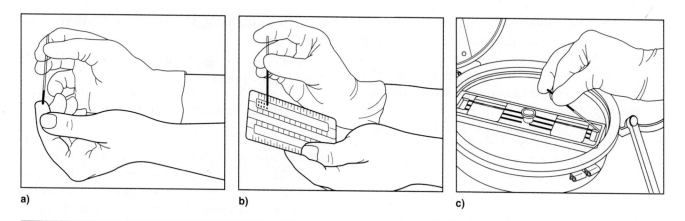

a)   b)   c)

FIGURE 14.1   **Steps in performing a hematocrit test. a)** Fill a heparinized capillary tube with blood.
b) Plug the capillary tube with clay. **c)** Load the tube in a microhematocrit centrifuge.

whose lab instructors have decided to simulate this procedure in a dry lab rather than perform it.

Materials for This Activity

Heparinized capillary tubes

Microhematocrit centrifuge

Microhematocrit reader

Seal-Ease or modeling clay

Sterile lancets

Alcohol prep pads

Band-Aids

Plastic centimeter ruler

Disposable gloves

"Sharps" container for sharp objects

Autoclave bag

Cleaning disinfectant or detergent solution

Paper towels

Container with 10% bleach solution

Freshly drawn animal blood (alternative choice)

Safety Precautions

- Please read the Review of Lab Safety Protocols section in Exercise 1.

- Remember that some infectious diseases are associated with the contact of blood. Protect yourself by following all instructions carefully. Perform only your own punctures and **handle only your own blood.** Wear gloves when you are in contact with slides containing other people's blood or animal blood and throughout this activity, including cleanup.

- Note that there are several types of lancet and methods of use. Please make sure you receive clear instructions on how to lance your fingertip to draw blood.

- Keep in mind that you are using a sharp instrument to draw your own blood. Concentrate and be alert.

Note: Your instructor will determine whether your own blood or animal blood will be used for this activity.

1. If you are using your own blood, obtain all of the supplies listed previously except for the animal blood, and return to your station.

   Clean one of your fingertips with the alcohol prep pad, and prick the fingertip with a lancet. When the blood is flowing freely, hold the unmarked end of the capillary tube to the blood drop, and fill the tube to the red line by capillary action (Figure 14.1a). Because blood clots quickly, the capillary tubes contain **heparin** to prevent **coagulation,** or clotting. If possible, prepare two samples. If the blood is not flowing freely, the end of the capillary tube will not be completely submerged in the blood during filling and air will enter. If this happens or you are not able to fill the tube to fully three-fourths, you will not be able to obtain an accurate reading, and it would be best to prepare another sample.

   Use a Band-Aid and, if necessary, talk to your instructor should you continue to bleed.

   If you are using the animal blood, you should wear gloves. Immerse the unmarked end of the capillary tube into the blood sample, and let it fill to the red line by capillary action.

2. Plug the blood-containing end of both capillary tubes by pressing it into the Seal-Ease or modeling clay (Figure 14.1b).

3. Give your two tubes to the instructor, who will load and operate the microhematocrit centrifuge (Figure 14.1c). Be sure to record the numbers of the grooves in which your tubes are placed. The centrifuge will be on for approximately five minutes.

   During the operation of a centrifuge, what causes the heavier portions to settle to the bottom of the tube?

   _____

   _____

4. Upon getting your tubes back, measure the percentages of RBCs, WBCs, and plasma. There are two ways of making these measurements. You may use the microhematocrit reader, or use a centimeter ruler to measure the height of each component or layer. The RBCs are the red bottom layer, the WBCs and platelets are the white middle layer, and the plasma is the yellowish fluid on the top layer. Record your height measurements in Box 14.1.

5. Cleanup. Place the used lancet in the "sharps" container. Place the blood-stained paper towels in the autoclave bag. Place the capillary tubes in the bleach jar. Spray some disinfectant on your counter area, and wipe it clean.

6. Go to Part 3: Calculating the Hematocrit Values.

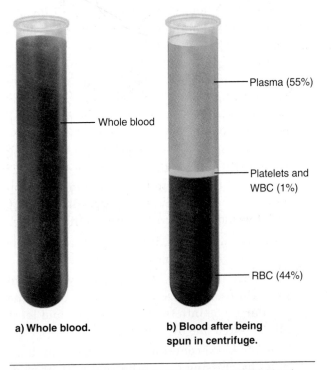

a) **Whole blood.**

b) **Blood after being spun in centrifuge.**

FIGURE 14.2 **Hematocrit test results. a)** Whole blood sample before centrifuging. **b)** Whole blood sample after centrifuging.

Part 2: Simulating the Hematocrit Test (Dry Lab)

Note: If you have completed Part 1, please go to Part 3: Calculating the Hematocrit Values.

Materials for This Activity

Plastic centimeter ruler

1. Let's imagine that a thin tube of blood is centrifuged for four minutes. Remember that blood clots quickly, so some **heparin** will be added to prevent **coagulation,** or blood clotting. As the tube is spun, the forces of gravity cause the heaviest part of the blood to settle at the bottom of the tube and the lightest part to move to the top. Thus, spinning a tube of blood will produce a gradient of the lightest to heaviest components, from the top to the bottom of the tube. This procedure is called a **hematocrit test,** and it represents a common lab test for blood.

2. Figure 14.2a shows the original tube of blood before spinning. Figure 14.2b shows the result after it has been centrifuged. Notice there are three major components of blood: a large portion at the top, a very small portion in the middle, and another large portion at the bottom.

3. If you held the centrifuged tube in your hand, you would notice that the top portion was an amber fluid, the middle portion was a white-colored solid substance, and the bottom portion was a deep-red solid substance. These three sections represent the packed cell volume of blood: plasma at the top, platelets and white blood cells in the middle, and red blood cells at the bottom.

4. Obtain a plastic metric ruler. First, observe the ruler. If each numbered centimeter has 10 smaller lines, each line is a millimeter. Some metric rulers have only five lines per centimeter, and each line would then be 2 mm. If necessary, refresh your memory of the metric measurements for length in Exercise 1.

5. Measure the total height (in cm or mm) of the column of blood in the capillary tube shown in Figure 14.2a. Measure from the bottom of the test tube to the top of the blood sample. Measure the height as accurately as possible. This means that your reading may not be a whole number, but it may include fractions or decimals.

6. Using the same units as in step 5 (cm or mm), measure the height of each element in the column of blood (Figure 14.2b). Due to the small

size, the middle portion will be hard to read; in fact, it will be in the millimeter range. These respective measurements represent the plasma, platelet and WBC, and RBC heights.

7. Record the four measurements under the Height column in Box 14.1.

8. Continue with Part 3: Calculating the Hematocrit Values.

Part 3: Calculating the Hematocrit Values

Materials for This Activity

Calculator

1. Using the height measurements you recorded in Box 14.1, calculate the percentage of each of the blood elements as follows:

$$\text{plasma}\% = \frac{\text{plasma layer height}}{\text{total height}} \times 100$$

$$\text{platelet and WBC}\% = \frac{\text{platelet and WBC layer height}}{\text{total height}} \times 100$$

$$\text{RBC}\% = \frac{\text{RBC layer height}}{\text{total height}} \times 100$$

2. Record the three calculations under the PCV% column in Box 14.1.

3. You have just calculated the **PCV%** for each of the three components of blood. Your values should be close to that of normal blood. The PCV% values for normal blood are as follows:

plasma%: 55%

platelet and WBC%: 1%

RBC%: 44% (also called the hematocrit)

4. The **hematocrit** is the PCV% for RBCs. This is a simple way of expressing your oxygen-carrying capacity. In the United States, average hematocrit values for males range from 42% to 52%; for females, the range is from 37% to 47%. If your hematocrit is lower than these values, it may mean that you are anemic, a disorder of insufficient RBCs. Speculate on the reason for higher hematocrit values in males.

Males are typically taller and bigger than the average female and needs more oxygen

5. Not all shifts from these reference values are due to disorders. If you moved to a city situated at a high elevation, for example, Denver, where

the air is "thinner" with less oxygen per volume of air, what immediate physical changes would you expect to experience?

They would get tired with little oxygen

After several weeks in this environment, what kind of shift in hematocrit would you expect to see?

The hematocrit would probably decrease due to lack of oxygen

Box 14.1 Hematocrit Calculations		
Blood Elements	**Height (cm or mm)**	**PCV%**
RBC	19 mm	
WBC and platelets	3 mm	
Plasma	50 mm	
Total	72 mm	**100%**

Blood Cells

The proper term for red blood cells is **erythrocytes.** There is only one type of erythrocyte. The function of erythrocytes is to transport gases. Earlier, you learned that the hematocrit, which is the erythrocyte percentage of blood, is a measure of the oxygen-carrying capacity of the body.

The proper term for white blood cells is **leukocytes.** There are five types of leukocytes. The general function of leukocytes is defense, and each type of leukocyte has a different way of defending the body.

Let's look at the actual structure of the blood cells by examining slides of normal and pathological blood.

ACTIVITY 2

Observing Erythrocytes and Leukocytes

Materials for This Activity

Compound light microscope

Prepared slides

Normal blood
Pathological blood

1. Obtain a microscope and a sample of each type of blood slide.

2. Start with the slide of normal blood. Using the coarse-adjustment knob, focus on the blood cells with the scanning-power lens. Now change to the low-power lens, and focus until you can see the individual cells. If necessary, change to the high-power (high-dry) lens, and focus carefully until you can see the details of the individual cells.

3. The numerous pinkish-red cells are **erythrocytes,** which appear lighter in the middle because the cells are thinner there. Scattered here and there are a few leukocytes, which look like transparent cells with purple lobes. Notice all of the leukocytes are similar in size, but the purple lobes of the nuclei are shaped differently.

   Why do you think there are so many more erythrocytes than leukocytes?

   More red than white

4. Observe the round, disklike appearance of the erythrocytes. In Box 14.2, draw a few erythrocytes as they appear on the slide. Remember to record the magnification of your samples.

5. Observe the differences between the five **leukocytes** illustrated in Figure 14.3. The agranular leukocytes are composed of monocytes and lymphocytes. What is the main structural difference between monocytes and lymphocytes?

   Nucleus

The granular leukocytes are composed of the neutrophils, eosinophils, and basophils. What are the main structural differences between neutrophils, eosinophils, and basophils?

Complexity of organelle

Examining your blood slide, find and draw a few of each type of leukocyte, and write the total magnification used in the space provided in Box 14.2. You may have to move the slide around a bit to find each type of leukocyte. What appears to be the *main* structural difference between the agranular and granular leukocytes?

Granular are dense and complex

6. Focus on a slide of pathological blood. Because each campus has a different collection of pathological slides, we will only make some general observations. Infections involve the proliferation of specific leukocytes, so depending on the type of disease, certain types of leukocytes will be more prevalent. For example, sickle cell anemia is a genetic disorder in which the red blood cells have a crescent or sickle shape, unlike the usual round, concave disk shape.

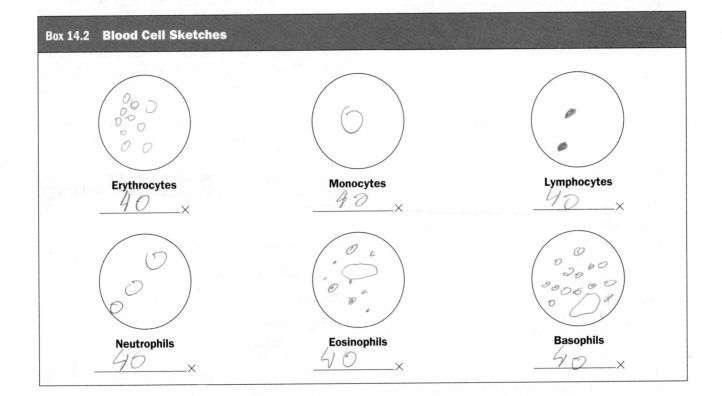

Box 14.2   Blood Cell Sketches

Erythrocytes _40_ ×

Monocytes _40_ ×

Lymphocytes _40_ ×

Neutrophils _40_ ×

Eosinophils _40_ ×

Basophils _40_ ×

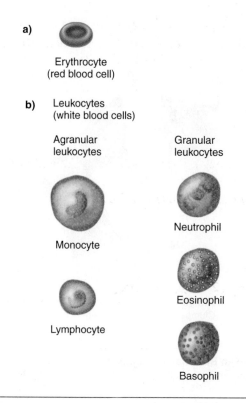

a)

Erythrocyte
(red blood cell)

b) Leukocytes
(white blood cells)

Agranular
leukocytes

Granular
leukocytes

Monocyte

Neutrophil

Lymphocyte

Eosinophil

Basophil

FIGURE 14.3 **Blood cells. a)** Erythrocytes. **b)** Leukocytes.

Compare the cells of the normal blood slide with the cells of the pathological blood slide. Describe some differences.

The white blood cell outnumber the red. The majority is purple than pink

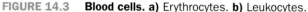

Blood Types

Successful transfusions of blood from one person to another depend upon the compatibility of the people's blood types. The right transfusion can save lives; the wrong one can result in severe illness or even death for the recipient.

Your blood type is determined by your genes, and it does not change during your lifetime. Before you can understand about blood type, you must learn some new terms: *antigen, antibody,* and *antigen-antibody complex.*

Antigen

The surfaces of your erythrocytes have specific proteins called "self" surface proteins, which allow your immune system to identify your erythrocytes

as your own cells (Figure 14.4a). The surfaces of erythrocytes from other blood types contain foreign surface proteins, called antigens, which allow your immune system to identify these erythrocytes as foreign cells (Figure 14.4a). There are different antigens, depending on your blood type.

Antibody

Your plasma contains specific plasma proteins called antibodies, which mount an attack on cells identified as foreign cells (Figure 14.4a). There are different antibodies, each one specialized to attack a particular antigen.

Antigen-Antibody Complex

Once the antigens have been identified, the antibodies mount an attack on foreign erythrocytes, cells, or chemical substances by binding to the antigen and forming an **antigen-antibody complex** (Figure 14.4b). Incompatible transfusions lead to the formation of an antigen-antibody complex, which then leads to agglutination (clumping together of RBCs) and hemolysis (rupture of RBCs). In essence, incompatible transfusions are dangerous because the RBCs are critically damaged.

Figure 14.5 shows the different antigens and antibodies found in each blood type. Notice that in the United States, the distribution of blood types in Caucasians, African Americans, and Native Americans differs significantly.

There are two classifications for blood type: ABO and RH. The **ABO blood type** is based on the type of antigens on the red blood cells. As indicated in Figure 14.5, blood type A has the A antigen on the red blood cell. Note, however, that the plasma in blood type A contains B antibodies. Look at blood type O, which is mistakenly called the "**universal donor.**" In this blood type, the red blood cell does not have either A or B antigens, but the plasma contains both A and B antibodies. This means that if you are type A, you cannot receive a large, whole-blood transfusion of type O blood. Why is this? (*Hint:* What is the effect of the type O antibodies on the type A red blood cells?)

It will dominate and kill your own blood cells

Type AB blood is also mistakenly called the "universal recipient." If you have type AB blood,

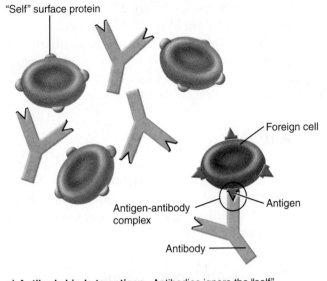

"Self" surface protein

Foreign cell

Antigen-antibody complex

Antigen

Antibody

a) Antibody binds to antigen. Antibodies ignore the "self" surface proteins but bind to the antigen of the foreign cell, forming an antigen-antibody complex.

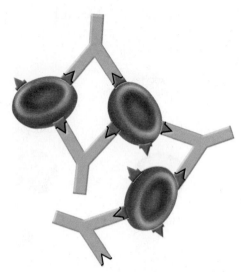

b) Antigen-antibody complexes clump together. Clumping effectively inactivates the foreign cells.

FIGURE 14.4 **The role of antibodies in recognizing and inactivating foreign RBCs. a)** The antibody recognizes the foreign RBC and binds to its antigen. **b)** The formation of the antigen-antibody complexes causes foreign RBCs to clump, thus rendering them inactive.

your red blood cells contain both A and B antigens. If you receive a large, whole-blood transfusion of type O blood, what is the effect? (*Hint:* What is the effect of the type O antibodies on the type AB red blood cells?)

It will fuse together with minimal effects

The Rh blood type is due to still another red blood cell surface antigen. It is called the Rh factor, because it was first discovered in rhesus monkeys. In the United States, 85% of the population is **Rh+.** This means that their red blood cells have the Rh factor and their plasma does not have Rh antibodies.

Normally, Rh− individuals do not have Rh antibodies in their plasma, but once there is an initial exposure to Rh+ blood, they will begin to produce Rh antibodies. This means that the first time an Rh− individual receives an Rh+ transfusion, there may be no danger. But the second Rh+ transfusion may cause a severe reaction. A special situation arises for an Rh− mother carrying an Rh+ fetus. The leakage of the fetal Rh+ blood into the mother's Rh− bloodstream may cause the mother to begin producing Rh antibodies. The result may be a form of fetal anemia, which is potentially life threatening.

In the following activity, we will determine blood type using anti-sera or chemical solutions with antibodies. The way an anti-sera reacts with blood indicates blood type.

ACTIVITY 3

Blood Typing

Materials for This Activity

Simulated blood (alternative choice for blood typing)

Anti-A, anti-B, and anti-Rh blood typing sera

Sterile lancets

Alcohol prep pads

Disposable gloves

Disposable slides for blood typing

Toothpicks

Wax marker

Band-Aids

Autoclave bag

Cleaning disinfectant or detergent solution

Paper towels

"Sharps" container for sharp objects

Container with 10% bleach solution

Safety Precautions

- Please read the Review of Lab Safety Protocols section in Exercise 1.

- Remember that some infectious diseases are associated with the contact of blood. Protect yourself by following all instructions carefully. Perform only your own punctures and handle only your own blood. Wear gloves when you are in contact with slides containing other people's blood, and throughout this activity, including cleanup.

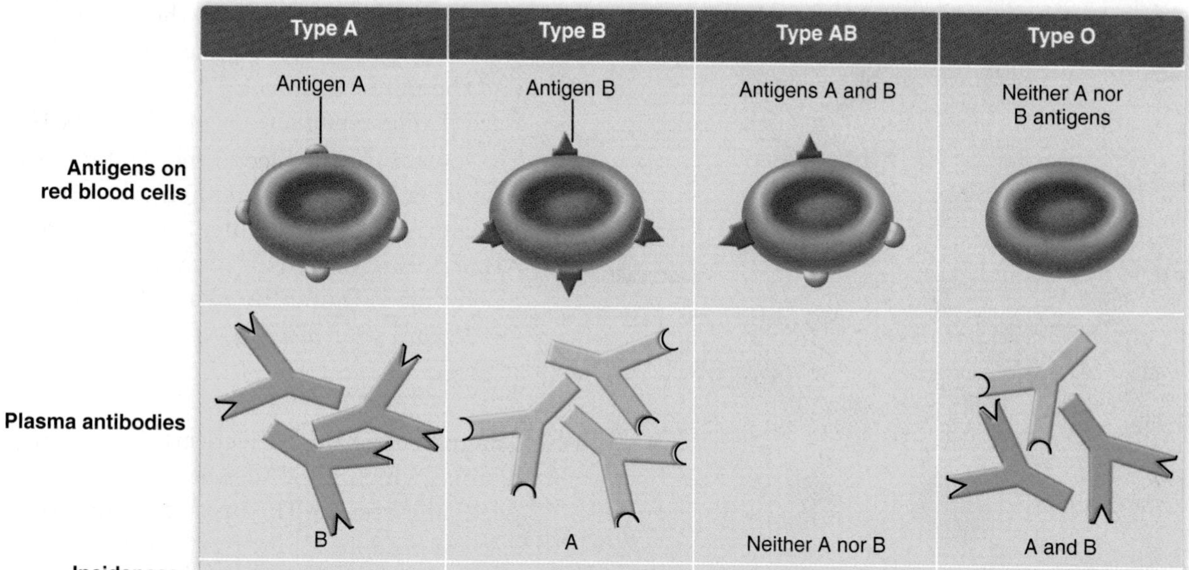

	Type A	Type B	Type AB	Type O
Antigens on red blood cells	Antigen A	Antigen B	Antigens A and B	Neither A nor B antigens
Plasma antibodies	B	A	Neither A nor B	A and B
Incidences:				
Caucasians	40%	11%	4%	45%
African Americans	26%	19%	4%	51%
Hispanics	31%	10%	2%	57%
Asians	28%	25%	7%	40%
Native Americans	8%	1%	0%	91%

FIGURE 14.5   Characteristics of ABO blood types.

- Note that there are several types of lancet and methods of use. Please make sure you receive clear instructions on how to lance your fingertip to draw blood.
- Keep in mind that you are using a sharp instrument to draw your own blood. Concentrate and be alert.

Note: Your instructor will determine whether your own blood or simulated blood will be used for this activity.

1. Obtain the indicated supplies from the supply area.

2. Cover your work area with paper towels, on which you can place blood-contaminated instruments and slides.

3. Mark slide 1 with a wax pencil as follows: Draw a line in the middle, mark the left upper corner "anti-A" and the right upper corner "anti-B." Place a drop of anti-A sera on the middle of the left side and a drop of anti-B sera on the middle of the right side. Place a toothpick on the left side and another toothpick on the right side.

4. Mark slide 2 with a wax pencil as follows: Write "anti-Rh" on the top of the slide. Place a drop

of anti-Rh sera in the middle of the slide. Then place a toothpick on the slide.

5. If you are using your own blood, thoroughly wash your hands with soap and dry them with paper towels. Clean your fingertips with an alcohol prep pad.

6. Pierce your fingertip with the lancet and wipe away the first drop of blood. Then place one drop of freely flowing blood on each of the three marked sections of the slides. Press your thumb over the fingertip to stop the blood flow. Then quickly mix each blood-antiserum sample with the designated toothpick. Dispose of the toothpicks and used alcohol prep pad in the autoclave bag. Dispose of the lancet in the sharps container.

Place a Band-Aid, if necessary, on your fingertip. If you continue to bleed, talk to your instructor.

If you are using simulated blood, use a medicine dropper to place one drop of blood on the left- and right-hand sides of slide 1 and a drop of blood on slide 2.

7. Observe the slide results after two minutes. **Agglutination,** or clumping, indicates an **antigen-antibody complex** has formed between

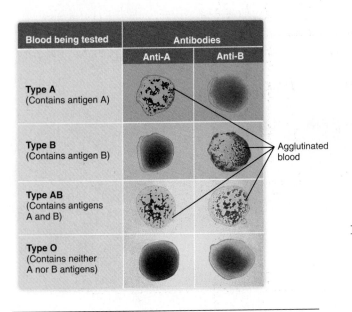

FIGURE 14.6   **Agglutination in ABO blood typing.**

the antiserum and erythrocytes. Use Figure 14.6 to help you interpret your ABO results. Type AB blood will agglutinate with both the anti-A and anti-B sera. Type A blood will agglutinate only with the anti-A serum. Type B blood will agglutinate only with the anti-B serum. Type O blood will not agglutinate with either anti-A or anti-B sera. Record your results in Box 14.3.

Sometimes, it is clear that agglutination has or has not occurred. It may be hard, however, to tell the difference between dried blood and agglutination. Ask your instructor to confirm your interpretations of agglutination.

8. Figure 14.6 does not illustrate agglutination for the Rh factor. Rh+ blood will agglutinate with the anti-Rh serum. Rh− blood will not. Agglutination in Rh+ blood is sometimes rather fine and hard to observe. If necessary, observe the slide under the microscope, or ask your instructor for assistance in determining agglutination.

The convention for labeling the Rh blood type is to use the plus symbol (+) to indicate Rh+ blood and the minus symbol (−) to indicate Rh− blood. Record your results as (+) or (−) in Box 14.3.

9. To determine your blood type, first determine if it is A, B, AB, or O, and then add (+) or (−) as appropriate. For example, if your blood type is AB and Rh+, you would record it as AB+. If your blood type is O and Rh−, you would record it as O−. Record your blood type.

Mr. Green; AB+

10. The instructor may select certain student slides against which all students may view and compare their results. The instructor may also focus certain student slides in a microscope to demonstrate the difference between agglutination and dried blood. It is useful to observe a range of slides and to see how blood agglutinations may look very different from slide to slide and between the naked eye and microscope.

Caution: Look without touching. Remember to use gloves if you handle any microscopes or slides containing other people's blood.

If you were able to view other slides, what were your observations about the different types of agglutination?

Mr. Smith; A+, Mr. Jones; B−
Ms. Brown; O−

11. Cleanup. Wear gloves. Place the used slides in the jar of bleach. Place gloves, toothpicks, and the blood-stained paper towels in the autoclave bag. Spray some disinfectant on your counter area and wipe it down. If the microscope was used, wipe it down. Remember that the next student to use the counter is relying on you for a clean, safe work area and instruments. ■

Box 14.3 **Blood Type Results**		
Results	**Observed**	**Not Observed**
Clumping with anti-A	*Mr. Smith* // *Mr. Green*	
Clumping with anti-B	*Mr. Green* // *Mr. Jones*	
Clumping with anti-Rh	*Mr. Smith* // *Mr. Green*	

THE CARDIOVASCULAR SYSTEM I: BLOOD

Critical Thinking and Review Questions

1. What are four functions of blood cells?

   Store hemoglobin, bind oxygen, fight bad bacteria, carry oxygen.

2. What are four functions of plasma?

   Transport medium, control cell, maintain blood volume, carry messengers.

3. What are three types of plasma proteins?

   Globulins, fibrinogens, albumins

4. What are two gases found in the plasma?

   Oxygen, carbon dioxide

5. Discuss the main difference between agranular and granular leukocytes.

   Agranular are made of monocytes, granular has neutrophils

6. Why do you think there are so many more erythrocytes than leukocytes?

   Erythrocytes transport oxygen and carbon dioxide. Leukocytes just defend against invasions.

7. Describe the hematocrit test procedure.

   A sample is centrifuged/spun, then the heavy components are at the bottom, light at the top.

8. What is the hematocrit? What does it measure?

   This is to measure red blood cells.

9. Why would you expect the hematocrit to be lower in an anemic individual?

   An anemic person has low red blood cells

10. What does the PCV% mean?

    Measures the volume of each blood component

11. What is coagulation? What prevents coagulation during the hematocrit test?

    Coagulation is when blood clots. Heparin is used
    to prevent coagulation.

12. Name any antigens and antibodies present in type A+ blood.

    A contains A antigens, B antibody

13. Name any antigens and antibodies present in type O– blood.

    O contains neither A or B. antyer
    A+B antibody

14. What does agglutination mean in blood typing tests?

    When antigens are mixed with corresponding
    antibodies.

15. Type O individuals used to be called universal donors because their blood could
    be donated to any other blood type. This label is now considered outdated because
    transfusion reactions can occur. Speculate on the reasons for the transfusion reaction.

    When O overpowers A/B blood, it will then clot

16. What might be the problems for a type A+ recipient who receives a type A– whole-blood
    transfusion?

    The body will accept the A blood, but not
    the Rh+. The body wont accept A–, but would accept A+
    clot on recipient blood

17. What might be the problems for a type AB recipient who receives a type O whole-blood
    transfusion?

    AB can have a small amount of type O, but if
    too much is transfused, the O will clot AB

EXERCISE
15

The Cardiovascular System II: Heart and Blood Vessels

Objectives

After completing this exercise, you should be able to

1. Describe the relationship of the heart to the lungs, rib cage, and diaphragm.
2. Identify the structures in the human heart.
3. Trace the blood flow through the heart.
4. Identify the structures of the sheep heart.
6. Demonstrate the procedure for measuring heart rate.
7. Compare and contrast arteries and veins.
8. Describe the reason(s) for measuring blood pressure and the method of doing it.
9. Demonstrate the use of a sphygmomanometer for measuring blood pressure.

Materials for Lab Preparation

Models and Charts

○ Torso model with removable heart
○ Human heart models (dissectible)
○ Sheep heart (preserved or fresh)
○ Anatomical charts of the systemic circuit
○ Models of the systemic circuit

Equipment and Supplies

○ Dissection equipment
○ Dissection tray
○ Disposable gloves
○ Autoclave bag
○ Paper towels
○ Blunt sticks
○ Disinfectant in spray bottle
○ Stopwatch or timer with second hand
○ Stethoscope
○ Sphygmomanometer
○ Alcohol prep pads

Introduction

This exercise continues our examination of the cardiovascular system, which consists of three components: blood, which we studied in Exercise 14, and the heart and blood vessels, which we will focus on in this exercise. The heart provides the pressure to pump blood through the network of blood vessels throughout our body. Because there are literally miles and miles of blood vessels, the extreme ends of the body, for example, the fingers and toes, would be deprived of blood without sufficient blood pressure. The blood vessels transport blood to and from each part of our body through arteries and veins.

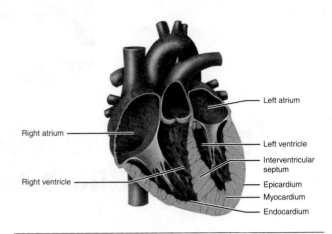

FIGURE 15.1   Layers of the heart wall. In this cross section, we can see the septum, which separates the right and left ventricles of the heart.

The Heart

The human heart, like that of any mammal, is primarily muscle. In Exercise 5, Tissues, we examined three kinds of muscle, one of which was cardiac muscle. Cardiac muscle is special because (1) it does not connect to bone like skeletal muscle, and (2) it contracts continuously to propel blood through the blood vessels to every region of the body.

The heart's natural position is in the chest, or thoracic cavity. A tough, fibrous sac, called the **pericardium,** protects the heart on the outside while anchoring it to surrounding structures. There are three layers to the wall of the heart: the **epicardium,** consisting of epithelial and connective tissue; a middle layer, the **myocardium,** which is thick and consists mainly of the cardiac muscle that forms the bulk of the heart; and the innermost layer, the **endocardium,** which consists of a thin endothelium resting on a layer of connective tissue (Figure 15.1).

The heart's structure is relatively simple. It consists of four chambers, two at the top of the heart called **atria** (singular *atrium*), and two at the bottom called **ventricles.** There is a muscular partition that separates the right and left sides of the heart, which is called the **septum** (Figure 15.1). When blood returns to the heart from the body's tissues, it enters the heart at the *right atrium* and then passes through a valve into the *right ventricle.* The right ventricle pumps blood through a second valve into the artery leading to the lungs. Blood returning from the lungs to the heart enters the *left atrium* and then passes through a third valve into the *left ventricle,* which pumps blood

through a fourth valve into the **aorta,** the body's largest artery (Figure 15.1).

In close proximity to the heart are the lungs (Figure 15.2). The lungs are intimately related to the heart, which sends blood to the lungs for oxygenation.

We will look at the position of the heart inside the thoracic cavity to see the physical relationships between the heart, lungs, trachea, and esophagus, and then we will examine the internal structure of the heart.

ACTIVITY 1

Locating the Heart and Examining Its Internal Structure

Materials for This Activity

Torso model with removable heart

Human heart models

Part 1: Locating the Heart

1. Using the torso model, locate the **heart,** which is near the center of the **thoracic cavity.** Notice that the right side of the heart rests on the diaphragm. The **lungs** surround most of the heart, which means the heart is well situated to send oxygen-poor blood to the lungs and to receive oxygen-rich blood in return (Figure 15.2).

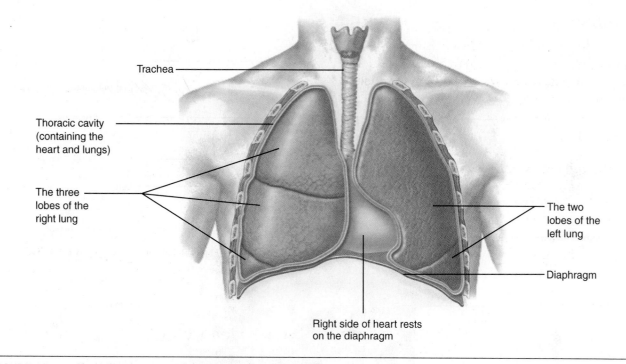

Trachea

Thoracic cavity
(containing the
heart and lungs)

The three
lobes of the
right lung

The two
lobes of the
left lung

Diaphragm

Right side of heart rests
on the diaphragm

FIGURE 15.2   The heart as it is situated in the thoracic cavity.

2. Remove the heart from the torso model. Notice that both the **trachea,** or wind pipe, and the **esophagus,** or food tube, are located near the major blood vessels on the upper portion of the heart. As you look at the tightly packed and entwined arrangement of the trachea, esophagus, and blood vessels of the heart, you can appreciate the efficient use of a small space.

3. With the heart removed, look at the point of the entry into the diaphragm for the esophagus. Notice that the stomach is underneath the heart, separated by the diaphragm. Can you see the basis for calling acid reflux "heartburn"?

4. Look for the cut rib bones in the **rib cage** on the edges of the thoracic cavity. Recall what you learned about the axial skeleton in Exercise 8.

What function does the rib cage play as it relates to the heart and lungs?

Ribs protect and keep structure

Part 2: Examining the Internal Structure of the Heart

1. Use the dissectible model of the heart to examine each of the internal structures illustrated in Figure 15.3. Notice that there are three large blood vessels. The one in the center, the **aorta,** is situated between the **superior vena cava** on the right side and the **pulmonary trunk** on the left side of the heart.

2. As mentioned earlier, there are four chambers of the heart. The upper chambers are the **right** and **left atria.** The lower chambers are the **right** and **left ventricles,** which are separated by the thick **interventricular septum.** Notice that the two lower chambers are more muscular. This is a consequence of their individual functions: The right ventricle pumps blood through a second valve into the artery that leads to the lungs; the left ventricle pumps blood through a fourth valve into the aorta. Because it requires much more effort to pump blood through the aorta to the entire body, the outer wall of the left ventricle is a lot thicker than that of the right ventricle.

3. The aorta arches over the pulmonary trunk. At the top of the **aortic arch,** there are three **aortic arteries,** which provide oxygenated blood for the head and arms. The aorta then descends to provide oxygenated blood for the lower part of the body. The aorta supplies the **systemic circuit,** which is the system of blood vessels for the entire body. We will discuss the systemic circuit in detail later.

4. The **inferior vena cava,** which brings blood to the heart from the inferior or lower part of the

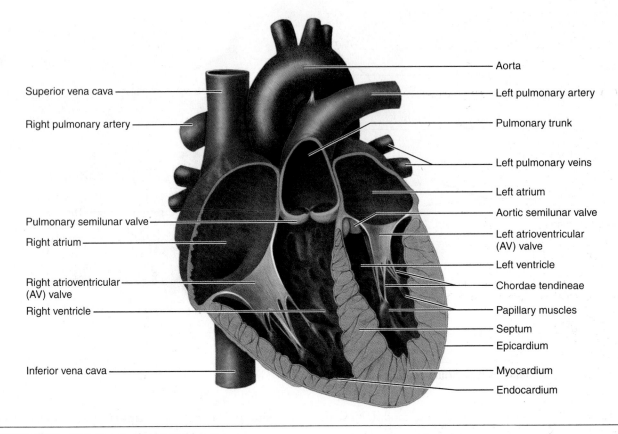

Superior vena cava

Right pulmonary artery

Pulmonary semilunar valve

Right atrium

Right atrioventricular
(AV) valve

Right ventricle

Inferior vena cava

Aorta

Left pulmonary artery

Pulmonary trunk

Left pulmonary veins

Left atrium

Aortic semilunar valve

Left atrioventricular
(AV) valve

Left ventricle

Chordae tendineae

Papillary muscles

Septum

Epicardium

Myocardium

Endocardium

FIGURE 15.3   An internal view of the heart.

body, is almost directly underneath the **superior vena cava,** which brings blood to the heart from the superior or upper part of the body. These two major veins are generally referred to as the venae cavae.

5. The pulmonary trunk separates into two branches, the **right and left pulmonary arteries,** which bring oxygen-poor blood to the lungs. These arteries supply the **pulmonary circuit,** which is the system of blood vessels in the lungs and which we will discuss later. The **right and left pulmonary veins** bring oxygen-rich blood from the lungs to the heart. Notice there are four pulmonary veins.

6. Identify the four heart valves. There are two semilunar valves to regulate the flow of blood from the heart. On the right, the **pulmonary semilunar valve** regulates the flow of blood into the pulmonary trunk. On the left, the **aortic semilunar valve** regulates the flow of blood into the aorta.

Two valves regulate the flow of blood from the atrial chambers to the ventricular chambers. On the right side, the blood flow is regulated by the **right atrioventricular (AV) valve,** also known as the **tricuspid valve.** On the left side,

the blood flow is regulated by the **left atrioventricular (AV) valve,** also known as the **mitral** or **bicuspid valve.**

7. The **chordae tendineae** anchor the AV valves to the **papillary muscles** of the ventricles. The contractions of these powerful ventricular muscles provide the blood pressure to transport blood from the heart to the extremities: the fingers and toes. ■

ACTIVITY 2

Tracing Blood Flow Through the Heart

1. Study Figure 15.4 a moment. Notice the arrows, which represent the flow of blood through the heart. Now, refer to Box 15.1, where you will see two sections: A and B. Fill in the blank beside each structure listed in Section A with the correct number from Figure 15.4. Then, list the structures through which blood passes as it travels through the pulmonary and systemic circuits in the appropriate blanks in Section A. There are 14 steps in the route, which begins (for

Box 15.1   Tracing Blood Flow Through the Heart

Section A

_7_ Pulmonary arteries

_5_ Pulmonary semi-lunar valve

_13_ Aortic arteries

_2_ Right atrium

_3_ Right AV valve

_4_ Right ventricle

_8_ Pulmonary veins

_14_ Aorta

_6_ Pulmonary trunk

_12_ Aortic semilunar valve

_9_ Left atrium

_10_ Left AV valve

_11_ Left ventricle

_1_ Venae cavae

Enter the numbers of the structures above that constitute the pulmonary circuit: _5–8_

Enter the numbers of the structures above that constitute the systemic circuit: _1–4, 9–14_

Section B

Oxygen-poor blood

1. Venae cavae
2. Right Atrium
3. Right AV valve
4. Right ventricle
5. Pulmonary semi-lunar
6. Pulmonary trunk
7. Pulmonary arteries

Oxygen-rich blood

8. Pulmonary veins
9. Left atrium
10. Left AV valve
11. Left ventricle
12. Aortic semilunar valve
13. Aortic arteries
14. Aorta

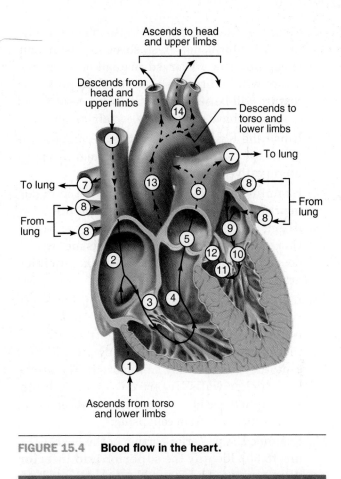

FIGURE 15.4   Blood flow in the heart.

ACTIVITY 3

Dissecting the Sheep Heart

Materials for This Activity

Sheep heart (preserved or fresh)

Dissection equipment

Dissection tray

Disposable gloves

Autoclave bag

Paper towels

Blunt sticks

Disinfectant in spray bottle

The sheep heart is used for dissection because, like the human heart, it has four chambers and is about the same size. Because this activity involves a dissection, please refresh your memory on the dissection protocol in Exercise 1.

our purposes) at the venae cavae ①. You'll first trace the flow of oxygen-poor (deoxygenated) blood through the right side of the heart and into the lungs. You'll continue as oxygen-rich (oxygenated) blood returns to the heart from the lungs and flows out of the heart to the head, torso, internal organs, and upper and lower limbs. Once you have assigned a number to each structure, identify the structures that constitute both the pulmonary and the systemic circuits.

2. In Section B of Box 15.1, enter the names of structures from Section A in their correct order, ①–⑭, to trace the pathway of blood.

Is this a correct definition? "Arteries carry oxygen-rich blood."

yes

1. Pair up with a lab partner. Supply yourself and your partner with dissection equipment, dissection tray, disposable gloves, and paper towels. Don your gloves, and obtain a sheep heart from the supply area.

2. Compare your sheep heart to the photos in Figure 15.5, which show fresh sheep hearts. If your sheep heart is preserved instead of fresh, the colors will be brownish rather than pink, the texture will be stiffer, and the blood vessels and heart may have collapsed in certain areas.

   Rinse the heart in cold water, and squeeze it gently over a sink a few times to remove trapped preservatives and blood clots.

3. Examine the external features of the sheep heart (Figure 15.5a–b). Observe the **pericardium,** if it is still present. Slit it open. Notice there is quite a bit of adipose tissue, which needs to be removed. The two earlike **auricles** are projections from the **atria.**

4. Identify the major blood vessels on the superior portion of the heart (Figure 15.5a). The **pulmonary trunk** is anterior to the aorta and the **pulmonary arteries** (if they are still present; they may already be removed). The **aorta** has thicker walls. Insert one or more blunt sticks into these blood vessels in the sheep heart to prevent them from collapsing.

5. Examine the posterior portion of the heart (Figure 15.5b). Identify the **superior and inferior venae cavae,** and insert blunt sticks into both veins in the fresh sheep heart. Notice that the insertion of blunt sticks provides additional firmness to the structure of the dissected heart organ.

   Compare the thickness and diameter of the aorta and venae cavae. Which one is larger? Which one has thicker walls?

   Venae is bigger, aorta is thicker

6. Insert a blunt stick into the superior vena cava, and use scissors to cut through to the **right atrium.** Observe the **right AV valve,** which is anchored with thin strands of **chordae tendineae** to the muscles of the ventricle (Figure 15.5c).

   How does the size of the right ventricle compare to the left ventricle?

   The right is larger

7. Continue the cut into the **right ventricle.** Identify the **papillary muscles** of the ventricle. The **ventricles** do much more work and are correspondingly much larger than the atria.

Compare the size of the atrial chamber to the ventricular chamber. Compare the thickness of the heart wall. Which chamber is larger? Which wall is thicker?

Atrium=bigger, Ventricle=thicker

8. Insert a blunt stick into the aorta, and use scissors to cut through to the **left ventricle.** Notice that the heart wall there is substantially thicker than that of the right ventricle.

9. Continue the cut from the left ventricle to the left atrium. Compare the **left AV valve** to the right AV valve. Notice that the right AV valve has three sections. How many sections does the left AV valve have?

   The left AV has two.

10. Properly dispose of the dissected sheep heart and blunt sticks in the autoclave bag. Clean and dry the dissection tools and tray. Spray and wipe down your counter with a disinfectant. ■

Heart Rates

Your heart is about the size of your fist. It is constructed of living cells, yet it can work continuously for 60 to 100 years without stopping for repairs. There are only about 5 liters of blood in the adult body, yet the heart can pump 5 to 25 liters of blood per minute! This means that it is possible to move your entire blood supply through the heart every minute!

Another amazing fact about your heart is that it never stops pumping blood. At rest, the typical heart rate is 75 beats per minute. During exercise, the heart rate can exceed 125 beats per minute. Over the course of a typical lifetime, your heart will probably beat more than 2 billion times!

The easiest way to measure the heart rate is by measuring the pulse rate of the radial artery. The pulse point for the radial artery is located on the side of the wrist, above the thumb. The pulse refers to the alternating surges of the pressure in an artery, which correlates to the contraction and relaxation of the left ventricle. In a normal heart, the pulse rate (number of surges per minute) correlates to the heart rate (beats per minute).

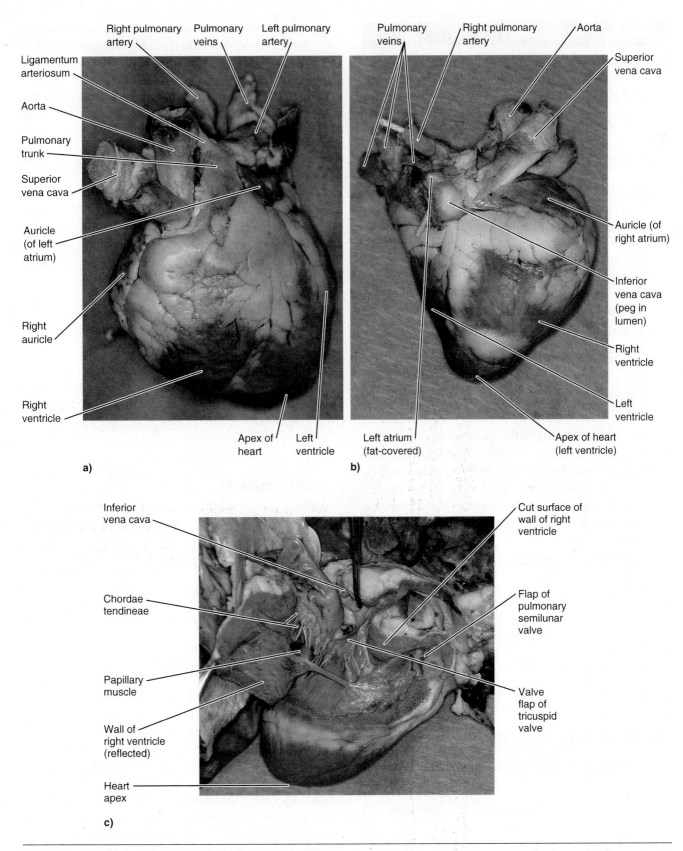

FIGURE 15.5   **Sheep heart. a)** Anterior view. **b)** Posterior view. **c)** Internal view.

ACTIVITY 4

Measuring Heart Rate

Materials for This Activity

Stopwatch or timer with second hand

1. Have your lab partner sit comfortably, with his or her arm and hand on the counter. Have your timer or stopwatch easily accessible. Place the fingertips of the first two fingers of one hand over the pulse point, and press gently. Remember not to press too hard.

2. Count the number of pulses in 15 seconds, and then multiply that number by four. For example, if you measure 20 pulses over 15 seconds, then 20 × 4 = 80 pulses per minute, or a heart rate of 80 beats per minute. Repeat the count, and verify your measurements. Record the heart rate in Box 15.2. This number will be referred to as the initial heart rate.

3. Ask your partner to do 10 jumping jacks or run in place for one minute. Measure the heart rate immediately after the exercise, and record it in Box 15.2. This will be the exercise heart rate.

4. Measure the heart rate three minutes after the exercise, and record it in Box 15.2. This will be referred to as the final heart rate.

5. Now, reverse your roles. Have your lab partner measure your heart rate under the three conditions.

   Checking the results in Box 15.2, answer the following.

   Use this formula to determine the percentage change over initial heart rate: Initial HR/Exercise HR × 100. Write the percentage in the space below.

   _____

Box 15.2	Heart Rate Measurements	
	Your Heart Rate	**Lab Partner's Heart Rate**
Initial heart rate		
Exercise heart rate		
Final heart rate		

What was the percentage change between the initial heart rate and the exercise heart rate?

_____

If there was a decrease, instead of the expected increase, try to determine the reason. You may repeat the measurement, and you may also explore the medical reasons for the decrease.

_____

_____

What was the percentage change from the initial to the final heart rate (Initial HR/Final HR × 100)? What was the change from the exercise to the final heart rate (Exercise HR/Final HR × 100)?

_____

_____

If the initial heart rate was approximately the same as the final heart rate, what does this information say about the health of the individual?

_____

At 75 beats per minute, how much time does a heartbeat take? (*Hint:* There are 60 seconds per minute. Calculate the time in seconds.)

_____

At 75 beats per minute, how many times will your heart beat in one day? (*Hint:* There are 60 minutes per hour, and 24 hours per day.)

_____

The Blood Vessels

There is an extensive network of blood vessels to deliver blood to all the tissues of your body. The arteries and veins that are located near the heart have the largest diameters. Remember that arteries are blood vessels that transport blood away from the heart, whereas veins transport blood back to the heart. As they extend to the extremities, arteries and veins begin to narrow and branch off into the arterioles and venules. The capillaries are the smallest and thinnest blood vessels, and they deliver the blood at the cell level. There is an inverse relationship between the diameter and length of blood vessels. The larger the diameter, the shorter the length.

Thus, the arteries and veins have the largest diameter and the shortest length, while capillaries have the smallest diameter and the longest length.

As pointed out earlier in this exercise, there are two major circulatory circuits in the human body. The **pulmonary circuit** is located in the lungs. It transports oxygen-poor blood through the lungs, where the blood gives up carbon dioxide and picks up a fresh supply of oxygen from the air we inhale. The **systemic circuit** is located throughout the body. The systemic **arteries** transport oxygen-rich blood from the heart to every part of the body (Figure 15.6). The arteries divide into arterioles, which divide again into the capillaries that serve the individual cells of the tissues. At the cells, the blood gives up oxygen and picks up a fresh supply of carbon dioxide, which is a product of the metabolic work of the tissues. The systemic veins transport this oxygen-poor blood back to the heart (Figure 15.6).

Blood Pressure and Its Relationship to the Cardiac Cycle

Keep in mind that the right side of the heart (the right atrium and the right ventricle) pumps blood to the lungs. The left side of the heart (left atrium and left ventricle) pumps blood to the head, torso, and upper and lower limbs. This pumping action consists of two distinct pulses: **contraction** and **relaxation.** Contraction in the two atria forces blood into the ventricles. The subsequent contraction of the two ventricles pumps blood into the pulmonary artery and the aorta. The repeated sequence of contraction and relaxation is known as the **cardiac cycle.**

As blood surges into the arteries, the artery walls are stretched by the higher pressure to accommodate the extra volume of blood. These same artery walls recoil as blood flows out of them. **Blood pressure** is the force that blood exerts on the wall of a blood vessel, and its maintenance is crucial to drive the flow of blood throughout the body and force the return of blood to the heart. **Systolic pressure** is the highest pressure of the cardiac cycle, which is reached when the ventricles contract to eject blood *from* the heart. The **diastolic pressure** occurs when both the atria and the ventricles relax and the blood pressure in the chambers of the heart is lower than the blood pressure in the veins.

Measuring blood pressure helps health care providers track trends in the blood pressure of an individual over time, and it is a good means for tracking changes in the cardiovascular system, as well as its overall efficiency.

Measuring Heart and Blood Pressure

Materials for This Activity

Stethoscope

Sphygmomanometer

Alcohol prep pads

Stopwatch or timer with second hand

The pressure of body fluids is measured in millimeters of mercury (mm Hg). A **sphygmomanometer** measures the changing blood pressure associated with the pulse.

We learned earlier that when measuring the pulse rate, we use the pulse point for the radial artery at the wrist. In this exercise, you will be introduced to a new pulse point: the elbow for the brachial artery. A stethoscope will be used to enable you to listen to the sounds in the artery as it experiences blood pressure changes.

1. Obtain a stopwatch, sphygmomanometer, and stethoscope. With an alcohol prep pad, clean the earpieces of the stethoscope, and let them dry. Examine the stethoscope sensors to see if they need cleaning as well. Examine the tubing of the sphygmomanometer to make sure there are neither tears nor holes in it. Also, be sure all the air is pressed out to avoid inaccurate readings. Return damaged equipment to your instructor.

2. Place your fingertips over the pulse point for the radial artery on the wrist. Using a watch, count the number of pulses in 15 seconds. Multiply your result by four. This will be the pulse rate. Normal pulse rates usually range from 60 to 100 and depend greatly on age and physical condition. For example, it is common for athletes to have pulse rates in the 50s and even lower. Record your pulse below.

   _____

3. Keep your fingertips lightly pressed on the pulse point to assess the pulse strength. Pulse strength is usually rated as absent, weak, diminished, strong, or bounding. The last two ratings are more common in healthy adults, but there are common factors such as mild dehydration that may affect this. How would you rate your pulse strength?

   _____

4. Have your lab partner sit comfortably and place his or her arm on the counter. Wrap the sphygmomanometer air cuff around his or her left upper

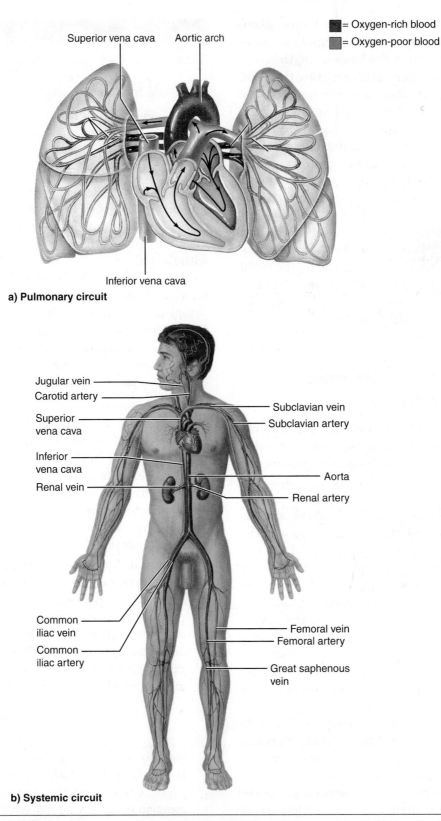

a) Pulmonary circuit

b) Systemic circuit

FIGURE 15.6   **Blood vessels of the pulmonary and systemic circuits. a)** Pulmonary circuit: Notice that the arteries are blue, because they carry oxygen-poor blood to the lungs. **b)** Systemic circuit (simplified): The opposite occurs here. The red arteries carry oxygen-rich blood to the tissues.

arm, just above the elbow. Place the stethoscope sensor over the pulse point for the brachial artery. Compare your setup to the depiction in Figure 15.7. Now place the earpieces of the stethoscope in your ears. Listen to the pulse in the brachial artery.

5. Close the valve of the sphygmomanometer's air bulb and pump the cuff to a pressure of about 180 mm Hg.

6. While listening to the artery and watching the mercury column, open the valve slightly and slowly release the air. This action will slowly release the pressure and cause the blood pressure in the artery to decrease.

As this happens, tapping sounds, the **heart sounds,** can be heard. The first heart sound is caused by the closure of the AV valves; however, that is not what you hear while taking blood pressure measurements. The **first heart sound** used for measurement is that of the brachial artery opening and closing under the cuff. Systolic pressure opens the artery, but because the diastolic pressure is lower than that in the cuff, the artery closes again after each pulse of blood passes. The **second heart sound** is actually no sound at all; it is the first sound going away because the pressure is now equal to that of the blood at diastole, and the blood is moving freely through the artery. Note the pressure of the first heart sound. This is the **systolic pressure.** Continue to decrease the pressure until the second heart sound. The second heart sound is caused by the closure of the semilunar valves. Note the pressure of the second heart sound. This is the **diastolic pressure.**

Blood pressure is expressed as systolic pressure *over* diastolic pressure. Average values are "120 over 80," which means 120 mm Hg/80 mm Hg. These two values are labeled as points ① and ② on the graph in Figure 15.7.

7. Let your lab partner rest a minute, and then measure the blood pressure again to verify your measurements. Record these numbers in Box 15.3. This recording will be referred to as the initial blood pressure.

8. Ask your partner to do 10 jumping jacks or run in place for one minute. Measure his or her blood pressure immediately after the exercise. These numbers will be referred to as the exercise blood pressure.

9. Measure his or her blood pressure three minutes after the exercise. These numbers will be referred to as the final blood pressure.

10. Now reverse your roles. Have your lab partner measure your blood pressure at rest, immediately after exercise, and three minutes after exercise.

11. Compare your respective values for initial blood pressure to those in Table 15.1. If you find that your blood pressure values are in stage 3 or 4, repeat your measurements again.

12. Refer to your measurements in Box 15.3 to answer the following questions. *Note:* Although exercise-related changes to diastolic pressure occur, this activity will focus on systolic pressure.

What was your initial systolic pressure? 120 / 81

What was your exercise systolic pressure? 136 / 72

What was the percentage change from your initial systolic pressure?

$$\frac{\text{Exercise systolic} - \text{Initial systolic}}{\text{Initial systolic}} \times 100$$

If there was a decrease instead of the expected increase, try to determine the reason.

What was the final blood pressure?

What was the percentage change from the initial systolic pressure?

$$\frac{\text{Final systolic} - \text{Initial systolic}}{\text{Initial systolic}} \times 100$$

Box 15.3	**Blood Pressure Measurements**	
	Your Blood Pressure	**Lab Partner's Blood Pressure**
Initial blood pressure	116 / 74	
Exercise blood pressure		
Percentage change		
Final blood pressure		

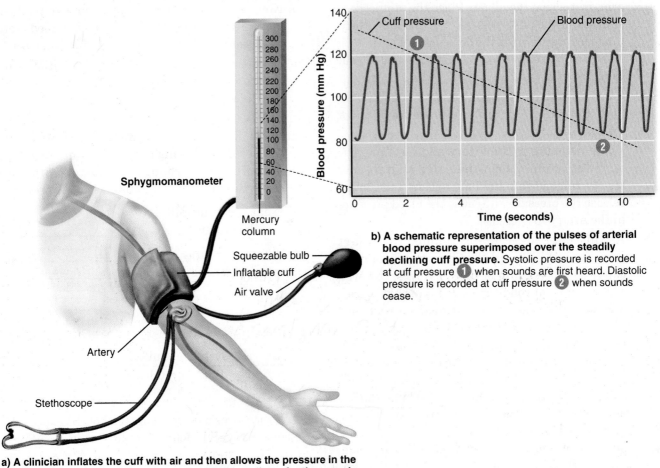

Sphygmomanometer

Mercury column

Squeezable bulb

Inflatable cuff

Air valve

Artery

Stethoscope

a) A clinician inflates the cuff with air and then allows the pressure in the cuff to fall gradually while using a stethoscope to listen for the sounds of blood movement through the artery.

b) A schematic representation of the pulses of arterial blood pressure superimposed over the steadily declining cuff pressure. Systolic pressure is recorded at cuff pressure ❶ when sounds are first heard. Diastolic pressure is recorded at cuff pressure ❷ when sounds cease.

FIGURE 15.7   Using a sphygmomanometer. a) The sphygmomanometer air cuff is wrapped around the upper arm over the brachial artery. **b)** In this schematic representation, blood pressure pulses are superimposed over declining cuff pressure. Point 1 indicates when sounds are first heard (systolic pressure); point 2 indicates when sounds cease (diastolic pressure).

Table 15.1 **Systolic and Diastolic Blood Pressures**

Systolic Pressure

Normal	129 and lower
High-normal (borderline hypertension)	130–139
Stage 1 (mild hypertension)	140–159
Stage 2 (moderate hypertension)	160–179
Stage 3 (severe hypertension)	180–209
Stage 4 (very severe hypertension)	210 and higher

Diastolic Pressure

Normal	84 and lower
High-normal (borderline hypertension)	85–89
Stage 1 (mild hypertension)	90–99
Stage 2 (moderate hypertension)	100–109
Stage 3 (severe hypertension)	110–119
Stage 4 (very severe hypertension)	120 and higher

What was the percentage change from the exercise systolic pressure?

$$\frac{\text{Exercise systolic} - \text{Final systolic}}{\text{Final systolic}} \times 100$$

_____

If the initial blood pressure was approximately the same as the final blood pressure, what does this statistic say about the health of the individual?

_____

What was your pulse rate and strength? _____

Let's say that your blood pressure was 180/120. Refer to Table 15.1. Is this a normal or hypertensive value?

_____

THE CARDIOVASCULAR SYSTEM II
Critical Thinking and Review Questions

1. Label the parts of the heart.

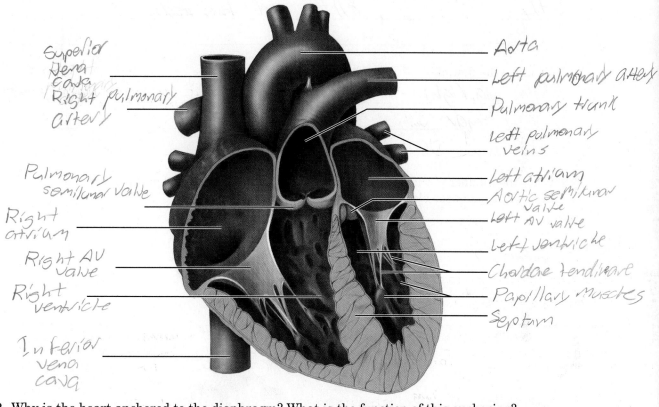

Superior Vena Cava

Right pulmonary artery

Pulmonary semilunar valve

Right atrium

Right AV Valve

Right ventricle

Inferior vena cava

Aorta

Left pulmonary artery

Pulmonary trunk

Left pulmonary veins

Left atrium

Aortic semilunar valve

Left AV valve

Left ventricle

Chordae tendineae

Papillary muscles

Septum

2. Why is the heart anchored to the diaphragm? What is the function of this anchoring?
   The heart constantly moves and can't be anchored to a uniformed structure.

3. Describe how the rib cage protects the heart.
   The ribs is like chest armor. The tough bones can withstand impacts to prevent damage to the heart.

4. How would a rib fracture affect the heart?
   Bone fragments could damage the heart and it the rib strikes the heart, it could cause even more damage

5. How are the AV valves anchored inside the heart? What is the function of this anchoring?

They are similar to strings. The AVs regulate blood flow from the atrial to ventricular chambers.

6. How does the right side of the heart differ from the left side of the heart? How do the atria differ from the ventricles?

The right is stronger, bigger, and more muscular since they are pumped all around the body.

7. Describe the flow of blood through the heart.

Vena cava, right atrium, right AV, right ventricle, right pulmonary semilunar valve, pulmonary artery, lungs, left atrium, left AV, left ventricle, semilunar aortic valve, aorta, body, repeat

8. Color the number in the white circle according to the oxygen level of the blood. Use blue for oxygen-poor blood and red for oxygen-rich blood.

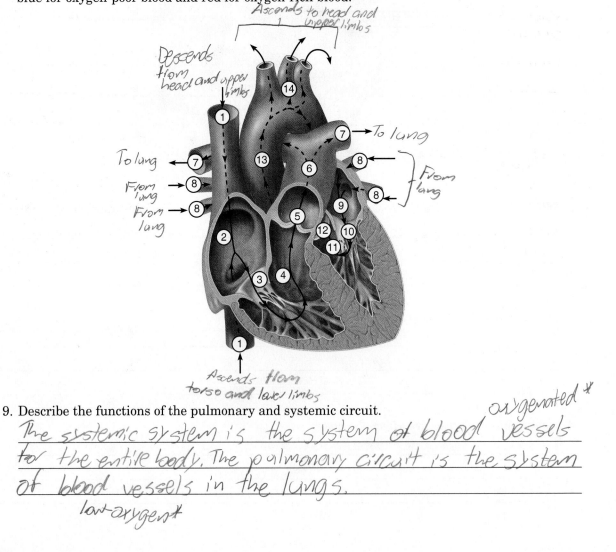

9. Describe the functions of the pulmonary and systemic circuit.

oxygenated *

The systemic system is the system of blood vessels for the entire body. The pulmonary circuit is the system of blood vessels in the lungs.

low-oxygent *

10. At 75 bpm, how many times will your heart beat in one day? In one week? When does your heart "take a break" and stop beating for a while?

    108,000 beats per day, 756,000 per week. The heart takes a break when you die.

11. Describe the events that make up the cardiac cycle.

    The hearts contract and relax to transport blood. It repeats to continue the cardiac cycle

12. What creates the first heart sound? The second heart sound?

    The contraction is the first sound/beat, and the relaxation is the second sound/beat.

13. What is blood pressure? What is the difference between diastolic and systolic blood pressure?

    Blood pressure is the pressure pushing on the walls of blood vessels. Systolic is the ejection of blood, diastolic is when the atria and ventricles relax.

14. If your blood pressure is "135/85," is this low, normal, or high? Explain your answer. (*Hint:* Refer to Table 15.1.)

    This is high-normal. It's not dangerous, but can be monitored for safety.

EXERCISE

16

The Respiratory System

Objectives

After completing this exercise, you should be able to

1. Describe the functions of the respiratory system.
2. Identify the structures in the upper and lower respiratory tract.
3. Trace the air flow from the nose to the alveoli.
4. Describe the functions of the tracheal walls.
5. Describe the functions of the bronchial walls.
6. Describe the process of breathing.
7. Use a bell jar to demonstrate breathing principles.
8. Measure chest and lung volume changes.
9. Calculate the components of total lung capacity.
10. Use a spirometer to measure lung volumes.
11. Compare the respiratory data between "impaired" and "healthy lung" groups.

Materials for Lab Preparation

Equipment and Supplies

○ Bell jar model of the lungs
○ Metric tape measure
○ Nonrecording spirometer (wet or dry)
○ Disposable mouth pieces for spirometers
○ Alcohol prep pads
○ Nose clips

Models

○ Human torso
○ Human half head
○ Human lung
○ Human alveolar sac
○ Human thoracic organs

Introduction

When someone chokes on a piece of food, he or she has less than one minute to remove the obstruction to the air supply before he or she suffers a loss of consciousness. This scenario points out our absolute dependency on constant air supply.

Our cells generally require constant supply of oxygen to drive their normal cell processes. Oxygen deprivation can be dangerous, and the brain cells are the most vulnerable. When they "shut down," many body activities become uncoordinated; so our body also "shuts down" by becoming unconscious. Carbon dioxide removal is also critical because as the carbon dioxide level in the blood rises, carbonic acid is formed, which acidifies the blood. As blood pH drops, many body functions are affected.

The main function of the respiratory system is to provide oxygen to cells and remove carbon dioxide, which is the waste product that cells release upon oxygen use. The exchange of these gases with the air is called **respiration.**

Four processes, collectively referred to as respiration, must occur in order for the respiratory system to fulfill this main function. We will be primarily concerned in this exercise with two of these processes: **pulmonary ventilation,** known as breathing, and **external respiration,** which involves gas exchange in the lungs.

Respiratory Anatomy

The respiratory system is divided into the upper and lower respiratory tracts (Figure 16.1). The **upper respiratory tract** consists of the nose, nasal cavity, and pharynx. The **lower respiratory tract** consists of the larynx, trachea, bronchi, and lungs, in which we find the bronchioles and alveoli.

Let's study the respiratory anatomy by tracing the route of inhaled air, starting with the nose and

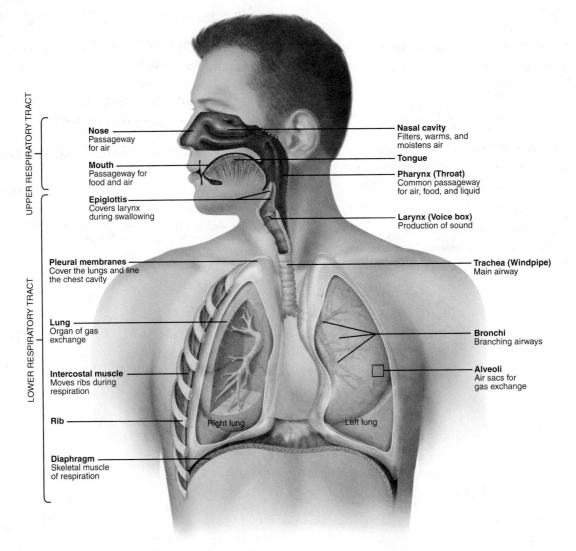

FIGURE 16.1 **Human respiratory system.**

ending at the lungs. The route for exhaled air follows the reverse direction, from the lungs to the nose.

ACTIVITY 1

Tracing the Route of Inhaled Air

Materials for This Activity

Models

Human torso
Human half head
Human lung
Human alveolar sac
Human thoracic organ

1. Using various models and referring to Figures 16.1–16.3, we will trace the route of inhaled air from the nose to the lungs. If the torso model is too heavy, bring your lab manual to that lab station. *Note:* Remember these models are used for many other lab exercises, so do not touch them with pencils or pens; use your fingers or the eraser ends of pencils.

2. During inhalation, nostril hairs filter particles from the air entering the **nose.** The air is then warmed and moistened in the **nasal cavity** before traveling down to the throat, or pharynx (Figure 16.1). The **pharynx** is the common area for air traveling from either the nose or the mouth. Therefore, air may also be inhaled through the **mouth,** but it would not be filtered,

warmed, and humidified as thoroughly as it would by the nose.

What is the potential harm of using the mouth to breathe all the time? (*Hint:* Review the functions of the nostril hair and nasal cavity.)

The mouth doesn't filter the air.

People who engage in water sports, such as water polo, may wear nose plugs. Thus, they are entirely dependent upon breathing through the mouth. What is the advantage of plugging up the nose? (*Hint:* What happens when you somersault in the water?)

No chemicals enter your nose and flush your nasal cavity

3. The cleaned, warmed, and moistened air now travels past the **epiglottis** (Figure 16.1). This little flexible flab of elastic cartilage is located at the opening to the larynx. It remains open when air is flowing into the larynx, but it temporarily covers up the larynx when it is pulled up as we swallow. Place your hand midway on the anterior surface of your neck and swallow. Can you feel the larynx rise as you do so? This "switching mechanism" routes food and drink into the esophagus and air into the trachea. If anything other than air enters the larynx, a cough reflex is triggered to expel the substance and prevent it from moving into the lungs.

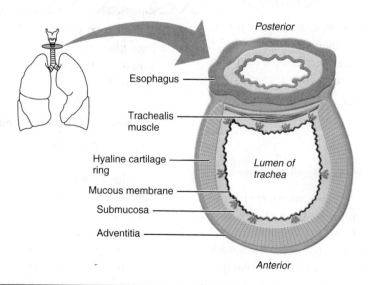

FIGURE 16.2 **Trachea and esophagus cross section.** Notice the cartilage ring is incomplete in the back to allow swallowing of large food amounts. The cartilage ring is complete in the front to strengthen the trachea and keep it open.

What else besides food might cause a blockage in your air supply?

Mucus dirt, etc

4. Within the **larynx,** or voice box, there are a pair of folds called **vocal cords,** which vibrate when air is expelled, thus giving us the ability to produce sound. When we are not talking, the vocal cords are generally relaxed and open. When we talk, the vocal cords are stretched tight by skeletal muscles. The tones of our speech depend on how tight the vocal cords are stretched.

In the event of a prolonged blockage of the upper respiratory tract, a tracheotomy is performed. This involves making an opening in the trachea and inserting a tube to assist in breathing.

How would a tracheotomy affect the tonal quality of our speech?

Some air leaks into vocal cords.

5. Inhaled air enters the **trachea,** or windpipe. The material of the tracheal walls performs some special functions:

- **moistening the air**
- **cleaning the air**

The lining of the trachea is made of ciliated, mucus-secreting pseudostratified columnar epithelium. (Refer to the histology in Exercise 5, Tissues.) This means that the inhaled air will continue to be moist, and clean, because dust particles, bacteria, and other debris will be trapped in mucus and propelled by the cilia away from the lungs and toward the throat.

When you "feel" the mucus pressure in the throat, what are your two choices about that mucus ball? (*Hint:* What do you do when you have a cold and there is a big ball of phlegm in your throat?)

You cough to clear your throat

- **swallowing food**
- **maintaining an open tracheal passageway**

The C-shaped cartilage rings provide two functions (Figure 16.2). The open part of the C is located posteriorly. This allows the esophagus to expand when a large food bolus is swallowed. The solid part provides firmness to the trachea wall so that the trachea is always open, regardless of the pressure changes during breathing.

What happens when you try to take a deep breath and swallow a big bite of food at the same time?

You would choke so your body stops yourself from breathing and eating at the same time

Based on what you've learned, why does this occur?

Your body has automatic reflexes and prevents yourself from choking

6. The inhaled air next passes into the bronchi. The trachea forms two major branches called the primary **bronchi,** which enter the right and left lung. Inside the lungs, the bronchi divide into smaller and smaller branches, and finally become the **bronchioles** (little bronchi). Notice the structural similarity to a tree, which is the basis for the term, **respiratory tree.**

Like the walls of the trachea, the material in the walls of the bronchi also perform some special functions:

- **maintaining an open bronchial passageway**

Hyaline cartilage reinforces the bronchial walls, which provides strength and firmness for maintaining an open bronchial passageway. However, as the branching continues, the amount of cartilage declines. Cartilage is absent in the bronchioles.

- **broncho-constriction and broncho-dilation**

Smooth muscles and autonomic nerves in the bronchial walls can influence the volume of air reaching a particular region of the lung. Air flow is regulated by constricting or dilating the airway lumen. This is similar to the control of blood flow by the smooth muscles in the arteries. There is an inverse relationship between cartilage and smooth muscle in the bronchial walls. The primary bronchial walls are mostly cartilage. In the bronchiole walls, there is a continuous layer of smooth muscle but cartilage is absent.

- **epithelial tissue becomes thinner**

As the bronchial branching continues, the epithelial cells gradually become thinner. In the smaller bronchi, simple columnar epithelium is present. In the bronchioles, simple cuboidal epithelium is found, while the alveolar walls consist of simple squamous epithelium.

7. The inhaled air is processed in the alveoli. The smallest bronchioles terminate in clusters of **alveoli,** or air sacs. Using Figure 16.3, trace the indicated blood flow from the blue pulmonary

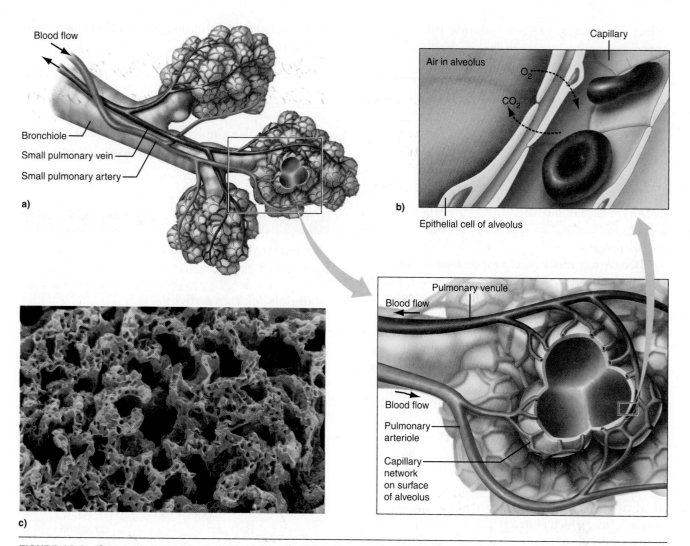

FIGURE 16.3 **Gas exchange between blood and alveoli. a)** Alveoli and pulmonary blood vessels.
b) Respiratory membrane—site of gas exchange. **c)** Alveolus and pulmonary capillary.

arteriole to the purple capillary, then to the red pulmonary venule. Clearly the blood has become oxygenated. We will discuss how the blood became oxygenated in the next section.

8. Trace the route of exhaled air, from alveolus to nose.

> *Nostrils, pharynx, glottis, trachea, bronchi, bronchioli, alveoli,*

In the respiratory tract, the inhaled air mixes with the exhaled air. Is this a problem?

> *Yes, exhaled air contains high CO₂*

External Respiration

Gas exchange in the lungs, also known as external respiration, is the process by which the deoxygenated blood in the pulmonary arteriole becomes oxygenated in the pulmonary venule. The inhaled air in the

alveoli provides air that is high in oxygen and low in carbon dioxide. Notice from Figure 16.3a and c that a capillary network surrounds the surface of each alveolus. This capillary brings in deoxygenated blood, which is low in oxygen and high in carbon dioxide.

The **gas exchange** takes place across the **respiratory membrane,** also called the **air-blood barrier,** which is made of the alveolar and capillary walls. Figure 16.3b shows oxygen moving from the alveolus into the capillary as carbon dioxide moves in the opposite direction. Gas exchange takes place by simple diffusion; in other words, each gas moves from areas of high concentration to areas of low concentration, across a permeable membrane.

Why is it important that the alveolar and the capillary walls are made of simple squamous epithelium?

> *It is more flexible*

What is the advantage of simple diffusion, as opposed to active transport, for gas exchange?

Diffusion is transport from high to low concentration

Breathing

Breathing, also called pulmonary ventilation, is the process of moving air through our lungs. We are usually aware of our breathing only when there is a problem, but as we will see, breathing is not a simple process.

Breathing consists of two phases: inhale and exhale (see Figure 16.5). (In earlier times, the equivalent terms were *inspire* and *expire*—to let the spirit in and out. Today, we use the term *expiration* to mean death.) To perform a normal **inhale,** we contract our inspiratory muscles (external intercostals and diaphragm), which **increases the volume of the thoracic cavity.** During this expansion, the elastic fibers in the bronchial walls are stretched. In a normal **exhale,** the inspiratory muscles are relaxed, these elastic fibers recoil, and the diaphragm, ribs, and sternum return to their original positions. As a result, the **thoracic volume is decreased.**

The next step is a bit more complex. Here, we need to understand two important principles: (1) **Boyle's law,** which states that the pressure of a gas is inversely proportional to the volume of the gas; and (2) When a **pressure gradient** exists, air will move from areas of high concentration to areas of low concentration.

Breathing is regulated by pressure gradients that form between the air in the atmosphere and the air in the airways. According to Boyle's law, during an **inhalation,** when the **chest volume increases,** the **air pressure in the airways decreases** below atmospheric pressure. The result is that the atmospheric pressure (area of high pressure) will force **air to move into the lungs** (area of low pressure).

What happens during an exhalation?

Chest volume decreases and forces air out

Remember that the lung is not directly attached to the chest. Due to the moisture and negative pressure—the so-called wet vacuum of the pleural cavity—the lung "sticks" to the chest cavity. As the chest volume changes, the lung volume changes correspondingly.

What happens to breathing when there is a chest puncture?

The volume would not be able to change and prevent breathing

We will use the bell jar model of the lungs to demonstrate the relationship between pressure and volume.

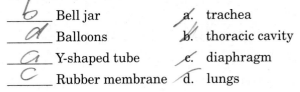

ACTIVITY 2

Demonstrating Air Flow into and out of the Lungs

Materials for This Activity

Bell jar model of the lungs

We have learned that in a closed container, increasing the volume will lead to a decrease in the air pressure. This condition establishes a pressure gradient, which results in air from the outside flowing into the lungs.

1. Obtain a bell jar model of the lungs and compare it to Figure 16.4. The bell jar model is a model for how air flow moves in and out of the lungs. Match the two columns:

   b Bell jar   *a.* trachea
   d Balloons   *b.* thoracic cavity
   a Y-shaped tube   *c.* diaphragm
   c Rubber membrane   *d.* lungs

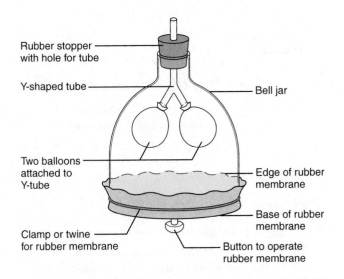

FIGURE 16.4 Bell jar model of the lungs.

Firmly hold the bell jar across the front of your body in such a way that the balloons are clearly visible. Work the bell jar by slowly pulling the button on the rubber membrane up and down. *Caution:* The rubber membrane may be punctured by long nails or can be torn by pulling the button too vigorously.

2. Cover the hole for the tube in the rubber stopper. Slowly pull the button up or down a few times. What are the changes in the shape and size of the balloons?

   It expands and contacts

3. Now uncover the hole for the tube in the rubber stopper. Slowly push the button up. Does the size of the balloons increase or decrease? Does the air flow into or out of the balloons? What is the equivalent action in our body? Does the lung volume increase or decrease? Do we inhale or exhale?

   The volume stays the same and nothing changes since there is a puncture in the bell jar

4. Still leaving the hole uncovered, slowly pull the button down. Answer the same previous five questions.

   _____

   _____

   _____

   _____

   _____

5. Return the bell jar to the proper storage area. ∎

ACTIVITY 3

Measuring Chest Volume Changes

Materials for This Activity

Metric tape measure

From the bell jar experiment, we have learned that expanding the volume of the closed container will cause air to flow into the balloons (lungs). In this activity, we will examine the volume changes of our own lungs by measuring the volume changes of the chest during three states: relaxed state, quiet breathing, and forced breathing. Notice the movement of the ribs and the diaphragm with inspiration and inhalation, as illustrated in Figure 16.5.

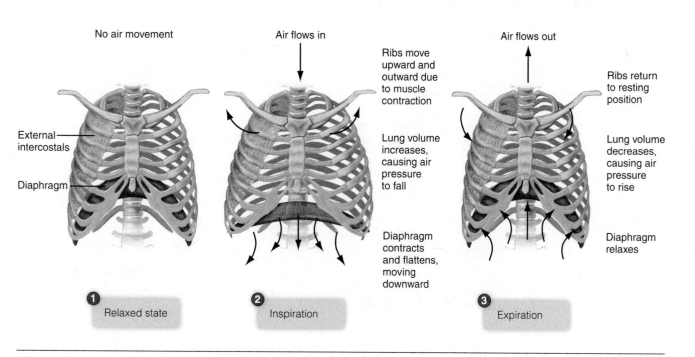

No air movement

Air flows in

Air flows out

Ribs move upward and outward due to muscle contraction

Ribs return to resting position

External intercostals

Lung volume increases, causing air pressure to fall

Lung volume decreases, causing air pressure to rise

Diaphragm

Diaphragm contracts and flattens, moving downward

Diaphragm relaxes

❶ Relaxed state

❷ Inspiration

❸ Expiration

FIGURE 16.5   Breathing.

1. In this exercise, you will measure the chest volume changes of your lab partner. Obtain a metric tape measure. Measure the chest circumference by placing the tape around your partner's chest, as high up under the armpit as possible. Make the following measurements in centimeters.

Quiet breathing

Hold breath after normal inhalation _____92_____

Hold breath after normal exhalation _____92.5_____

Forced breathing

Hold breath after forced inhalation _____92_____

Hold breath after forced exhalation _____89_____

2. Repeat your measurements to verify your results.
3. Subtract your quiet breathing normal *exhalation* measurement from your quiet breathing normal *inhalation* measurement and record. _____
4. Subtract your forced breathing *exhalation* measurement from your forced breathing *inhalation* measurement and record. _____
5. Considering that a larger difference indicates a larger volume change, what can you conclude about the relationship between chest volume changes and air movement?

More air movement = more chest volume ∎

Lung Capacity

Depending on age, gender, and physical condition, there can be great variation in the chest volume changes that were measured for normal and forced breathing. These variations make sense, considering the lungs are geared to accommodate personal oxygen needs, which are different for each individual.

Measurement of these **lung volumes** is useful because it allows medical personnel to track the changes in your lung capacity during the course of a disease or in the diagnosis of a respiratory disease.

There are four individual components to your **total lung capacity (TLC)** (Figure 16.6):

1. Inspiratory reserve volume (IRV)
   Volume of forcibly inhaled air after a normal inhale
2. Tidal volume (TV)
   Volume of air in a normal inhale or normal exhale
3. Expiratory reserve volume (ERV)
   Volume of forcibly exhaled air after a normal exhale
4. Residual volume (RV)
   Volume of air in the lungs at all times

In summary,

$$TLC = IRV + TV + ERV + RV$$

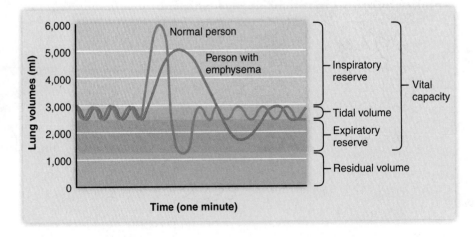

FIGURE 16.6   Components of total lung capacity.

Vital capacity (VC) represents the sum total of IRV, TV, and ERV. Notice it includes every lung volume except for RV, which is always part of the lung. Vital capacity refers to the total exchangeable air of the lungs. There are two ways of calculating VC:

$$VC = IRV + TV + ERV$$
$$VC = TLC + RV$$

ACTIVITY 4

Measuring Lung Volumes

Materials for This Activity

Nonrecording spirometer (wet or dry)

Disposable mouthpieces for spirometers

Alcohol prep pad

Nose clips

1. Obtain a wet or dry spirometer (Figure 16.7), mouthpiece, and alcohol prep pad. Upon return to your station, immediately write your name on your mouthpiece to avoid using the wrong one. If you obtained a dry spirometer, untwist the top and clean out the interior chambers with an alcohol prep pad. Replace the top of the dry spirometer. If you obtained a wet spirometer, prepare a clear work space in case of water spillage.

2. Take a look at your spirometer. They are called nonrecording spirometers because no record is kept of the data. After each use, you need to reset the gauge back to zero. Note that the volumes are to be measured in milliliters or liters. The typical values quoted for each measurement are based on a young, healthy male.

3. Your job is to provide the following instructions to your partner, record the data, and reset the gauge of the spirometer. As the subject, your lab partner should handle the mouthpiece of the spirometer, inserting it into his or her mouth.

4. Count the number of normal breaths in one minute.

Normal respiratory rate: ___*12*___

A typical value for a normal respiratory rate is about 12 breaths per minute.

5. Practice exhaling through the mouthpiece, without exhaling through the nose. If this is not possible, clean and use the nose clips.

6. Measure TV as follows: Inhale a normal breath, place the spirometer in your mouth, and exhale normally into the mouthpiece. Take three measurements, and calculate the average of the three measurements. Remember to set the gauge to zero before making the next measurement.

First TV   *400*

Second TV   *200*

Third TV   *200*

Average TV   *266.6 = .266*

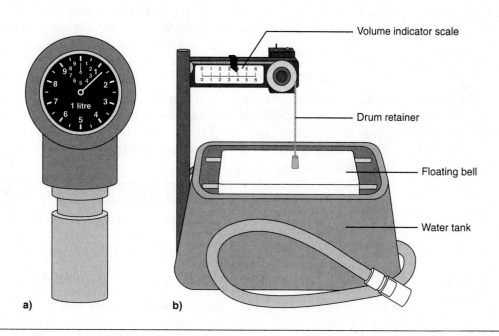

a)   b)

FIGURE 16.7   **Spirometers. a)** Dry spirometer. **b)** Wet spirometer.

Typical values for TV range from 300–500 ml. About one-third of the air stays in the respiratory tract and two-thirds reaches the alveoli to become involved in gas exchange.

7. Calculate the **minute respiratory volume (MRV)** as follows:

$$MRV = TV \times \text{respiratory rate}$$

MRV _3,199_

Typical values for the MRV range are determined as follows:

300 ml × 12 breaths per min = 3,600 ml or 3.6 l

500 ml × 12 breaths per min 6,000 ml or 6 l

Range: 3,600–6,000 ml, or 3.6–6 l

8. Measure ERV as follows: Take a few normal inhalations and exhalations. After your last inhalation, place the spirometer in your mouth and forcibly exhale. Take three measurements, and calculate the average of the three measurements. Remember to set the gauge to zero before making the next measurement.

First ERV _2200_

Second ERV _1900_

Third ERV _1800_

Average ERV _= 1,966_

Typical values for ERV are about 1,300 ml or 1.3 l.

ERV is greatly reduced in respiratory disorders, such as emphysema. Because energy must be used to deflate the lungs, exhalation is also exhausting.

9. Measure VC as follows: Take a few normal inhalations and exhalations. Bend over and exhale as completely as possible. Raise yourself and inhale as completely as possible. Quickly insert the mouthpiece and exhale forcibly. Take three measurements, and calculate the average of the three measurements. Remember to set the gauge to zero before making the next measurement.

First VC _2400_

Second VC _2500_

Third VC _2400_

Average VC _2,433 = 2,433_

The typical range for VC is 3,600–4,800 ml, or 3.6–4.8 l.

10. Calculate IRV as follows:

$$IRC = VC - TV - ERV$$

Average IRV _1201_

The typical range for IRV is 2,000–3,000 ml, or 2–3 l. ■

Respiratory Disorders

Let's apply our knowledge of the respiratory system to understanding some major respiratory disorders. Cigarette **smoking** is directly related to emphysema and chronic bronchitis. In **emphysema,** the alveolar walls are damaged and the lungs lose their elasticity. This means exhaustion (other muscles are used for breathing), a barrel chest (air is trapped in the alveoli, which flattens the diaphragm), and damage to the pulmonary capillaries (due to alveolar wall damage). In **chronic bronchitis,** there is excess mucus production and inflammation in the trachea and bronchial tubes. The result is airway obstruction and difficulty in breathing and gas exchange.

About 1 in 10 people in North America suffer from **asthma.** The symptoms are probably familiar to most people: coughing, labored breathing, wheezing, and a tight chest. The causes of asthma are quite diverse. Triggers include cold air, exercise, allergens, dust mites, and animal fur. Treatment focuses on preventing attacks by isolating the main cause, such as specific allergens, and reducing exposure. Medications (e.g., antihistamines), bronchodilators, and corticosteroids to reduce inflammation are also used as necessary.

Clearly, smokers and asthmatics have impaired lungs. What type of people would have healthy lungs? In general, people who exercise regularly and have a balanced diet and a sensible lifestyle will have healthy lungs. In this section, we will compare the performance of these two groups, "impaired" versus "healthy" lungs.

ACTIVITY 5

Comparing Respiratory Data

Materials for This Activity

Lung volume data from Activity 4

1. The instructor will divide the class into two groups: Group A has "impaired" lungs, Group B

has "healthy" lungs. Some suggestions are: asthmatics versus non-asthmatics, smokers versus non smokers, people who perform regular cardiorespiratory activities versus people who do not.

2. On the board, the instructor will put up two tables for students to record the individual data for the A and B groups.

3. Obtain the averages of your personal data from the previous activity. Reorganize the data as follows:

MRV	3,199
IRV	1,966
TV	.266
ERV	1,966
VC	2,433

4. Record your individual data on the board under Group A or B.

5. Calculate the average for each category on the board. The average is the sum of the values divided by the number of values.

Average Group A data:

MRV	3,2
IRV	1,5
TV	.35
ERV	1,11
VC	33

Average Group B data:

MRV	9,1
IRV	.52
TV	.32
ERV	.94
VC	1,576

6. How do your data compare to the appropriate group—Group (A) or B?
Compare your average to the group average. Are your statistics higher or lower?

MRV	lower
IRV	higher
TV	lower
ERV	higher
VC	lower

7. The range is the simplest measure of the dispersion of the data. Many public reports, for example, are presented in the form of ranges. Weather forecasts typically state the high and low temperature projections for each day of the week. Stock reports track the highs and lows of stock prices over a certain time period.

Calculate the range of statistics for each category on the board.
Range of Group A data:

MRV	1,776
IRV	1,9
TV	,4
ERV	,27
VC	2,1

Range of Group B data:

MRV	4,3
IRV	2,03
TV	,55
ERV	1,76
VC	1,6

8. How do your data compare to the appropriate group—Group A or (B)? Compare your data to the group range. Are you in the low, middle, or high end?

MRV	Lower
IRV	Lower
TV	Lower
ERV	Higher
VC	Lower

9. How do the Group B data compare to the Group A data?

Compare the averages. Are the nonsmoker numbers higher or lower than the Group A numbers?

MRV	higher
IRV	higher
TV	higher
ERV	higher
VC	higher

If the data are not what you expected, what might be some reasons?

Activity levels can affect lung health.

Compare the ranges. Is the middle of the Group B range higher or lower than the middle of the Group A range?

MRV	Higher
IRV	higher
TV	higher
ERV	higher
VC	lower

If the data are not what you expected, what might be some reasons?

Activity levels

THE RESPIRATORY SYSTEM
Critical Thinking and Review Questions

1. Label the parts of the human respiratory system.

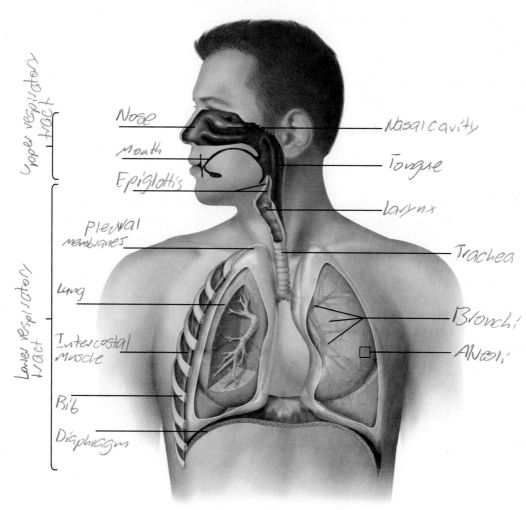

Upper respiratory tract

Nose

Mouth

Epiglottis

Pleural membranes

Lung

Intercostal muscle

Lower respiratory tract

Rib

Diaphragm

Nasal cavity

Tongue

Larynx

Trachea

Bronchi

Alveoli

2. Nose, alveolus, bronchi, bronchiole, pharynx, larynx, nasal cavity, trachea.

   a. Use the above terms to trace air entering, then leaving the body. Write the path below.

   Nose, nasalcavity, tongue, epiglottis, larynx, trachea, bronchi, alveoli

   b. Rank the oxygen level in each stage from highest to lowest.

c.   Rank the carbon dioxide level in each stage from highest to lowest.

_____

_____

3. Form and function are related. Describe how the "form" of the tracheal wall is related to its four functions.

_____

_____

_____

4. Describe how the "form" of the bronchial wall is related to two functions.

_____

_____

_____

5. The respiratory membrane is made of two thin membranes from the alveolus and capillary walls. What is important about the thinness of the membranes?

Makes gas exchange efficient

_____

_____

6. Draw and label an alveolus surrounded by a capillary network.

7. You smell something delicious, and you want to inhale deeply to enjoy the smell more. Explain how you can increase the chest volume more than usual. What other muscles might be involved?

Your diaphragm changes the pressure in the cavity to expand the lungs.

_____

8. Explain how increasing the chest volume will force air to flow into the lungs.

   The pressure pulls air into the lungs due to pressure change

9. If the rubber membrane of the bell jar model lung were accidentally punctured by sharp nails, what would be the effect on the two balloons?

   The balloons wouldn't be able to expand since pressure wouldn't be able to change

10. If the chest volume decreased during an inhalation instead of increasing, what might be some reasons?

    There may be a puncture in the cavity to prevent pressure change

11. What is the relationship between the components of TLC?

    All involve inhalation

12. What is the relationship between the components of VC?

    This is the sum total of the measurements.

13. What are two ways for MRV to increase?

    Activity, stress

14. Describe the damage to specific parts of the respiratory system in these disorders:

    a. emphysema

    Alveolar walls are damaged

b. chronic bronchitis

Excess mucus production

c. asthma

Tight chest

15. Describe the reasons for expecting larger lung volumes from "stronger" lungs as opposed to "weaker" lungs.

Larger lung volumes are able to have more oxygen in the lungs and be able to keep the body oxygenated with strong lungs.

EXERCISE 17

The Digestive System and Nutrition

Objectives

After completing this exercise, you should be able to

1. List, identify, and describe the major organs of the digestive system.
2. Describe the functions of the major organs of the digestive system.
3. List the major enzymes of the digestive system, and describe their basic functions.
4. Explain the basic concept of metabolism and how it is connected to nutrition.

Materials for Lab Preparation

Equipment and Supplies

- ○ Compound light microscope
- ○ Fetal pig
- ○ Dissection equipment
- ○ Dissection tray
- ○ Disposable gloves
- ○ 1-liter beaker filled about halfway with tap water heated to 38°C
- ○ 1-liter beaker filled about halfway with tap water heated to 70°C
- ○ 1-liter beaker filled about halfway with ice water
- ○ Tap water
- ○ Distilled or deionized water
- ○ 250-ml flask
- ○ Five large test tubes
- ○ pH paper
- ○ China marker
- ○ Standard personal weighing scale

Prepared Slides

- ○ Small intestine

Models

- ○ Small intestine
- ○ Human torso

Solutions

- ○ Phenol red solution
- ○ Dropper bottle with vegetable oil
- ○ Dropper bottle with 2% sodium carbonate
- ○ Dropper bottle with 1% bile solution (which has been adjusted to a pH of 7.5–8.0)
- ○ Dropper bottle with 1% pancreatin or pancreatic lipase solution

Introduction

There are three major parts to the study of the digestive system: digestion, absorption, and metabolism (how the absorbed nutrients are used by the body). Indirectly related to all three is the concept of nutrition. As always, it is important to first learn the "parts" of the system before we can understand what they do and why being aware of nutrition issues is so important. It is also important to understand the role played by enzymes in the digestive system.

The structures of the digestive system are usually divided into two sections: the organs of the **gastrointestinal (GI) tract** and the **accessory organs** located outside it (Figure 17.2). After the mouth, the GI tract consists of the pharynx (throat), esophagus, stomach, small intestine, large intestine, rectum, and anus. The main accessory organs are the pancreas, liver, gallbladder, and salivary glands (Figure 17.2). We will examine each of these structures and the tissues they are made from in the next section.

Digestive System Anatomy and Basic Function

The Layers of the GI Tract

All of the tubular structures of the organs composing the GI tract have the same basic tissue organization. There is an outer **serosa**, a thick layer of muscle called the **muscularis,** an inner layer of connective tissue called the **submucosa,** and a membrane with epithelial cells lining the inside called the **mucosa** (Figure 17.1).

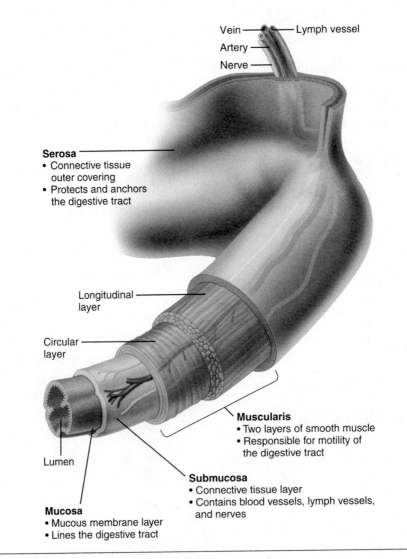

Vein — Lymph vessel
Artery
Nerve

Serosa
• Connective tissue outer covering
• Protects and anchors the digestive tract

Longitudinal layer

Circular layer

Muscularis
• Two layers of smooth muscle
• Responsible for motility of the digestive tract

Lumen

Submucosa
• Connective tissue layer
• Contains blood vessels, lymph vessels, and nerves

Mucosa
• Mucous membrane layer
• Lines the digestive tract

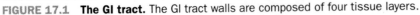

FIGURE 17.1 **The GI tract.** The GI tract walls are composed of four tissue layers.

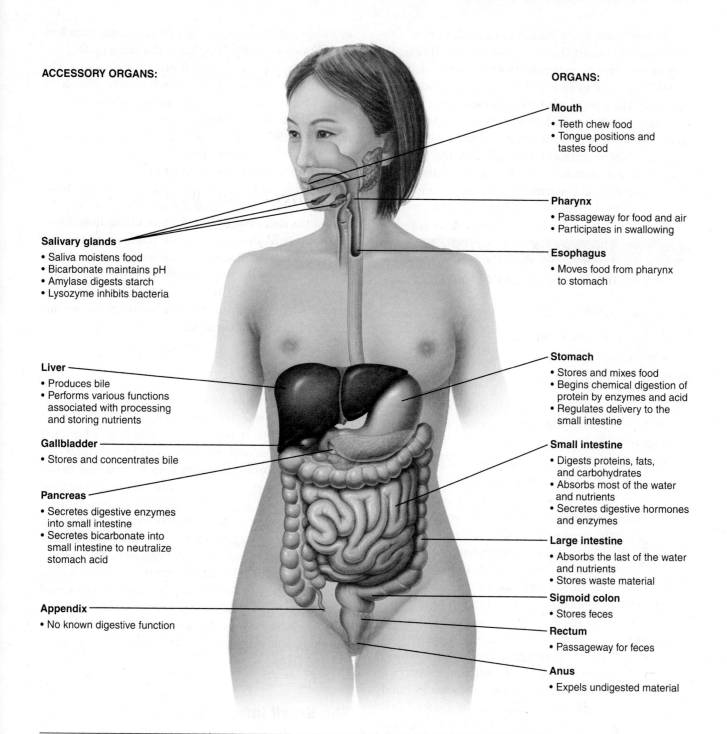

ACCESSORY ORGANS:

Salivary glands
- Saliva moistens food
- Bicarbonate maintains pH
- Amylase digests starch
- Lysozyme inhibits bacteria

Liver
- Produces bile
- Performs various functions associated with processing and storing nutrients

Gallbladder
- Stores and concentrates bile

Pancreas
- Secretes digestive enzymes into small intestine
- Secretes bicarbonate into small intestine to neutralize stomach acid

Appendix
- No known digestive function

ORGANS:

Mouth
- Teeth chew food
- Tongue positions and tastes food

Pharynx
- Passageway for food and air
- Participates in swallowing

Esophagus
- Moves food from pharynx to stomach

Stomach
- Stores and mixes food
- Begins chemical digestion of protein by enzymes and acid
- Regulates delivery to the small intestine

Small intestine
- Digests proteins, fats, and carbohydrates
- Absorbs most of the water and nutrients
- Secretes digestive hormones and enzymes

Large intestine
- Absorbs the last of the water and nutrients
- Stores waste material

Sigmoid colon
- Stores feces

Rectum
- Passageway for feces

Anus
- Expels undigested material

FIGURE 17.2 **Organs and accessory organs of the digestive system.** The digestive organs, from the mouth to the small intestine, share a common function: getting nutrients into the body. The rectum and anus provide the means for expelling waste. Accessory organs aid digestion and absorption.

These tissues are modified in each part of the digestive tract to fit the special needs and functions of each organ. For example, the stomach has three layers of muscle so that it can churn the food being digested. The rest of the tract has two layers so that food can simply be pushed along the length of the tract. The mucosa of some areas produces less mucus in places where nutrient absorption is the highest priority, and it produces a great deal of mucus where protection from acid or friction is the main concern.

ACTIVITY 1

Observing the Tissue Layers of the GI Tract

Materials for This Activity

Compound light microscope
Slide of the small intestine
Model of the small intestine

1. Observe a slide and/or model of the small intestine and identify the four tissue layers shown in Figure 17.1.

   Note the outer *serosa* (which is often mostly removed during slide preparation and may appear as one small area of adipose tissue). Next, observe the two layers of smooth muscle in the *muscularis*. The deep layer is circular, while the superficial layer runs longitudinally.

2. Observe the connective tissue of the *submucosa* and note the blood vessels, nerves, and other structures embedded in this layer. The submucosa provides most of the blood supply for the other layers and protects the delicate nerves and vessels that pass through its connective tissue.

3. Finally, observe the *mucosa* and note its surface epithelium, connective tissue, capillaries, and lacteals (lymphatic capillaries). Its major functions are secretion (e.g., of hormones, mucus, and enzymes), absorption of digested food, and protection against bacteria. Into which structure in the mucosa would water-soluble nutrients be absorbed?

_____

The Mouth, Pharynx, and Esophagus

You probably already know that the mouth, pharynx, and esophagus are structures that allow us to chew land swallow food. But how are they specialized for this job? All three are lined with *stratified*

squamous epithelium to protect the lining from friction. You may recall that this is the same epithelium as the skin; however, there is one big difference. The skin cells are filled with keratin for extra protection and prevention of water loss. Because water loss is not really an issue in these structures, and a softer, more flexible lining is desirable, the lining of these structures is **nonkeratinized.**

Another aspect common to all three structures of the upper GI tract is muscle tissue. Skeletal muscle is found in and around the mouth, including the tongue, in order to chew and manipulate food. Skeletal muscle is also found in the pharynx and upper portion of the esophagus so that we may voluntarily initiate swallowing. The muscle tissue of the esophagus gradually transitions to smooth muscle, which explains why initiating swallowing is voluntary but the continuation of the swallowing process is not.

The Stomach

The saclike appearance of the stomach suggests that it plays a role of a food reservoir (Figure 17.2). It is also a major site of digestion, as solid food is liquified by the enzymes and acid produced by the stomach. A pair of **sphincter** muscles help to hold food in the stomach during this part of the digestive process. The cardiac (or gastroesophageal) sphincter, near the top of the stomach where the esophagus and stomach connect, prevents the acidic food matter from moving back up into the esophagus. The pyloric sphincter, where the small intestine and stomach connect, prevents food from moving into the small intestine until it is fully liquified. What common ailment is caused when the cardiac sphincter allows a small amount of material to leak into the esophagus?

_____

The Small Intestine

The first portion of the small intestine, which connects to the stomach, is called the **duodenum.** Ducts from the pancreas, liver, and gallbladder also connect to this busy crossroads of the digestive tract. Bicarbonate ions from the pancreas neutralize the acid from the stomach. Bile from the liver and gallbladder and enzymes from the pancreas are also added to the duodenum so that complete digestion of our food can occur during the long trip through the small intestine.

The remainder of the small intestine is divided into two parts, the **jejunum** and **ileum.** Not only is food digested down to simple molecules in these

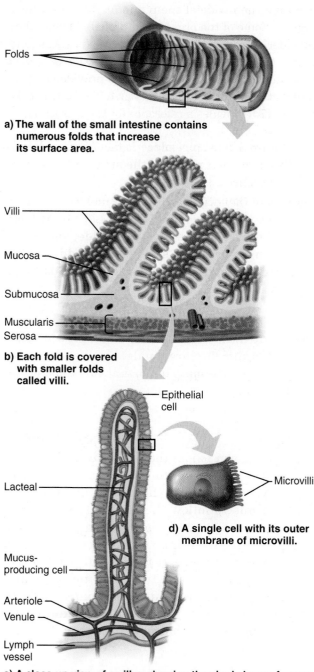

a) **The wall of the small intestine contains numerous folds that increase its surface area.**

b) **Each fold is covered with smaller folds called villi.**

d) **A single cell with its outer membrane of microvilli.**

c) **A close-up view of a villus showing the single layer of mucosal cells covering the surface and the centrally located lymph vessel (lacteal) and blood vessels.**

FIGURE 17.3 **The walls of the small intestine. a)** The numerous folds of the small intestine increase its surface area. **b)** Each fold is covered with microscopic folds called villi. **c)** This view of a single villus shows a single layer of mucosal cells, a lacteal (lymph vessel), and blood vessels. **d)** This detail view shows a single cell with its outer membrane of microvilli.

structures, but absorption of nutrients is a major function as well. Numerous folds and microscopic folds, called villi, increase surface area for absorption (Figure 17.3a–b). For the most part, water-soluble nutrients such as glucose are absorbed

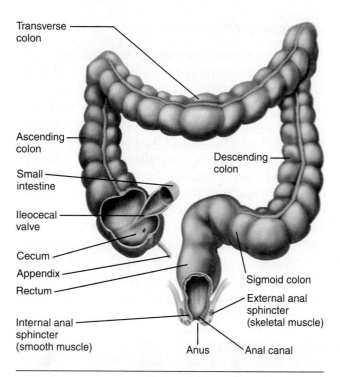

FIGURE 17.4 **The large intestine.** The large intestine absorbs the last of the remaining water, ions, and nutrients in the digestive tract and stores the remaining near-solid waste until it can be expelled.

into the capillaries, and fat-soluble nutrients are absorbed into the lacteals, which are lymphatic capillaries in the mucosa of the villi (Figure 17.3c).

The Large Intestine

The ileum of the small intestine plugs into the large intestine slightly above its starting point. The short blind alley of large intestine below this entry point is called the **cecum.** In humans, the cecum has a wormlike "tail" called the **appendix** (Figure 17.4).

The **colon** begins just above the cecum, and it is named for the direction in which the digestive wastes move: the **ascending colon** on the right side of the abdominal cavity, the **transverse colon** stringing across from right to left in the upper portion of the abdominal cavity, the **descending colon** down the left side, and the squiggly **sigmoid colon** connecting to the rectum in the pelvic cavity. We refer to the material in the large intestine as waste because essentially all of the nutrients and most of the water should have been absorbed from the food before it left the small intestine.

It is important to note that a significant amount of water and some salts are absorbed by the large intestine, a process that helps to solidify the wastes. Bacteria in the large intestine also provide a service to us by producing vitamins such as vitamin K, which plays an important role in blood clotting.

Accessory Organs

Although not considered part of the digestive *tract,* the pancreas, salivary glands, liver, and gallbladder play critical roles in the digestive process (Figure 17.2). The **pancreas** and **salivary glands** produce enzymes and other materials needed by the digestive system. As mentioned previously, the pancreas helps to neutralize the acid from the stomach by secreting bicarbonate ions, and it produces enzymes that help digest many kinds of food molecules. The salivary glands produce mucus and amylase, an enzyme that helps to break down starches into sugars. The **liver** produces bile, which is then stored and concentrated in the **gallbladder.** Although it is not an enzyme, bile plays an important role in fat digestion by breaking up or emulsifying large fat globules into much smaller ones.

ACTIVITY 2

Identifying Human Digestive System Organs

Materials for This Activity

Human torso models

Using Figures 17.2 and 17.4 as references, trace the pathway of food through the GI tract by identifying each of the following structures on the human torso models.

- Esophagus
- Stomach
- Pancreas
- Duodenum
- Jejunum
- Ileum
- Cecum
- Ascending colon
- Transverse colon
- Descending colon

ACTIVITY 3

Observing the Major Digestive Organs in the Fetal Pig

Materials for This Activity

Fetal pig
Dissection equipment
Dissection tray
Disposable gloves

Because your fetal pig should already be open from previous lab exercises, you will now simply need to

examine the inside of the pig in order to find these organs. Some of the organs will probably be familiar, considering they were pointed out previously as landmarks to help you find other structures. Please review the safety procedures for dissection provided by your instructor and those that appeared in Exercise 13. Figure 17.5 provides a reference for the dissection.

1. Obtain a fetal pig, place it faceup on a dissection tray, and reattach the string as described in Exercise 13.
2. Locate the large dark **liver,** which covers much of the upper abdominal cavity. Lift up the lobes of the liver on the *pig's* right side, and locate the small, greenish, saclike **gallbladder.** What secretion do the liver and gallbladder add to the digestive tract?

_____

3. Lift up the lobes of the liver on the *pig's* left side, and locate the large, pale, saclike **stomach.** What are the most important functions of the stomach?

_____

4. Follow the stomach toward the pig's right as it tapers down to join the **duodenum.** Then, follow the duodenum to where it hooks back to the pig's left. This is the start of the **jejunum.** Continue to follow the small intestine until it attaches to the large intestine. The latter half of the small intestine, which attaches to the large intestine, is the **ileum.** Because the jejunum and ileum are coiled together and difficult to differentiate, most sources refer to them collectively as the **jejuno-ileum** in the fetal pig. What is the main function of the small intestine?

_____

Why do you suppose that the small intestine is the single longest portion of the digestive tract?

_____

5. At the junction of the large and small intestines, you should see a small, thumblike pouch called the **cecum.** The thicker intestine above the cecum is the **colon,** which is typically coiled together in the pig. For this reason, it is usually referred to as the *spiral colon* in the fetal pig. The colon straightens out to run along the back of the pelvic cavity and form the **rectum.**

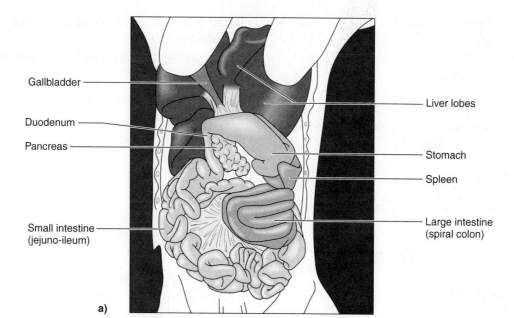

Gallbladder

Duodenum

Pancreas

Liver lobes

Stomach

Spleen

Small intestine
(jejuno-ileum)

Large intestine
(spiral colon)

a)

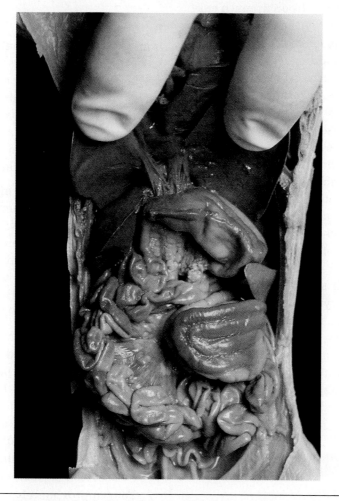

b)

FIGURE 17.5 **The digestive organs of the fetal pig. a)** This diagram illustrates the photograph in part **b)**.

6. Using your fingers and/or a blunt probe as needed, lift up (do *not* cut and remove) the liver on the pig's left side (your right). Then, lift up the saclike stomach. You should be able to see the **pancreas,** which you first observed in Exercise 13, The Endocrine System. If you did not locate the pancreas previously, use forceps to break and remove the thin, plasticlike material under the stomach. You should now be able to see the pancreas. What structure is the pancreas attached to on its right side (remember, the *pig's* right side)?

_____

Enzymes

Enzymes are the helper molecules, made of protein, that make many chemical reactions occur much easier and faster. They are used by the digestive system to break the large food macromolecules down to their more easily absorbed building blocks.

The enzymes of the digestive system are generally named for the substance they break down. For example, proteases break down proteins, lipases break down lipids, and amylases break down starches such as amylose. Within these general categories, there are enzymes with specific jobs. Pepsin is a protease produced by the stomach, which is mainly responsible for breaking big proteins down into smaller ones. Other proteases, such as trypsin produced by the pancreas, then break down proteins further.

Because they are made of delicately folded proteins, enzymes require fairly specific environmental conditions to maintain their shape and function effectively. Although some enzymes are more tolerant than others, a slight change in temperature or pH can render an enzyme useless. Not surprisingly, each enzyme is designed for the environment in which it must work. For example, pepsin works best in the acidic conditions we find in the stomach, while trypsin only works at higher pHs found in the small intestine.

ACTIVITY 4

Observing Enzyme Function

Materials for This Activity

1-liter beaker filled about halfway with tap water heated to 38°C

1-liter beaker filled about halfway with tap water heated to 70°C

1-liter beaker filled about halfway with ice water

Tap water

Distilled or deionized water

250-ml flask

Five large test tubes

Phenol red solution

pH paper

China marker

Dropper bottles containing the following:

- Vegetable oil
- 2% sodium carbonate
- 1% bile solution (which has been adjusted to a pH of 7.5–8.0)
- 1% pancreatin or pancreatic lipase solution

In this experiment, you will be testing the performance of an enzyme, pancreatic lipase, under a variety of environmental conditions. You will be preparing five test tubes: one serves as a control and each of the remaining tubes is subjected to a different variable. The lipid that pancreatic lipase will be breaking down is called a **triglyceride.** When broken down, the triglyceride is separated into a glycerol molecule and three fatty acid molecules.

1. Add about 120 ml of distilled water to the 250-ml flask. The exact amount is not important. Test your solution with pH paper. The pH of distilled water should be approximately 7.0. Add 1–2 drops of sodium carbonate to the flask and swirl the flask to mix the solution. Test your solution with pH paper. If the solution has a pH between 7.5–8.0, proceed to step 2.

   If the pH of the solution is lower than 7.5, repeat the step of adding sodium carbonate and testing the solution with pH paper until the pH is in the 7.5–8.0 range. *Note:* If the pH of your solution is already above 8, your source of distilled water may not be pure and has an abnormally high pH. You may substitute tap water or bottled water if necessary, as long as you end up with a pH in the slightly alkaline range (7.5–8.0).

2. Pour approximately equal amounts of the solution you prepared in step 1 into each of the five large test tubes. Label the tubes 1, 2, 3, 4, and 5. Add five drops of phenol red solution to each test tube. Phenol red is somewhat like liquid

litmus paper; it turns a bluish color at a high pH and an orange color at lower pHs. Because you made your solution slightly alkaline in step 1, your tubes should all be about the same color— a fuchsia or magenta color.

3. Add five drops of pancreatin or pancreatic lipase solution to each test tube.

4. Add nothing to test tube 1; add five drops of vegetable oil to test tubes 2, 3, 4, and 5. Add three drops of bile solution to test tube 3. Propose a reason why you added nothing further to test tube 1. (What function is test tube 1 serving in this experiment?)

_____

5. Place test tubes 1, 2, and 3 in the beaker serving as the 38°C water bath; place test tube 4 in the beaker serving as the 70°C water bath; place test tube 5 in the beaker serving as the ice-water bath.

6. Every 10 minutes for one hour, check to make sure the temperature has not changed more than two to three degrees from the starting temperature. If it has, place the 38°C or 70°C beaker on a hot plate for a minute or two in order to return it to its starting temperature. The ice-water bath is unlikely to require adjustment in one hour's time, but you may add more ice if necessary. Use this 10-minute check as an opportunity to remix the contents of the test tube by swirling or shaking the test tube from side to side. (Your instructor may demonstrate the best procedure for your test tubes.)

7. After one hour, compare the color changes, if any, that occurred in your test tubes. Use test tube 1 as your point of comparison, considering it should not have changed color during this experiment.

Recalling that the enzyme should break down the vegetable oil into glycerol and fatty *acids,* a change in pH should occur in any tube in which the enzyme was working.

Recalling that phenol red is like liquid litmus paper, a color change to pinkish-red or red is a strong positive result. A slight color change to pink would still be considered a positive reaction, but a weaker one.

Record your results in Box 17.1. *Note:* It is helpful to hold the tubes against a white background when judging your color changes. A dark background makes it difficult to notice slight color changes.

8. The bicarbonate ions, which are secreted by the pancreas along with enzymes such as pancreatic lipase, raise the pH of the small intestine to approximately the same alkaline range used in this experiment. However, the temperature of the human body in the abdominal cavity is more likely to be in the 37–38°C range. Which were the only test tubes to give significant positive reactions?

_____

Which test tube demonstrated the strongest reaction?

_____

What ingredients were initially placed in the test tube of your previous answer?

_____

_____

Explain your results.

_____

_____

9. The temperatures experienced by test tubes 4 and 5 should have been too extreme for the enzyme to work. Propose a reason why it might be possible to get a slight color change in these test tubes anyway. (*Hint:* Review your procedure for this activity for potential sources of experimental error.)

_____

_____

Box 17.1	Results of Enzyme Activity as Measured by Color Change
Test Tube	**Color Change (insignificant, slight, moderate, substantial)**
1	
2	
3	
4	
5	

Speculate as to why either a high fever or hypothermia can be lethal.

_____

_____

_____ ■

Metabolism and Nutrition

Metabolism mainly describes how the nutrients we absorb are used by the body. These nutrients can be used to build molecules, cells, and tissues, or to repair existing ones. Most of the molecules we absorb can also be used for energy. Because humans are warm-blooded creatures, an enormous amount of the nutrient molecules we take in are simply used to maintain our body temperature and keep our organs functioning.

Basal metabolism, or **basal metabolic rate (BMR),** is the measure of the energy required for essential life-support functions such as generating body heat. A calorie is a very small heat measurement unit and is simply the amount of energy needed to raise 1 gram of water 1°C. Because this tiny unit would be impractical in discussing human nutritional needs, the kilocalorie (1,000 Calories) is used. We use the capitalized term, **Calorie,** when referring to the larger kilocalorie in discussions of nutrition.

For the average-sized person, the Calories of energy required for basal metabolism would be approximately 2,000 Calories per day. This may vary greatly depending upon numerous factors such as gender, size, and muscle density. The total number of Calories we expend depends on our BMR plus our daily activities. A sedentary person will clearly expend a number of Calories that is not substantially higher than his or her BMR, while a triathlete in training will expend many more Calories in a day. A simple fact of weight gain versus weight loss is that excess Calories are stored as fat, and reducing body weight requires burning more Calories than we consume.

It is an unfortunate fact of nutrition that fatty foods tend to be significantly higher in Calories, ounce for ounce, than other food sources. They also tend to taste better to most people—probably a leftover adaptation from a time in human history when food was far less abundant; so the more Calories, the better.

ACTIVITY 5

Learning About Metabolism, Nutrition, and You

Materials for This Activity

Standard personal weighing scale

In this activity, we will examine a person's caloric needs, compare that to a proposed menu, and examine the amount of additional activities this person would need to undertake to avoid weight gain.

1. To calculate your BMR, it is necessary to determine your weight in kilograms and adjust the number if you are female, because females burn slightly fewer Calories pound for pound than males. Weigh yourself, divide the number of pounds by 2.2, and record this number.

   __118.18_____

   If you are female, multiply this number by 0.9, and record it.

   _____

2. Multiply the number you obtained in step 1 by 24 and record it.

   __2836.32_____

   This is your BMR, or the number of Calories you burn every day just to maintain body temperature and organ function.

3. Use Table 17.1, or other references provided in the lab, to choose items for a hypothetical menu for typical daily food intake. *Note:* This activity will not work if you make food choices that do not truly represent your food intake on most days. Enter these items and their number of Calories in Box 17.2. Add the Calories in Box 17.2 to calculate total Calories for the day. Record this total.

   __1180_____

4. Subtract the BMR you recorded in step 2 from the Calorie total you recorded in step 3. If this is a negative number, you may have forgotten to include all the items you would truly eat in a typical day, or you may have planned a menu that you would follow if you were on a diet. If you are certain that you did plan a representative menu and still obtained a negative number,

Table 17.1 **Common Foods and Their Estimated Caloric Content**

Meal	Food Item	Calories
Breakfast foods	Cereal with milk	200
	Pancakes (2) with syrup	450
	Doughnut	300
	Bagel, medium (with margarine or cream cheese, add 40)	250
	Nonfat yogurt	150
	Medium muffin (with margarine, add 50)	250
	Fruit, medium size (apple, orange)	80
	French toast (2 slices) with syrup	450
	Toast, one slice, with margarine and jam	170
	Scrambled eggs (2)	200
	Glass of milk	150
	Orange juice	100
	Black coffee (with milk, add 25)	5
Lunch and dinner items	Sandwich*	400
	Fish (fried, add 150)*	200
	Medium hamburger (fast food, add 150)*	350
	Steak*	350
	Chicken breast (fried, add 100)	150
	Pizza, 1 slice (each meat topping, add 50)*	250
	Pasta with tomato sauce	250
	Soup*	150
	Potato, baked or mashed (with margarine, add 50)	150
	Mixed vegetables (with margarine, add 50)	50
	Garden salad (with dressing, add 40)*	100
	French fries (fast food, add 150)	200
	Average soft drink (non-diet)	130
Dessert and snack items	Cake*	300
	Potato or corn chips*	150
	Popcorn*	70
	Apple (or similar) pie slice*	300

Note: The listed Calorie numbers are simplified, rough estimates provided only for the purposes of this activity. They should not be used for dietary planning. The estimates will vary greatly depending on the exact preparation, portion size, and ingredients used.

*The Calories in these items are especially variable depending upon the exact preparation and ingredients. For example, some species of fish are much higher in fat content than others, and a lean steak can easily be half the Calories of a fatty one.

this result suggests that your current eating habits *may* result in weight loss over time. If you obtained a positive number (the expected outcome), proceed to steps 5 and 6.

5. There is no need for concern if the number of Calories you calculated from your menu is only 10–15% higher than your BMR. Unless a person's lifestyle is very sedentary, it is likely that he or she will burn at least 10% more Calories above his or her BMR. But if the difference between the BMR and caloric intake from the proposed menu is great, he or she would need to reconsider dietary choices or increase his or her activity level in order to avoid weight gain over time.

6. Walking and most light activities burn Calories at a rate of about 250 Calories per hour for a person of average size. Jogging and more

Box 17.2	Proposed Menu and Calorie Totals	
	Food Items	**Calories**
Breakfast		
	Total	
Lunch	Chicken Breast	150
	Garden Salad	140
	Total	290
Supper	Steak	350
	Vegetables	100
	Salad	140
	Total	590
Snack	Pie	300
	Total	300
	Total for all food items	

intense physical activities burn approximately 600 Calories per hour or considerably more depending on a person's size and the intensity of the activity performed.

How many hours of walking would you need to do to burn the Calories above your BMR? _____

How many hours (or minutes) of jogging would you need to do? _____

It is important to conclude by mentioning that numerous factors may affect your metabolic rate and caloric needs; your doctor or other health professional is in the best position to assess your actual dietary needs. Also, Calories are not the only consideration in a proper diet. Other factors, such as vitamins, fiber, and fat content are important too. The purpose of this activity was simply to get you off to a good start in thinking about the Calories you burn versus the Calories you consume. ■

THE DIGESTIVE SYSTEM AND NUTRITION
Critical Thinking and Review Questions

1. What are the four tissue layers of the GI tract wall?

   _____

   Summarize the importance of each layer.

   _____

   _____

   _____

   _____

2. What are the four accessory organs of the digestive tract, and what does each do?

   _____

   _____

   _____

   _____

   _____

3. Fill in the following table to describe the location of each organ.

Organ	Description of Location
Salivary glands	
Esophagus	
Liver	
Gallbladder	
Pancreas	
Appendix	
Descending colon	

4. Match the organ to its description or function.

   ____ Stomach   a. stores and concentrates bile
   ____ Duodenum   b. blind pouch near the beginning of the colon
   ____ Jejunum and ileum   c. most absorption occurs here
   ____ Cecum   d. first part of the small intestine
   _a_ Gallbladder   e. food reservoir

5. What are villi, and why are they important to the function of the small intestine?

_____

_____

6. Match the digestive secretion to its function or description.

____ Protease   a.   breaks down starches

____ Amylase   b.   not an enzyme; breaks up large fat globules

____ Lipase   c.   not an enzyme; neutralizes acid from the stomach

____ Bile   d.   breaks down proteins

____ Bicarbonate ions e.   breaks down lipids

7. Do enzymes require relatively stable environmental conditions, or do they tend to work over broad ranges of temperature and pH?

_____

Explain your answer.

_____

_____

8. In Activity 4, you experimented with using pancreatic lipase at both normal and abnormally high temperatures. Recall your study of the scientific method at the beginning of this book and review your answer to the previous question. Formulate a hypothesis about pH and the effectiveness of enzymes.

_____

_____

_____

Predict what would happen if you conducted an experiment similar to Activity 4, using a mixture of pepsin and protein in two different test tubes—one adjusted to a pH of 2, and the other adjusted to a pH of 8.

_____

_____

_____

_____

9. What is *basal metabolism*?

_____

Why is it important to determine *basal metabolic rate (BMR)* as part of a Calorie-counting activity?

_____

10. One pound of fat has a caloric equivalent of about 3,500 Calories. If a person is on a diet and exercise program whereby he or she burns 500 more Calories a day than he or she consumes, how long will it take for him or her to lose 1 pound?

_____

Do you think that diets that encourage the loss of 5 pounds per week or more are safe? Explain.

_____

_____

_____

_____

11. Label the following diagram.

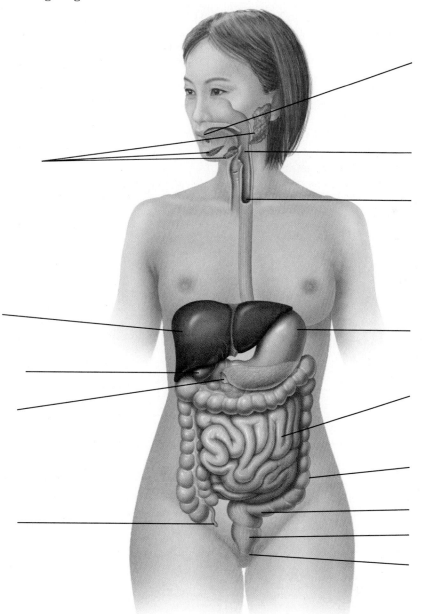

The Urinary System

Objectives

After completing this exercise, you should be able to

1. List, identify, and describe the major organs of the urinary system.
2. List the main parts of the nephron and describe their functions.
3. Explain how urine is formed and the factors that affect its composition.

Materials for Lab Preparation

Equipment and Supplies

- ◯ Dissection equipment
- ◯ Fetal pig
- ◯ Disposable gloves
- ◯ Urine specimen containers
- ◯ Clinistix or other suitable testing strips that test urine pH and specific gravity
- ◯ Drinking cups (8-ounce and 12-ounce)
- ◯ Salted pretzels
- ◯ Distilled water
- ◯ Baking soda (sodium bicarbonate)
- ◯ Caffeinated coffee, tea, or diet soda
- ◯ Cleaning disinfectant or detergent solution

Models

- ◯ Human torso
- ◯ Nephron

Introduction

The digestive tract absorbs nutrients, as well as substances that we cannot metabolize. Also, metabolism produces waste molecules when we use nutrients for energy or other cellular functions. Urea, a nitrogen-containing waste, is an example of a waste product of amino acid metabolism. These wastes must be removed from the blood so that they do not build up to toxic levels. The kidneys and the urinary system not only provide this crucial service of waste removal, but they also contribute significantly to body fluid homeostasis. In this exercise, we will first examine the organs of the urinary system, and then we will focus on the process of urine formation.

Organs of the Urinary System

The Kidneys

The kidneys are bean-shaped organs about the size of your fist, and they are located near the posterior abdominal wall (Figure 18.1a). Like some other organs we have encountered in the study of the human body, the kidneys have an outer **cortex** and an inner **medulla.** The cortex is where most of the blood is filtered and enters the microscopic kidney tubule system, called the **nephron.** The medulla mainly contains tubules and ducts involved in urine formation and concentration (Figure 18.1b). The tiny collecting ducts of the medulla (Figure 18.1c) eventually dump the urine they produce into the **renal pelvis,** which collects and funnels urine into the ureter (Figure 18.1b).

The **ureters** from each of the two kidneys lead down to the **urinary bladder,** a saclike organ that stores urine. Both the ureters and urinary bladder have smooth muscle in their walls. In the ureters, the muscle produces wavelike contractions that push urine toward the bladder. In the urinary bladder, the smooth muscle contracts as part of a reflex, which helps to push urine through the **urethra** to the outside of the body (Figure 18.1a).

As you can see in Figure 18.2, the urethra is longer in males because it passes through the penis. In this figure, you can also see the two sphincter muscles that control urination. The **internal urethral sphincter,** which is near the base of the bladder, is involuntary. The **external urethral sphincter,** which is 1 to 2 inches farther down the urethra, is voluntary.

Observing the Organs of the Urinary System

Materials for This Activity

Fetal pig

Dissection equipment

Disposable gloves

Human torso models

Although your fetal pigs should be open from previous exercises, you may need to cut farther down in the pelvic cavity to view the urethra. See Figure 18.3 for views of the male and female urinary systems of the fetal pig. Review the safety information provided by your instructor and the precautionary information for performing dissections presented in Exercise 13.

1. Move the digestive organs over to one side of the pig's abdominopelvic cavity, and look for the bean-shaped **kidney** located near the posterior abdominal wall. You may need to break and remove the plasticlike connective tissue that is covering the kidney to see it clearly. Repeat this process to find the kidney on the other side. Speculate as to why animals have two kidneys, as well as duplicates of other important organs.

   _____

   _____

2. Find the **ureter,** which will look like a piece of thin, wavy string emerging from the middle of the kidney. Follow each ureter down to its point of attachment to the urinary bladder. Does this tube use simple gravity feed, or does it use muscle to push urine into the bladder?

   _____

3. Find the **urinary bladder,** which looks more like a long, muscular tube than a saclike organ. It is located between the two umbilical arteries, which helps to make the urinary bladder relatively easy to find.

4. Use a scalpel to cut through the ventral body wall, slightly off center. Find the start of the urethra at the base of the urinary bladder, and follow it to where it reaches the surface of the skin. The urethra is the portion of the urinary system that is most different in males and females.

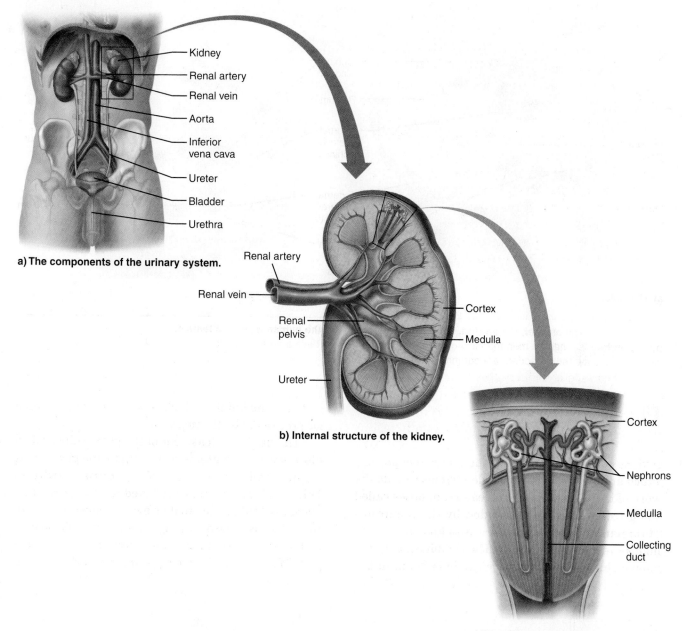

a) **The components of the urinary system.**

b) **Internal structure of the kidney.**

c) **The cortex and medulla of the kidney are composed of numerous nephrons.**

FIGURE 18.1 The human urinary system. a) Structures of the urinary system within the body. **b)** Internal structure of the kidney. **c)** Nephrons within the cortex and medulla in relation to the collecting duct, through which urine produced by the nephron passes.

5. Using Figures 18.1 and 18.2 as references, find the following structures on the human torso models.

- Kidney
- Renal cortex
- Renal medulla
- Renal pelvis
- Ureter
- Urethra
- Urinary bladder

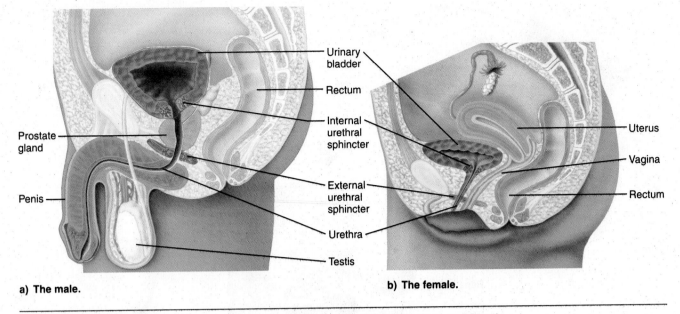

a) The male.

b) The female.

FIGURE 18.2   **The bladder, the urethra, and associated organs in the human male and female.**
a) The urethra is longer in men than in women. **b)** Women generally have a smaller bladder
capacity because their bladders are compressed by the uterus.

The Nephron and Urine Formation

A kidney typically contains about 1 million nephrons, which are the microscopic tubule systems that do the work of the kidney. A knot of leaky capillaries called the **glomerulus** is surrounded by the nephron's **glomerular capsule,** which is also known as Bowman's capsule (Figure 18.4). The combination of the glomerulus and glomerular capsule is found in the cortex of the kidney, which is where blood is filtered for processing by the nephron.

The fluid that leaks out of the glomerulus, called **glomerular filtrate,** is picked up by the glomerular capsule and funneled into the **proximal tubule.** This tubule is sometimes referred to as the proximal *convoluted* tubule due to its wavy appearance (Figure 18.4). This is a hardworking tubule, with cells loaded with mitochondria to provide energy for active transport. Most valuable molecules are reabsorbed by the

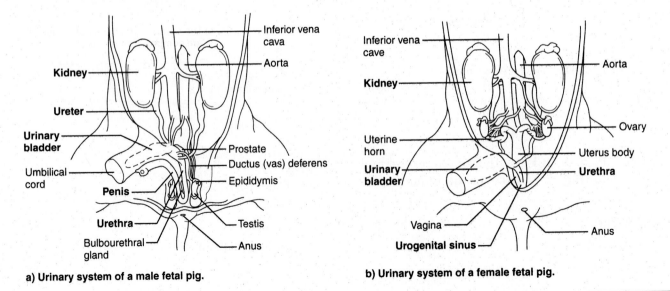

a) Urinary system of a male fetal pig.

b) Urinary system of a female fetal pig.

FIGURE 18.3   **Male and female fetal pig urinary systems.**

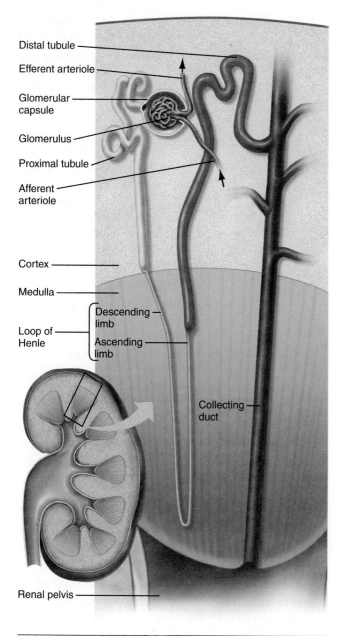

Distal tubule

Efferent arteriole

Glomerular capsule

Glomerulus

Proximal tubule

Afferent arteriole

Cortex

Medulla

Loop of Henle
- Descending limb
- Ascending limb

Collecting duct

Renal pelvis

FIGURE 18.4 **The tubular structure of a nephron.** In this view, a portion of the glomerular capsule is cut away to expose the blood vessels that serve the nephron.

proximal tubule, and most of the water lost from the glomerulus is reabsorbed by the tubule as well.

Some nephrons have a hairpinlike structure called the **loop of Henle,** which dips down deep into the medulla (Figure 18.4). The selective permeability of the loop of Henle contributes to the medulla's salt gradient: a feature important to producing a concentrated urine. Consult your textbook for more details on the work of the loop of Henle.

The **distal tubule,** also known as the distal *convoluted* tubule, is a wavy tubule somewhat similar in appearance to the proximal tubule (Figure 18.4).

Like the proximal tubule, this tubule reabsorbs some important molecules. It also *secretes* molecules into the urine.

The **collecting duct,** or collecting tubule as it is sometimes called close to its point of attachment to the distal tubule, is the place where regulation of water reabsorption occurs (Figure 18.4). Collecting ducts are sensitive to a hormone, **antidiuretic hormone (ADH),** which was first discussed in Exercise 13, The Endocrine System. Without ADH, water remains trapped inside the collecting duct and is eventually delivered to the bladder. This action results in the production of a high volume of dilute urine. When ADH is present, water easily passes out of the collecting duct into the kidney tissue of the medulla, where it is picked up by capillaries. Thus, much less water is dumped into the pelvis for eventual delivery to the bladder. This action results in the production of a lower volume of concentrated urine.

ACTIVITY 2

Observing the Tubules of the Nephron

Materials for This Activity

Model of the nephron

Using Figure 18.4 as a reference, locate the following structures on a model of the nephron.

- Glomerulus
- Glomerular capsule
- Proximal tubule
- Loop of Henle
- Distal tubule
- Collecting duct

Which tubules seem to have a wavy appearance?

ACTIVITY 3

Conducting a Urine Formation Experiment

Materials for This Activity

Urine specimen containers

Clinistix or other suitable testing strips that test urine pH and specific gravity

Disposable gloves

Drinking cups

Salted pretzels

Distilled water

Baking soda (sodium bicarbonate)

Caffeinated coffee, tea, or diet soda

Disinfectant solution

In this experiment, you will experience firsthand the process of urine formation and the factors that may affect urine formation. Normal urine is mostly water with small amounts of dissolved waste materials (e.g., urea) as described earlier. Its pH tends to be slightly acidic, as the kidneys are one of the places excess $H+$ can be excreted.

Substances that we eat or drink can affect the urine we produce as we dump wastes and excess materials into the urine. Similarly, *not* drinking water has a great impact on urine formation. We lose a little water every time we exhale, and if the water is not replaced by drinking, our body fluids become hypertonic (more concentrated).

Consuming salt is a fast way to mimic the effect of not drinking water for several hours because it quickly makes our body fluids hypertonic, just as water loss without replacement would do. This triggers ADH release and our thirst drive, mechanisms for maintaining homeostasis. Drinking a large amount of distilled water makes our body fluids hypotonic (more dilute), which inhibits ADH release. Sodium bicarbonate (baking soda) molecules, like many drugs and most other water-soluble substances, are simply dumped into the urine as a waste product. Finally, caffeine increases blood pressure in the glomerulus, increasing glomerular filtration rate. The following experiment will test the response of the urinary system to each one of these factors.

For this experiment, it will be necessary to test your urine after consuming one of four designated items. The general rule is not to consume food or drink in the lab, as some chemicals and materials handled or stored may contaminate foodstuffs. Your instructor will be able to advise you of the best way to consume your designated food or drink.

When handling urine or any body fluids, it is imperative that proper precautionary and sanitary procedures are followed. Urine samples should be disposed of in a restroom toilet. Any containers or items that have come in contact with urine should be properly disposed of or soaked for one hour in disinfectant solution and washed before re-use. Your instructor can advise you of the proper way to handle, test, and dispose of your samples. The single most important precautionary step is to handle *only* your own sample. Wear gloves, and use disinfectant solution to wipe down all work areas on which you have handled urine.

Note: For best results, it is important that you do not consume anything other than a small quantity of water for two to three hours before you begin this experiment. If you have recently consumed anything that could interfere with this experiment (such as salty foods or caffeinated beverages), you should empty your bladder and wait one hour before starting this experiment.

1. Work in groups of four, and decide which one of the four items each person in the group will consume. It is *not* advisable for a person with high blood pressure or other cardiovascular or urinary system disorders to consume either the caffeinated beverage or pretzels.

2. Consume one of the following *within five minutes* time:
   - One serving of pretzels (see the food label for serving size) and no more than 3–4 ounces of tap water
   - Three 8-ounce cups of distilled water
   - Two cups of distilled water with a tablespoon of baking soda mixed in each
   - Twelve ounces of caffeinated coffee, tea, diet soda, or diet "energy drink"

Note: Particularly for baking soda solution, consume small amounts at a time over the five minute period, and stop if your stomach becomes upset.

3. Obtain a specimen container, and obtain a sample of your urine immediately after consuming the item in step 2. Test this sample for pH and specific gravity, and enter that information as your baseline data in Box 18.1. Obtain and record the information for the other three test items in Box 18.1 as well.

4. Half an hour after the start of the experiment, attempt to obtain another sample. Test your sample, this time recording the volume as well. If you produce more urine than the sample container can hold, do your best to estimate the amount. If you cannot produce a sample at the ½-hour point (which is likely for at least one of the test items), write "no data" in the ½-hour column boxes and wait for the 1-hour interval.

5. One hour after the start of the experiment, obtain another sample. Test your sample for specific gravity, pH, and volume. Again, if you produce more urine than the sample container can hold, do your best to estimate the amount. Repeat at 1½ and 2 hours from start time and enter your data in Box 18.1.

Box 18.1	Urine Test Data					
Test Item	**Parameter**	**Baseline**	**½ Hour**	**1 Hour**	**1½ Hours**	**2 Hours**
Pretzels	Specific gravity					
	pH					
	Volume					Total Vol.
Distilled water	Specific gravity					
	pH					
	Volume					Total Vol.
Baking soda	Specific gravity					
	pH					
	Volume					Total Vol.
Caffeine	Specific gravity					
	pH					
	Volume					Total Vol.

6. Evaluate your data. In the following space, describe the changes to specific gravity and pH, and calculate the volume generated per hour by adding the volumes recorded during the experiment.

_____

_____

_____

_____

Which of the substances caused an increase in volume?

_____

Which of the substances caused an increase in pH?

_____

Which of the substances caused the specific gravity to decrease (approach 1.000)?

_____

Salted pretzels made the subject's body fluids hypertonic and caused the release of ADH. The resulting more concentrated urine should have a higher specific gravity, and a very low volume of urine should be produced per hour. Did your group's salt/hypertonic fluid data match the expected results?

_____

Summarize what was happening in the subject's kidneys during this experiment.

_____

_____

Distilled water made the subject's body fluids hypotonic and inhibited the release of ADH. The resulting more dilute urine should have a lower specific gravity (approaching 1.000), and a high volume of urine should be produced per hour. Did your group's distilled water/hypotonic fluid data match the expected results?

_____

Summarize what was happening in the subject's kidneys during this experiment.

_____

_____

Bicarbonate ions dumped into the urine as a waste product tend to absorb hydrogen ions and raise pH above 7.0. Thus, a pH that increased at least one point above baseline would be a positive

test for bicarbonate ions in the urine. Was your group's bicarbonate ion test positive?

_____

Caffeine increased glomerular filtration rate and overloaded the nephrons' mechanisms to reabsorb substances, including water. Considering that not just water reabsorption was affected, urine output should have been relatively high; but you may find little change in specific gravity or pH. Did your group's caffeine data match the expected results?

_____

Compare your results to the expected results and explain.

_____

_____

Would water-soluble drugs also pass into the urine, and could they be detected by an appropriate test?

_____ ∎

THE URINARY SYSTEM
Critical Thinking and Review Questions

1. List the structures through which urine passes on its way to the outside of the body, starting with the medulla of the kidney.

   _____

   Which structures rely on fluid pressure for the movement of urine?

   _____

   Which structures rely on muscle action for the movement of urine?

   _____

2. Match the organ to its description or function.

   ____ Urinary bladder   a. stores urine
   ____ Ureter   b. carries urine to the outside of the body
   ____ Urethra   c. collects urine and funnels it into the ureter
   ____ Renal pelvis   d. carries urine to the bladder

3. Match the structure to its description or function.

   ____ Glomerulus   a. carries urine through the medulla of the kidney
   ____ Proximal tubule   b. wavy tube found farther from the glomerular capsule
   ____ Distal tubule   c. actively reabsorbs most nutrients
   ____ Glomerular capsule   d. collects glomerular filtrate
   ____ Collecting duct   e. ball of leaky capillaries

4. List the structures of the nephron where filtration occurs.

   _____

   _____

   List the structures of the nephron where reabsorption occurs.

   _____

   _____

5. What would cause ADH release?

   _____

   What effect does ADH release have on urine production?

   _____

   _____

6. Summarize the effect of each of the following substances on urine production:
   Excess salt consumption

   _____

   _____

   Excess water consumption

   _____

   _____

   Excess bicarbonate consumption

   _____

   _____

   Excess caffeine consumption

   _____

   _____

7. Label the following diagrams.

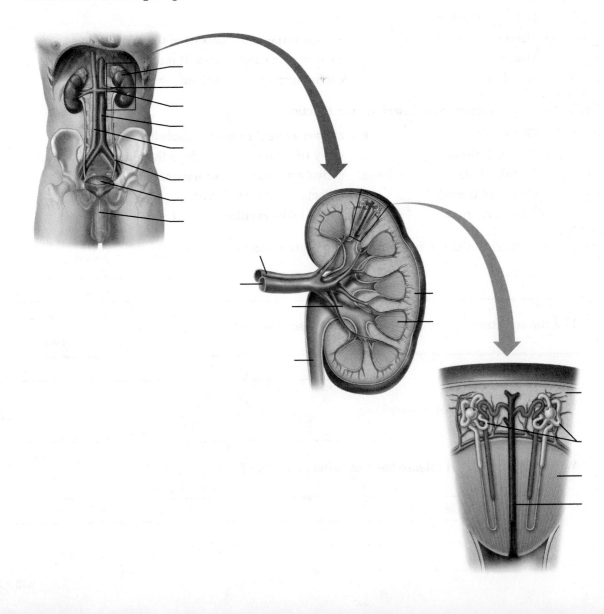

EXERCISE 19

The Reproductive System

Objectives

After completing this exercise, you should be able to

1. Describe the function of the ovarian and uterine cycles in human reproduction.
2. Describe the major anatomical features of the female reproductive system.
3. Trace the egg route from the ovary to the uterus.
4. Describe the major anatomical features of the male reproductive system.
5. Trace the sperm route from the testes to the urethra.
6. Identify the major anatomical features of the fetal pig reproductive system, and relate this to the human reproductive system.
7. List and describe the major STD-causing organisms.
8. Describe the basic methods for prevention of STDs and unitended pregnancy.

Materials for Lab Preparation

Equipment and Supplies for Dissection

- ◯ Fetal pigs (male and female if possible)
- ◯ Dissection equipment
- ◯ Dissection tray
- ◯ Bone cutters
- ◯ Disposable gloves
- ◯ Autoclave bag for biologically hazardous material

Equipment and Supplies for Condom Experiment

- ◯ Latex condoms
- ◯ Natural membrane condoms (e.g., lambskin)
- ◯ Threads
- ◯ 0-ml or 25-ml graduated cylinders
- ◯ Dye solution in squeeze bottles (dye solution: clothing dye dissolved in warm water so as to be sufficient to show as colored water)
- ◯ 200-ml beaker
- ◯ Paper towels

Models

- ◯ Human female pelvis (full, half, or dissectible)
- ◯ Human female reproductive organs
- ◯ Human pregnancy series (if available)
- ◯ Human male pelvis (full, half, or dissectible)
- ◯ Human male reproductive organs

Introduction

The human reproductive system consists of the tissues and organs required to create a new human being. Both male and female reproductive systems have specific functions that play complementary roles to achieve the goal of reproduction. The male reproductive role is to produce sperm and deliver them to the female reproductive tract. The female reproductive role, initially, is to produce unfertilized eggs. The sperm and egg may combine in the female reproductive tract to produce a fertilized egg. If fertilization occurs, the female uterus then provides an environment in which the embryo, later called a fetus, can safely develop.

Female Reproductive System

The reproductive structures of the female may be considered in terms of primary organs (the two ovaries) and secondary organs (such as the uterus). In this exercise, we will be concerned principally with the monthly menstrual cycles of the female reproductive system—the ovarian cycle and the uterine cycle—and the internal structures that contribute directly to human reproduction.

The Ovarian Cycle

As you learned in Exercise 13, The Endocrine System, the ovaries are organs that produce the hormones estrogen and progesterone. They also produce egg cells called **oocytes.** Interestingly, the process of producing oocytes begins before birth and results in a million immature egg cells that stay intentionally stalled partway through meiosis for many years. By puberty, this number declines to about 150,000 in *each* ovary, still more than enough to supply the 400–500 oocytes that actually may be used in an average female's lifetime.

Each immature primary oocyte is found in a small **follicle,** a structure with an oocyte surrounded by a ball of smaller, hormone-producing cells. Stimulation by the pituitary hormones, follicle-stimulating hormone (FSH) and luteinizing hormone (LH), causes the follicle to grow (Figure 19.1). These primary follicle cells divide, enlarge, and produce the hormone **estrogen,** while the oocyte completes the first of two meiotic divisions. As the follicle grows more over the next few days, the two or more layers of follicle cells (now called granulosa cells) produce estrogen, fluid, and a protective coating around the oocyte called the zona pellucida. The follicle is called a secondary follicle at this point.

The granulosa cells make more fluid over the next several days and produce a large fluid-filled chamber in the follicle called an antrum. This mature follicle is often termed a **Graafian follicle** (named for Dutch anatomist Dr. de Graaf). A mid-cycle surge in LH from the pituitary causes the follicle to burst, shooting the oocyte out of the ovary, still surrounded by the zona pellucida and whatever cells stick to it. This is called **ovulation.**

The majority of the follicle cells stay behind, and the damaged follicle quickly "heals," becoming a hormone-producing structure called the **corpus luteum.** The hormones estrogen and progesterone are produced by the corpus luteum. In addition to having important implications for preparing the uterus for possible pregnancy, estrogen and progesterone are important for secondary sex characteristics in females. Receptors for these sex hormones are found in the breast, numerous bones, and other structures associated with feminine characteristics.

ACTIVITY 1

Ovaries and the Ovarian Cycle

Materials for This Activity

Model of the ovary

Microscope slide of mammalian ovary (cat is frequently available)

Microscope

Part 1: Using the model of the ovary, find the small follicles containing the immature, primary oocyte and about one layer of follicle cells. Next, find the secondary and Graafian follicles, noting the difference in size and the amount of fluid they contain. Then, locate the ovulated follicle and the corpus luteum. Note the significantly different shape of the corpus luteum. The model may also have a structure called the corpus albicans, the small "scar tissue remnant" of the corpus luteum.

Suggest a reason for the different colors of the corpus luteum ("yellow body") and corpus albicans ("white body").

_____

_____

Make a sketch of this model on following page and label the stages of follicle development.

Ovary: Stages of the ovarian cycle

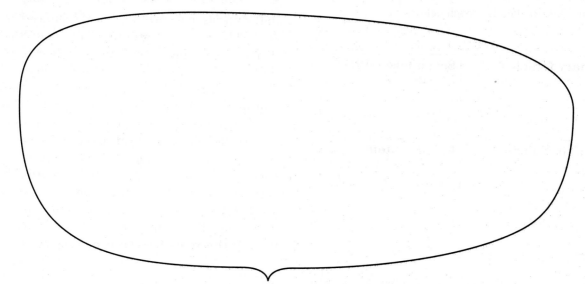

Part 2: Focus the microscope slide using your lowest power objective lens. Oocytes are some of the largest cells found in the human body, and the larger follicles with oocytes should be easily identifiable at lower magnifications. The smaller primary follicles may require that you increase magnification to 100× (using your 10× objective lens). Using Figure 19.1 as a guide, find good examples of a small primary

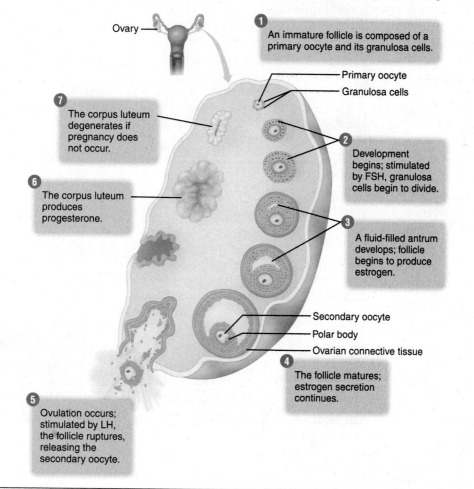

FIGURE 19.1 **The ovarian cycle.** Approximately one dozen follicles start this process each month, but generally only one completes it. For any particular primary follicle, the events take place in one location, but for clarity the events are shown as if they migrate around the ovary in a clockwise fashion.

follicle, secondary follicle, Graafian follicle, and corpus luteum. Make a sketch of each below.

Primary follicle	Secondary follicle
Graafian follicle	Corpus luteum

Uterine Cycle

In the uterus, a monthly series of events occurs that is also linked to the maturation of the egg. The function of the uterine or menstrual cycle is to provide safe housing and nutrition for the developing fetus.

There are also three phases to the monthly uterine cycle (Figure 19.2). The **menstrual phase** consists of the sloughing off of the uterine lining, also called the **endometrium.** This process is accompanied by bleeding for three to five days and is referred to as **menstruation.** The **menstrual period** is the period of visible menstrual flow, which is a discharge consisting of shed tisues and blood flowing through the vagina and out the vaginal opening. During the **proliferative phase,** which takes

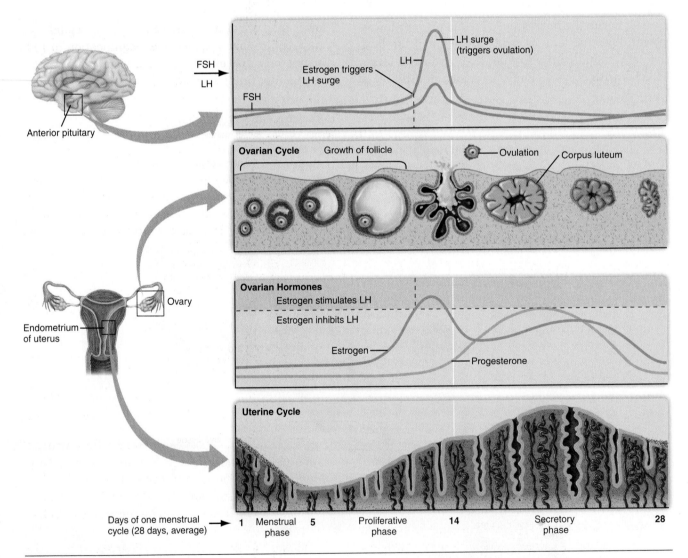

FIGURE 19.2   Ovarian and uterine cycles. The ovarian cycle is the monthly changes in the ovary that are linked to oocyte maturation. Ovulation is triggered by surges of FSH and LH. The uterine cycle is the monthly changes in the uterine lining, which responds to rising levels in estrogen and progesterone.

approximately 9 days and is triggered by rising levels of **estrogen,** the endometrium is rebuilt and blood vessels again proliferate. In the **secretory phase,** which takes approximately 14 days and is stimulated by **progesterone,** the endometrium becomes even thicker and begins to secrete nutritious substances that are capable of sustaining the embryo while it begins to implant in the endometrium.

During which phase of the uterine cyles is the endometrium the thinnest? The thickest?

_____

_____

During which phase of the uterine cycle is the best time for a developing embryo to implant in the endometrium?

_____ ■

ACTIVITY 2

Female Reproductive Anatomy

Materials for This Activity

Models

Human female pelvis (full, half, or dissectible)
Human female reproductive organs
Human pregnancy series (if available)

Let's learn about the female reproductive anatomy by tracing the path of an egg as it is released from the ovary and travels through the female reproductive tract. Use Figure 19.3 and Figure 19.4 together with the female reproductive models in your lab.

1. The **ovaries,** which are the primary female reproductive organs, produce eggs, called **oocytes,** and secrete the two major female sex hormones, **estrogen** and **progesterone.** Look at how this paired organ is located on both sides of the **uterus** and is held in place, within the peritoneal cavity, with several strong ligaments.

2. After ovulation, the egg cell is released from the ovary into finger-like projections over the ovary called **fimbriae.** Look at how the fimbriae are draped over the ovary. The fimbriae contain thousands of cilia, which beat in unison to capture the oocyte and transport it to the next location.

If the oocyte is not caught by the fimbriae, what are the chances that the oocyte will get to the oviduct?

_____

_____

3. The oocyte is next moved into the **oviduct,** also called the **fallopian tube,** or **uterine tube.** Notice that sperm is already present in the oviduct in Figure 19.3. **Fertilization** is the successful union of the egg and sperm, and the result is called a **fertilized egg.** Ideally, fertilization occurs in the upper third of the oviduct. Remember the sperm has traveled a long way to get to this location.

Trace the sperm path within the female reproductive system, starting with the vaginal opening and ending with the oviduct.

_____

_____

_____

4. The fertilized egg moves through the oviduct, over the course of three to six days, to the **uterus.** The uterus is a hollow chamber in which the fertilized egg is nourished as it changes into an embryo and then into a fetus.

Look at how the uterus is located between the large intestine and the urinary bladder. Give some reasons for frequent urination and constipation as the fetus grows bigger.

_____

_____

_____

5. There are two layers in the walls of the **uterus.** The **endometrium,** the inner layer, is composed of epithelial and connective tissues. It is rich and thick in nutrients and will eventually become the **placenta,** which provides shelter, nourishment, and waste material removal for the growing fetus. The **myometrium** is the outer layer, and it is composed of thick layers of smooth muscles, called the **uterine muscles.**

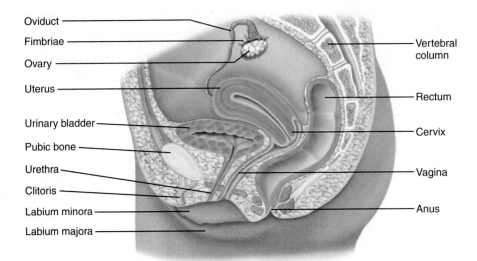

a) Sagittal section showing the components of the system in relation to other structures.

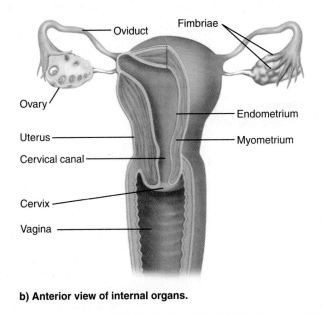

b) Anterior view of internal organs.

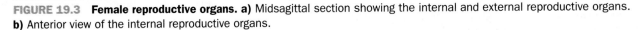

FIGURE 19.3   **Female reproductive organs. a)** Midsagittal section showing the internal and external reproductive organs. **b)** Anterior view of the internal reproductive organs.

If the embryo implants in the oviduct, this is called an ectopic pregnancy. Compare the structure of the uterus and the oviduct. Explain why an ectopic pregnancy is more likely to spontaneously abort than a regular pregnancy.

_____

_____

_____

6. During **labor,** the uterine muscles provide powerful uterine contractions to first expel the fetus through the **cervix, vagina,** and out of the **vaginal opening.** Later, the contractions continue in order to detach the **placenta** from the uterus. To prevent post partum bleeding and infection, the placenta and attached fetal membranes, commonly known as the **afterbirth,** must be completely removed. Compare the relatively empty uterus in Figure 19.4 and the almost full-term fetus in the uterus in Figure 19.5. Notice the extent of the "stretch" of the uterus and how completely protected the fetus is inside the uterus. There is a limit to this stretch, of course, and at a certain size, the uterus will be torn and damaged.

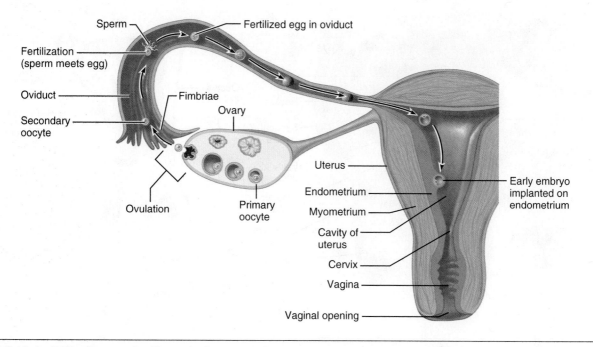

FIGURE 19.4 **Path of an oocyte.** The path of an oocyte begins with ovulation from the ovary. The oocyte then travels from the oviduct to the uterus.

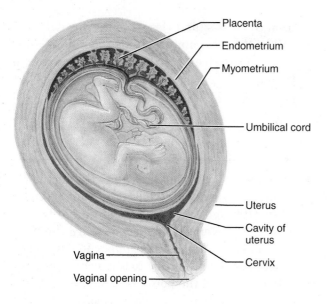

FIGURE 19.5 **A full-term fetus inside the uterus.**

If a small woman carries a large baby, what are the dangers to the mother?

_____

_____

_____

Male Reproductive System

The male reproductive system produces sperm, which must travel through the female reproductive system to potentially fertilize the egg released by the female ovaries. **Testosterone,** the male sex hormone, is produced in the testes (singular: testis). This hormone produces secondary sexual characteristics such as facial and body hair and a deeper voice. In the context of reproduction, testosterone stimulates sperm formation by stimulating growth of the male reproductive structures—ducts, glands, and the penis.

ACTIVITY 3

Male Reproductive Anatomy

Materials for This Activity

Models

Human male pelvis (full, half, or dissectible)
Human male reproductive organs

Let's learn about the male reproductive anatomy by tracing the path of a sperm as it is released from the testes and travels through the male reproductive tract. Use Figure 19.6 together with the male reproductive models for this lab.

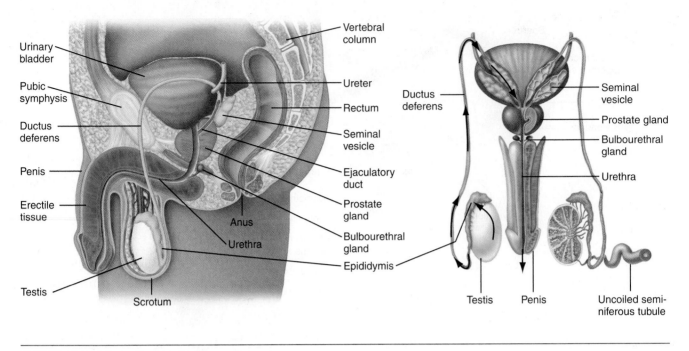

FIGURE 19.6   The male reproductive system.

1. The **testes** (singular: testis) lie in the **scrotum,** a sac of skin and other membranes located outside the abdominopelvic cavity. The external location is both good and bad. The advantage is a cooler temperature, which is ideal for sperm production, or **spermatogenesis;** the disadvantage is the exposed location.

2. The testes are composed of the **seminiferous tubules** where the actual sperm formation takes place. These tubules are tightly packed, and they total about 100 meters or 100 yards in length. Here, **spermatogonia** will undergo a series of cell divisions, finally becoming **immature sperm.**

   What is the advantage of such long lengths in the seminiferous tubules? (*Hint:* greater surface area.)

   _____

   _____

3. Ducts emerging from the testes merge to form the epididymis, a structure about 1½ inches long outside of and attached to each testis. Inside, the epididymis is a coiled tube that would be about 20 feet long if stretched out. This long coiled tube ensures proper "inventory control"— first in, first out. The sperm at the farthest point

from the testis (and closest to the vas deferens) will be the most mature. It also contains tissues that can resorb excess, dead, or damaged sperm. Why wouldn't a sac-like structure work as well as a storage area for sperm? _____

   _____

   _____

4. The **vas deferens,** also known as the ductus deferens, is a heavily muscled tube about 18 inches long. Although sperm is also stored in the vas deferens, movement of sperm toward the penis prior to ejaculation is its primary function. This substantial length is due to the meandering trip it takes around the pelvic cavity. The vas deferens emerges from the scrotum, loops around the front of the pubic bone, circles around the top of the bladder, and hooks into the ejaculatory duct at the back of the bladder (Figure 19.6). The last portion of the ductus deferens widens into a portion referred to as its **ampulla.**

   Speculate as to why the ductus deferens has (proportionally speaking) smoother muscle in its wall than the intestines.

   _____

   _____

5. When a male is sexually stimulated and ejaculates, there are smooth muscles in the walls of the epididymis and vas deferens that contract propelling the mature sperm into the **ejaculatory duct.**

6. **Semen** is sperm combined with gland secretions from the **seminal vesicle** and **prostate gland.** The **bulbourethral glands,** also known as **Cowper's glands,** produce a mucus that neutralizes the acidic urine in the urethra and provides lubrication during sexual intercourse.

7. During a sexual climax, rhythmic smooth muscle contractions will propel the sperm through the 20-cm (about 7-in.) long **urethra.** Most of the urethra is located in the **penis,** which is the male organ of sexual intercourse. Its reproductive function is to deliver the sperm internally to the female as close to the cervix as possible.

Notice that the urethra is used for both urine and semen movements. Name one disadvantage of this "common duct."

Fetal Pig Reproductive System

We will dissect the male and female fetal pig in order to identify the major fetal pig reproductive structures and to compare the major differences between the fetal pig and human reproductive systems.

ACTIVITY 4

Dissecting a Fetal Pig

Materials for This Activity

Fetal pigs (male and female if possible)
Dissection equipment
Dissection tray
Bone cutters
Disposable gloves
Autoclave bag for biologically hazardous material

Note: The following instructions assume that the abdominal cavity has been opened up in previous dissection exercises.

1. Obtain the listed supplies and return to your lab area. For a few moments, study Figure 19.7a if you have a female pig or Figure 19.7b if you have a male pig. The focus of this dissection is the reproductive structures in the pelvic portion, but there are other tubular structures in the area (e.g., the vena cava, colon, urinary bladder, and umbilical vein). Review these ducts so you can differentiate between these ducts and those of the reproductive system.

2. If you are using a female fetal pig, perform the following steps:
   a. Push the digestive organs far to the pig's right side and identify the **uterus** in the pelvic cavity. The pig's uterus is "Y" shaped, with long arms extending right and left called the horns of the uterus. Careful! It is easy to confuse these uterine horns with the oviducts or fallopian tubes of the human reproductive tract. But the horns are long extensions of the *uterus* that allow pigs to have litters of a dozen or more piglets.
   b. Near the lateral ends of the horns of the uterus, find the small rounded **ovaries.**
   c. Unlike the human female, the fetal pig's **uterine tube** is a very short tube squiggling its way across the ovary to the end of the horn of the uterus right nearby.
   d. Cut through the front of the pubic bone with a strong scissors or bone cutter, and follow the bottom of the uterus to where it connects to the **vagina.** Note that the vagina remains a separate tube for a very short distance before merging with the urethra to form the **urogenital sinus.**

How does the lower end of the fetal pig reproductive tract compare with the human females?

Speculate as to how this setup might be of some advantage to the female pig.

If you are using a male fetal pig, perform the following steps:
   a. Identify the urethral orifice just under the umbilical cord. Carefully cut through the skin of the orifice to identify the **penis,** which lies within a fold of skin.

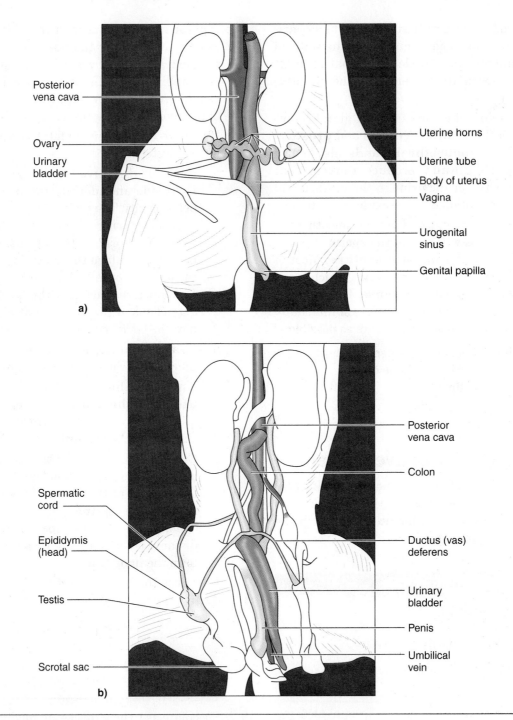

FIGURE 19.7   Fetal pig reproductive system. a) Female fetal pig reproductive organs.
b) Male fetal pig reproductive organs.

b. Follow the penis toward the back of the fetal pig where it joins with the **urethra.** A pair of glands, the bulbourethral glands, are at this intersection.

c. Follow the urethra toward the top of the fetal pig, until you can see the right and left **ductus deferens,** encapsulated in the **spermatic cord.**

d. The paired **seminal vesicles** are located near the beginning of the ductus deferens.

e. The **prostate gland** is located between the bladder and **urethra,** but it is difficult to find.

f. Follow the spermatic cord toward the bottom of the fetal pig as it enters the **scrotum.**

g. Carefully cut through the scrotal sac and identify the **testis,** a round structure, and the **epididymis,** a fine tube running alongside the surface of the testis.

How does this fetal pig structure compare with the male human structure?

_____

_____

_____ ■

Sexually Transmitted Diseases

Sexually transmitted diseases (STDs) are transmitted by various kinds of sexual contact (i.e., genital, oral-genital, and/or anal-genital). It is speculated that worldwide STD rates are increasing because younger segments of the population are engaging in sexual contact, the availability and improvement of birth control measures have removed some fear of pregnancy and spreading disease, and people assume many STDs have been eradicated or simply won't spread to them (the "it-could-never-happen-to-me" mind-set). When used properly, many birth control and barrier methods can help prevent the spread of disease, but with sexual contact no method is flawless.

Most disease microorganisms enter our body through a warm, moist entry point, such as the digestive tract, respiratory tract, or reproductive organs. Certain organisms, like the HIV virus for AIDS, travel in the body fluids.

STD-causing organisms include viruses, bacteria, fungi, and more. **Gonorrhea, syphilis,** and **chlamydia** are caused by **bacteria. AIDS, hepatitis B, genital herpes,** and **genital warts** are caused by **viruses. Yeast infections** are caused by a **fungus.** In terms of size, bacteria are considerably larger than viruses. (Refer to Table 16.4 in your main text, Sexually Transmitted Diseases, for information about the names of the organisms, symptoms, and complications.)

Preventing STDs and Unintended Pregnancies

There are a variety of ways of preventing STDs and unintended pregnancies, from simple barrier methods to behavioral ones. The latter include things like careful selection of a partner, communication, and choosing the timing and kinds of sexual contact that are less likely to result in pregnancy or disease transmission. Figure 19.8 shows many of the available methods of birth control.

One of these birth control methods, the male condom, is acknowledged to be useful as a birth control method *and* barrier against transmission of STDs. When used properly, a condom forms a physical barrier against the transmission of sperm to the female reproductive tract, as well as preventing the contact of STD organism-containing body fluids from either partner with the other. However, there are different kinds of male condoms, and there are questions regarding whether all provide equal protection against STDs such as HIV. In the next activity, we will compare the effectiveness of two different kinds of condoms.

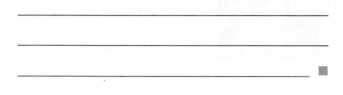

ACTIVITY 5

Comparing Condoms

Materials for This Activity

Latex condoms

Natural membrane condoms (e.g., lambskin)

Threads

10-ml or 25-ml graduated cylinders

Dye solution in squeeze bottles

200-ml beaker

Paper towels

1. Obtain one of each type of condom, two beakers, and the other listed supplies.
2. Fill each condom with the same volume (20–30 ml) of the dye solution, tie them tightly with the thread, and rinse them thoroughly.

   What is the purpose of rinsing the condoms?

   _____

3. Place each filled condom in a separate beaker of water and observe them for leakage and any other changes over the course of one hour, three hours, and overnight.

   What were the results after one hour? (Note all changes you observe in either condom.)

   _____

   _____

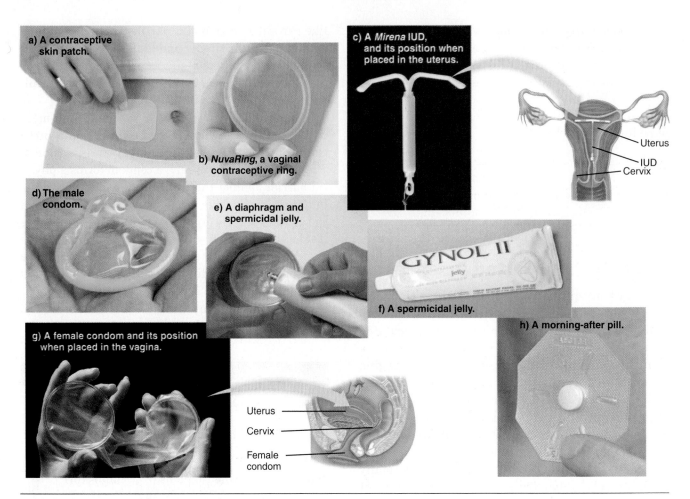

a) A contraceptive skin patch.

b) *NuvaRing*, a vaginal contraceptive ring.

c) A *Mirena* IUD, and its position when placed in the uterus.

Uterus
IUD
Cervix

d) The male condom.

e) A diaphragm and spermicidal jelly.

GYNOL II
jelly

f) A spermicidal jelly.

h) A morning-after pill.

g) A female condom and its position when placed in the vagina.

Uterus
Cervix
Female condom

FIGURE 19.8 **Some birth control methods.**

What were the results after three hours? (Note all changes you observe in either condom.)

_____

_____

What were the results overnight? (Note all changes you observe in either condom.)

_____

_____

4. The natural membrane condom should have leaked pretty quickly and swelled up overnight. Is this consistent with the changes you observed? Is this consistent with what you expected to see?

_____

Water and dye moved out of the natural membrane condom in the early stages, and water likely moved into the condom in the later stages. How can you tell if dye has moved into the condom?

_____

_____

Based on your observations, what are your conclusions at this point about the potential safety drawbacks of natural membrane condoms?

_____

_____

_____

5. The latex condom should not have leaked or swelled up overnight. Thus, water and dye should not have moved in or out of the latex condom. Is this consistent with your observations? _____

At this point, what are your conclusions about the potential safety of latex condoms?

_____

_____

_____

6. Despite the leakage observed in the natural membrane condom, the condom does provide birth control because sperm in semen is too large to pass through the natural openings in the membrane during intercourse. How many of your conclusions regarding the natural membrane condom in the previous activity were based on the swelling of the condom while in the water? Does this new information change your thoughts on the safety of natural membrane condoms?

_____

_____

7. Let's say an STD-causing bacteria, such as the *Neisseria gonorrhoeae* (gonorrhea) and *Treponema pallidum* (syphilis) were the same size as the sperm, and that STD-causing viruses such as HIV (AIDS), Herpes simplex virus (genital herpes), and HPV (genital warts) were the same size as the dye. (Research does indicate that some viruses can pass through the natural membrane condoms.)

Based on these size assumptions, describe the potential protection against STDs provided by the natural membrane condom.

_____

_____

Describe the potential protection against STDs provided by the latex condom.

_____

_____ ▪

ACTIVITY 6

Internal Search for Information on Condoms

1. Visit the Centers for Disease Control and Prevention (CDC) Web site at *http://www.cdc.gov* and search for information about the safety guidelines and recommendations about latex condoms and natural membrane condoms.

   a. In what way(s) does your experiment support the recommendations given?

   _____

   _____

   _____

   b. Based on your research on the CDC Web site, what are the ways you can increase the safety and reliability of either latex or natural membrane condoms?

   _____

   _____

   _____

   c. What other types of condoms are available aside from the latex and animal membrane condoms you used in your experiment? Why is this availability important?

   _____

   _____

   _____

   d. Does the CDC Web site provide additional information about other safer sex options? Are there other methods considered more safe or effective for birth control and the prevention of STDs? (See Figure 19.8 for other options to consider.)

   _____

   _____

   _____ ▪

THE REPRODUCTIVE SYSTEM

Critical Thinking and Review Questions

1. What are the two parts of the menstrual cycle, and what are their respective roles in reproduction?

_____

_____

_____

2. What is the name and function of each of the three phases in the uterine cycle?

_____

_____

_____

3. What hormones are released from the female pituitary, and what is their effect?

_____

_____

4. What hormones are released from the ovaries, and what is their effect?

_____

_____

5. Trace the route of the egg from the ovary to the uterus.

_____

_____

_____

6. Describe the two tissue layers of the uterus. What are their functions?

_____

_____

7. Trace the route of the sperm from the testes to the urethral opening.

_____

_____

_____

8. Describe the advantages and disadvantages of the external location of the testes.

_____

_____

9. Compare and contrast the main differences in the reproductive tract of the female fetal pig and human.

_____

_____

_____

10. Compare and contrast the performance of latex and natural membrane condoms regarding birth control and STD prevention.

_____

_____

_____

11. Label the female reproductive organs.

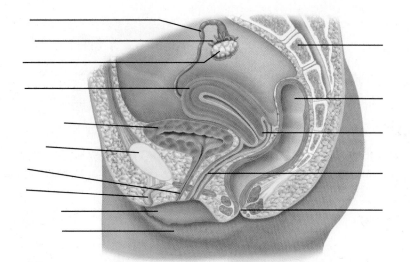

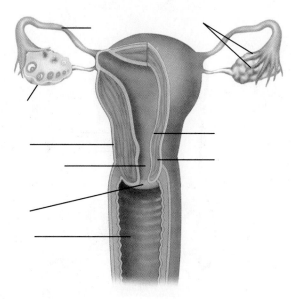

12. Label the male reproductive organs.

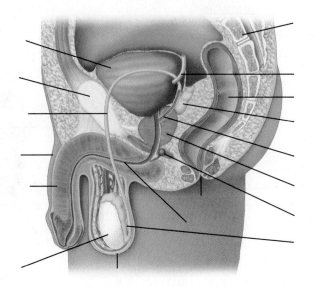

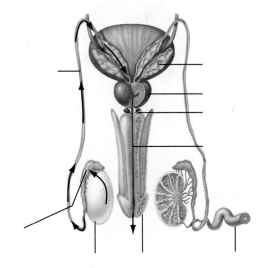

EXERCISE 20

Genetics

Objectives

After completing this exercise, you should be able to

1. Describe the process of cell division (mitosis and meiosis).
2. Describe the basic principles of inheritance.
3. Explain the connection between genetics, DNA, and chromosomes.
4. Draw and complete Punnett squares, and calculate simple genetic probabilities.
5. Describe the concept of linkage.
6. Interpret pedigree charts.

Materials for Lab Preparation

- ○ Meiosis models
- ○ Two coins
- ○ Two small boxes
- ○ Four prepainted tongue-depressor (or Popsicle) sticks
- ○ Whitefish blastula microscope slide
- ○ Microscope

Introduction

The study of the transfer of hereditary material from parents to offspring is the science of genetics. For asexual reproduction by mitotic division, the genetic makeup of the offspring would be fairly predictable, considering offspring are typically exact copies of the parent. Conversely, sexual reproduction allows genes from different organisms of the same species to mix, creating far more variety among offspring. However, a new form of cell division that cuts the chromosome number in half was necessary for sexual reproduction to work. **Meiosis** is that form of cell division that produces **gametes,** or **sex** cells, with one-half the usual number of chromosomes. The two gametes fuse to form a new cell with one full set of chromosomes. In this exercise, we will study meiosis and then the different patterns of inheritance made possible by the fusion of gametes.

Mitosis

Before discussing how sex cells are produced, and how they are related to the study of genetics, it is important to have a basic understanding of how cells divide in the first place. The process of cell division in human cells is called **mitosis.** Because the usual purpose of mitosis is to produce two daughter cells from one mother cell, mitosis usually involves three aspects: (1) copying of the DNA; (2) portioning of the DNA into separate areas where new nuclei will form around it; and (3) splitting the cell in two.

It is important to note that while all cells go through the first two steps, some do not go through the third. For example, the megakaryocyte cells in bone marrow that produce platelets do not split and just become large, multinucleated cells. However, this is the exception, not the rule, as most cells undergo mitosis in order to divide.

A cell that is to undergo mitosis will receive some internal or external stimulus during **interphase,** the part of the cell cycle when it is otherwise busily doing its job. The DNA is replicated during the **S-phase** ("synthesis" phase) of interphase, and the cell then begins preparations for mitosis.

Mitosis begins with **prophase,** when the copies of the DNA molecules wrap themselves around proteins and coil up into **chromosomes.** The nuclear membrane disintegrates as the cell prepares to move the chromosomes. A **mitotic spindle** forms for the purpose of attaching to chromosomes and pulling them to opposite sides of the cell (Figure 20.1).

During **metaphase,** spindle fibers pull the chromosomes to the center or **equator** of the cell. **Anaphase** occurs when the spindle fibers snap the connection between double chromosomes or **sister chromatids,** and pull them to opposite sides of the cell. Finally, **telophase** is the reversal of prophase, where the chromosomes uncoil, the nuclear membrane re-forms, and the cell *usually* splits in two. The process of cell splitting is called **cytokinesis.**

ACTIVITY 1
Observing the Stages of Mitosis

Materials for This Activity

Whitefish blastula microscope slide
Microscope

View a slide of the whitefish blastula, which goes through the process of mitosis somewhat similarly to human cells. The entire blastula (with hundreds of cells) may look like one cell under low power, so make sure to go up to at least 400× magnification to ensure you are viewing individual cells within the round blastula. Select examples of each phase of mitosis from your slide. Draw an example of each phase in Box 20.1. Refer to Figure 20.1 as necessary. ∎

Meiosis

Meiosis makes the process of reproduction and inheritance significantly more complicated than asexual reproduction. Depending upon the number of chromosomes in the cells of the life form and how much variation there is within the species, there can be thousands or millions of genetically different offspring that are possible from each sexual reproduction event. Basically, meiosis starts out similarly to mitosis. After the chromosomes double, the cell prepares to divide up these chromosomes—one-half going to each new cell. The difference is that the cells divide a second time in meiosis, which then cuts the chromosome number in half (Figure 20.2). See your textbook for more information on mitosis and meiosis.

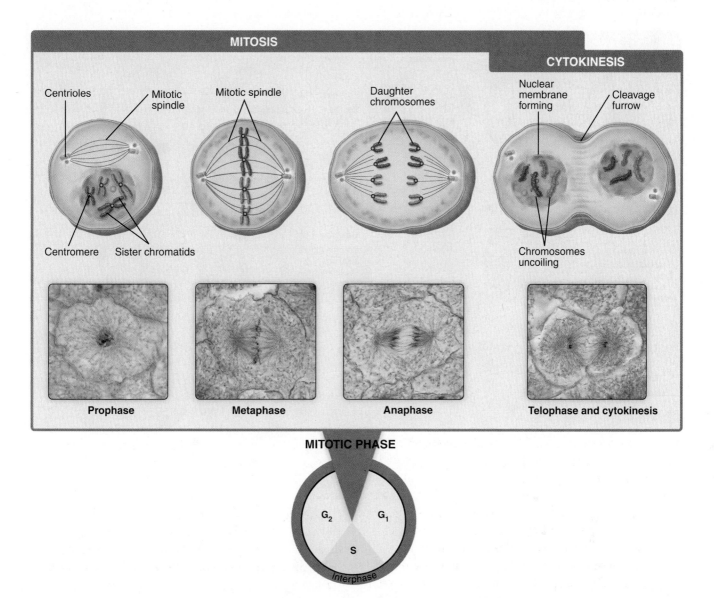

Prophase   Metaphase   Anaphase   Telophase and cytokinesis

FIGURE 20.1   Mitosis. Mitosis comprises most of the mitotic phase of the cell cycle. The duplicated chromosomes become visible in prophase, line up in metaphase, separate in anaphase, and uncoil and are surrounded by nuclear membranes in telophase.

Box 20.1 **Stages of Mitosis**			
Prophase	**Metaphase**	**Anaphase**	**Telophase**

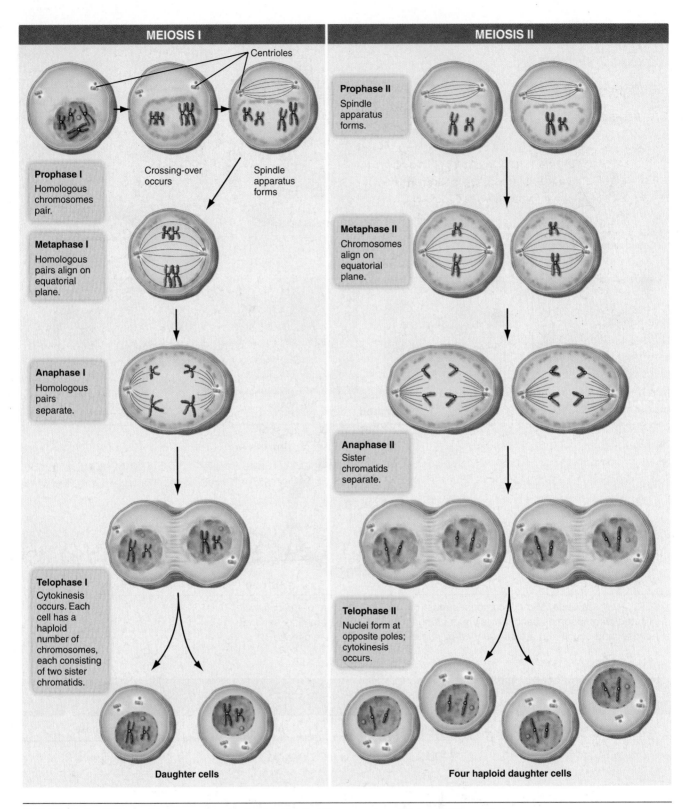

| MEIOSIS I | MEIOSIS II |

Centrioles

Prophase I
Homologous chromosomes pair.

Crossing-over occurs

Spindle apparatus forms

Metaphase I
Homologous pairs align on equatorial plane.

Anaphase I
Homologous pairs separate.

Telophase I
Cytokinesis occurs. Each cell has a haploid number of chromosomes, each consisting of two sister chromatids.

Daughter cells

Prophase II
Spindle apparatus forms.

Metaphase II
Chromosomes align on equatorial plane.

Anaphase II
Sister chromatids separate.

Telophase II
Nuclei form at opposite poles; cytokinesis occurs.

Four haploid daughter cells

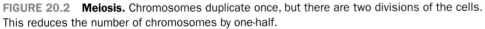

FIGURE 20.2 **Meiosis.** Chromosomes duplicate once, but there are two divisions of the cells. This reduces the number of chromosomes by one-half.

ACTIVITY 2

Reviewing the Stages of Meiosis

Materials for This Activity

Meiosis models

Review the models that demonstrate meiosis, and follow the number and arrangement of the chromosomes through each stage. Make a sketch of each stage in Box 20.2. As you make your sketches, consider how this form of cell division could produce much more variation than asexual reproduction. ■

Basic Genetics

Gregor Mendel, often called the "father of genetics," was probably one of the earliest researchers to use the scientific method. First, he observed pea plants and determined that they had several clearly identifiable traits (e.g., flower color, pea shape, size of plant). He then wondered if there was a measurable pattern to the inheritance of these traits. He formed a generalization based upon his observations that purebred plants would produce offspring exactly like themselves through numerous generations. This seemed true—purebred plants with purple

flowers *never* produced offspring with white flowers and vice versa. This "revelation" alone was of little value to science as farmers already knew at least this much, but his next step is what set Mendel on the road to some important discoveries. He decided to collect observations about a cross between opposite purebred strains.

Crossing a purple-flowered and white-flowered pea plant, or a tall and short plant, was a simple yet effective test to see what would happen when two dissimilar varieties were crossed. The experiment did not result in pink-flowered or medium-sized offspring. All offspring of such purebred crosses had purple flowers and were tall. From this experiment, he formed a hypothesis, later referred to as the **law** or **principle of dominance and recessiveness.** Apparently, whatever was responsible for causing these traits did not simply blend them together; instead, one trait could have the ability to completely overshadow (dominate) the other (the recessive trait).

Mendel then crossed these offspring among themselves. Once again, no blending occurred (no pink flowers or medium-sized plants), but a few white flowers and a few short plants were produced. Whatever was responsible for causing these traits in the hybrids did not become lost or assimilated in

Box 20.2   Stages of Meiosis

Prophase I	Metaphase I	Anaphase I	Telophase I

Prophase II	Metaphase II	Anaphase II	Telophase II

the dominant trait. He referred to this phenomenon as the **law** or **principle of segregation.** Mendel did not know about genes and DNA, so we will now leave his story for another time and move forward into modern genetics.

Genes are the segments of DNA on a chromosome responsible for producing a particular trait, such as hair color. However, not all hair color genes are identical. Each variety of a gene for a particular trait is called an **allele.** For example, everyone has hair color genes, but some have blond alleles for that gene, some have brown alleles, and so on.

The **phenotype** is the observable trait expressed, such as blue or brown eyes. The **genotype** describes the alleles present in the offspring. For example, people can have freckles because they have two identical alleles of the freckles gene (FF). Or they may have no freckles because they have two identical alleles of the nonfreckles gene (ff). There is a third possibility: people can have freckles because they have one of each allele (Ff). Because having freckles is dominant, they only need to have *one* freckles allele to display that phenotype. Because we bring two of these alleles together to form a single cell or "zygote," the suffix **zygous** is used to describe the genotype. When describing genotype in words (not letters as in "FF," "Ff," or "ff"), the terms **homozygous** (same alleles) or **heterozygous** (different alleles) are used to describe purebred and mixed alleles respectively. For example, "FF" means homozygous dominant (with freckles); "Ff" means heterozygous dominant (with freckles); and "ff" means homozygous recessive (without freckles).

How would you describe the genotype of Mendel's pea plants that had purple flowers, but had one purple allele and one white allele (Pp)?

_____

How would you describe the white flowering plant that had two white alleles (ww)?

_____

Using the Punnett Square

A **Punnett square** is a visual demonstration of simple genetic probabilities in chart form. It is a simple graphical device where each of the parent's alleles that could end up in the gametes are placed along the top and side of a box. The box is divided into columns and rows, depending upon the number of possible

Table 20.1   Punnett Square Demonstrating Two Individuals Who Are Heterozygous for Freckles

Parent 1 → Parent 2 ↓	F	f
F	FF	Ff
f	Ff	ff

F = freckles; f = no freckles

gametes for each trait to be studied. In each empty box, you simply enter the letters representing the alleles for that intersection as you look up to the top of the column and then to the side of the row. This method can be demonstrated with a Punnett square showing a cross of two people who are "hybrids" (heterozygous) for freckles (Table 20.1). Note that three out of the four boxes have at least one capital "F" and would represent a person with freckles. Only one box has two lowercase "f's" and would represent the recessive, nonfreckled person. This 3:1 **ratio** is typical of a cross of two hybrid genotypes.

ACTIVITY 3

Setting Up and Completing the Punnett Square

Recall Mendel's purple pea plants from his F1 generation (the first generation after crossing his purebred purple and purebred white flowering plants). All the plants were Pp, or heterozygous dominant for flower color. After meiosis, all the pollen and eggs each plant would produce would have either the dominant P allele, or the recessive p allele.

1. For parent 1, place the capital P in the empty top left cell in Box 20.3, and the lowercase p in the empty top right cell. Repeat this for parent 2 in the two cells on the left.

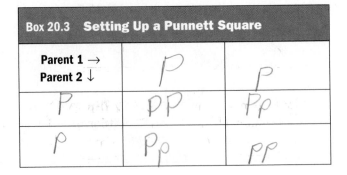

Box 20.3   Setting Up a Punnett Square

Parent 1 → Parent 2 ↓	P	P
P	PP	Pp
p	Pp	PP

2. Fill in the four empty cells by simply looking for the letter (allele) donated by the parent directly above the cell, and the letter (allele) donated by the other parent directly to the left of the cell.

3. Considering only one capital P (purple) allele is necessary for a purple flower, how many cells had a purple flowering plant? _____3_____

How many had a white flowering plant?

_____1_____

Did you obtain the 3:1 ratio typical of a cross involving two hybrid genotypes?

_____yes_____ ■

ACTIVITY 4

Exploring Genetic Probabilities Using the Punnett Square

Materials for This Activity

Two coins

Part 1: Coin-Toss Experiment

Although coins may seem quite different from chromosomes, flipping two coins is somewhat analogous to what happens to chromosomes during meiosis. Just as a person and his or her mate will randomly contribute *one* of their pairs of each chromosome, each of the two coins may randomly land on *either* heads (H) or tails (h). Consider the purple flowering plant from the previous activity, and imagine its purple allele is like heads, and its recessive allele is like the tails side of the coin. Thus, you can complete the Punnett square for heads and tails in Box 20.4.

Box 20.4 **Coin-Toss Punnett Square**		
Coin 1 → Coin 2 ↓	**H**	**h**
H	H H	H h
h	H h	h h

Work in pairs, and *carefully* flip two coins simultaneously. (You and your partner each flip a coin at the same time.) Record your data for 20 tries in Box 20.5 and on the laboratory blackboard.

Box 20.5 **Coin-Toss Work Sheet**		
	Your Data	**Entire Class Data**
Both heads (HH)	||||	38
One head, one tail (either Hh or hH)	|||| ||||		66
Both tails	||||		37

Just like the freckles example, one-fourth of the boxes in the coin-toss Punnett square (Box 20.4) would be HH, one-fourth would be hh, and half should be mixed (either Hh or hH). Your data should approximate this distribution as well.

Did your results come within 5% of the expected results (25%–25%–50%)?

_____yes_____

If not, did the data for the entire class come closer?

_____yes_____

Explain any potential sources of experimental error.

Human error, flipping incorrectly

Part 2: Hairy Hands Experiment

Having hair on the back of the hands is a dominant trait, compared to the absence of hair on the back of the hands. Thus, a person would have to have both recessive alleles (hh) to have hairless hands, considering even one of the dominant alleles would produce hair on the back of the hands. Fill in the Punnett square presented in Box 20.6, which represents the

Box 20.6 **Hairy Hands Punnett Square 1**		
	H	**h**
H	H H	H h
h	H h	h h

mating of two people who have hairy hands but are "hybrids" in that they carry the recessive allele for hairless hands. People carrying a recessive allele without displaying the trait are called **carriers.**

Your Punnett square should match the coin-toss Punnett square (see Box 20.4).

Of the four boxes, how many have at least one capital H? _____3_____

How many have two lowercase h's? _____1_____

Is it possible for two people with hairy hands to produce a child without hair on the backs of his or her hands?

_____yes_____

Would it be possible to produce a child without hair on the backs of his or her hands if one of the parents had two dominant alleles?

_____No_____

Prove your answer by completing the Punnett square presented in Box 20.7.

Box 20.7 **Hairy Hands Punnett Square 2**		
	H	**H**
H	HH	HH
h	Hh	Hh

Does your Punnett square match your prediction? _____yes_____

Explain your answer.

the square visually showes the prediction

Probabilities

Although Punnett squares provide great ways of visually demonstrating the possible outcomes of a genetic cross, **probabilities** are often better ways of determining results for more complicated crosses. Probabilities are expressed as a fraction of *one*, like the "whole" pie in a pie chart. The simple rule as it applies to genetics is that probabilities of events that must happen simultaneously (like mixing of egg and sperm in mating/crossing) are *multiplied*. Nonsimultaneous events and alternate probabilities are *added*.

For a simple example, let us work through the coin-toss "cross" (Hh × Hh) as a probability. What would be the probability of obtaining a toss where both coins are tails (hh)? The chance of coin 1 landing as tails is 50%, or one-half, and the chance of coin 2 landing as tails is also 50%, or one-half. According to the procedure, getting both coins to land as tails is something that must happen at the same time, so we must multiply these two numbers together. Thus, one-half *multiplied* by one-half would be *one-fourth,* which nicely matches the one out of four boxes that were "hh" in your coin-toss Punnett square.

There are *two* ways to get a coin toss that is mixed: One toss could be Hh (coin 1 = heads, coin 2 = tails), and another toss could be hH (coin 1 = tails, coin 2 = heads). The probability of each of these different tosses would be the same one-fourth that we calculated in the previous example. But because there are two independent, nonsimultaneous ways that this can happen (Hh or hH), we *add* them: $1/4 + 1/4 = 2/4$ or $1/2$.

How many of the four boxes in your coin-toss Punnett square represent a mixed toss?

_____2_____

Does this match the probability calculated here?

_____yes_____

Explain your answer.

the amount of mixed toss was about half

ACTIVITY 5

Calculating Probabilities

Without using a Punnett square, answer the following questions using the probabilities technique explained in this section. (You may compare your results to a Punnett square to check your work.)

1. What is the probability of tossing two coins and getting two heads?

$\frac{1}{4}$, 25%

2. What is the probability of a hairy-handed father who is a carrier for the hairless allele (Hh) and a mother who has no hair on the back of her hands (hh) producing a child with hairy hands? ■

25%

ACTIVITY 6

Calculating Probabilities Using Two Methods

1. Being able to roll one's tongue is a dominant trait, and being unable to roll the tongue is a recessive trait. Using both the probability method and then a Punnett square, determine the probability of two carrier parents (Rr × Rr) producing a child who cannot roll his or her tongue.

$\frac{1}{4}$, 25%

	R	r
R	RR	Rr
r	Rr	rr

2. Having unattached earlobes is a dominant trait, and having attached earlobes is a recessive trait. Using both the probability method and then a Punnett square, determine the probability of a carrier parent with unattached earlobes and his or her mate with attached earlobes (Uu × uu) producing a child with unattached earlobes. ■

$\frac{1}{2}$, 50%

	U	u
u	Uu	uu
u	Uu	uu

Sex Determination, Linkage, and Pedigree Charts

A different kind of inheritance pattern involves the only mismatched chromosome pair in sexually reproducing organisms, namely the **X** and **Y chromosomes.** The Y chromosome apparently carries the genes needed to produce a male, and it is smaller than its X-chromosome partner. A person with two X chromosomes is female, and a person with an XY combination is male (Table 20.2).

However, the sex chromosomes do have genes for traits other than sex determination. These "other genes" are said to be **linked** to the sex chromosome because they are obligated to travel with it when the gametes come together to produce a fertilized egg. Most **sex-linked** traits are linked to the X chromosome because the smaller Y chromosome has very few genes that are not related to sex determination.

This XY combination creates a potential problem for males, who do not have the safeguard of a complete second sex chromosome with an allele which would protect them against expressing a (usually recessive) mutant gene. The truncated Y chromosome is missing many of the alleles found on the X chromosome. So, in a male, there is only a single allele on his X chromosome, with no matching allele on the Y chromosome. This is why X-linked genetic disorders occur far more frequently in males. When a male receives a "diseased" X chromosome from his mother, the Y chromosome from his father has no matching, healthy allele to compensate for it.

ACTIVITY 7

Using Pedigree Charts to Determine Patterns of Inheritance

Part 1

Patterns of inheritance can be studied by looking at a person's genetic history using a device called a pedigree chart (Figure 20.3). Commonly referred to as a "family tree," a pedigree chart has boxes and circles that designate males and females, respectively, and its rows indicate each generation. Because these charts are typically used by medical professionals or geneticists to trace a genetic disease or trait of interest, a solid box or circle is used to designate a person with the phenotype in

Table 20.2 **Punnett Square Demonstrating Sex Determination by the X and Y Chromosomes**

Parent 1 → Parent 2 ↓	X	X
X	XX Female	XX Female
Y	XY Male	XY Male

question, and a lighter or colorless box or circle is used to indicate a person without the trait. Carriers, who are heterozygous for the allele, are either shown with a different color or a box or circle that is half solid.

Figure 20.3 displays a chart typical of an X-linked disorder such as color blindness. Using the terminology explained previously, describe the individual labeled as No. 1 in the figure.

a male w/a phenotype in question

Describe the individual labeled as No. 2 in the figure.

a female w/a heterozygous

Describe the individual labeled as No. 3 in the figure.

a male without the trait

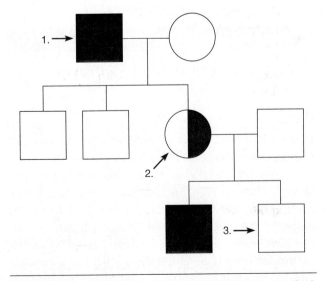

FIGURE 20.3 **A simple pedigree chart displaying a typical X-linked disorder.**

Part 2

Review the pedigree chart in Figure 20.4. Complete the chart by adding the following offspring to the couple in the chart: two sons—one color-blind and one with normal vision; one daughter.

Is the mother or the father the source of the color blindness gene?

father

Consider your answer to the previous question, and darken in half of the box for the appropriate parental unit chosen.

What is the probability that the daughter is a carrier?

50%

It is important to note that only some linked genes are on the sex chromosomes. The other 22 pairs of chromosomes contain thousands of genes, many of which are linked and transmitted together to the offspring. ■

ACTIVITY 8

Analyzing Patterns of Inheritance Using the Experimental Cross Method

Materials for This Activity

Two small boxes

Four prepainted tongue-depressor (or Popsicle) sticks

Work in pairs, and obtain two small boxes labeled "experimental cross male" and "experimental cross female." These boxes each contain two tongue-depressor sticks that have been painted on each side according to the following color schemes:

- **Sex (side one):** Blue = Male alleles (Y chromosome); Pink = Female alleles (X chromosome)
- **Color blindness (side two):** Red = Normal color vision allele; Black = Recessive color blindness allele

Both of the sticks in the female box are pink on one side, but on the other side, one is red and the other is black. This box represents a female who is a *carrier* for color blindness ($X^B X^b$). In the male box, one stick is pink on one side and red on the other side. The other stick is blue on one side and unpainted on the other side. This box represents a healthy male ($X^B Y$).

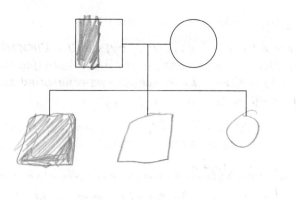

FIGURE 20.4 **A blank pedigree chart.**

Why is the blue stick unpainted on its reverse side?

To represent the healthy side of the male allele

1. *Without looking,* randomly take one stick out of each box. This combination represents the first of your hypothetical offspring. First, determine the sex of your child. Two pink sticks would be a female, and one pink and one blue stick would be a male. Then, determine if the child has color blindness. Recall that color blindness is recessive, so even one normal allele, represented by the red stick, results in a child with normal color vision. Record the results of this cross in Box 20.8. For example, if you were to pick a pink stick with red on side 2 from one box and a blue stick unpainted on side 2, you would write X^HY; **Normal male** in the empty box under **Child 1**.

Box 20.8	**Experimental Cross Results**	
Child 1	**Child 2**	**Child 3**
Colorblind mal	color blind fem	Male norm
Child 4	**Child 5**	**Child 6**
Color blind mal	Male normal	Male norm
Child 7	**Child 8**	**Child 9**
norm male	norm fem	Male colorb
Child 10	**Child 11**	**Child 12**
Male colorb	Norm fem	Norm fem

2. Return the sticks to their respective boxes and shake the contents. Repeat the process described in step 1 11 more times, tallying the results in Box 20.8.

Did you obtain any color-blind females? *yes*

Propose a reason why this result occurred.

Though it is rare, there is still a slim chance to get color blind females.

In the space provided below, draw a pedigree chart for your family.

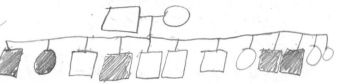

3. Next, complete the Punnett square in Box 20.9. This Punnett square will provide you with the *theoretical* ratios for this experimental cross.

Based on the Punnett square, what proportion of your females *should be* color-blind? *0%*

What proportion of your males *should be* color-blind? *25%*

4. Based upon your *experimental* results in Box 20.8, what proportion of your females were color-blind? *8%*

Does this match the *theoretical* results of your Punnett square in Box 20.9? *No*

Explain your answer.

I shouldn't have had any color blind females, and I

Box 20.9   Color Blindness Punnett Square

	X^B	Y
X^B	$X^B X^B$	$X^B Y$
X^b	$X^B X^b$	$X^b Y$

Based upon your *experimental* results, what proportion of your males were color-blind?

25%

Does this match the *theoretical* results of your Punnett square?

This matches

Explain your answer.

This is an exact match. In my punnett square, it also predicted 25%

5. Next, pool your data with that of the other groups in the class.

Does your percentage of color-blind offspring more closely match the Punnett square results?

Yes

If so, propose a reason for this result.

My data was extremely close to the Punnett square, but similar to other students but some people had less or more color blind children

GENETICS

Critical Thinking and Review Questions

1. What advantage does sexual reproduction have over asexual reproduction?

   In asexual reproduction, cells are copied, but with sexual reproduction, then disorders can be cancelled.

2. Compare and contrast meiosis and mitosis.

   Mitosis and meiosis both include the division of cells, but mitosis is for repair, meiosis is for sex cells

3. Match the term to its best description.

   _c_ Alleles   a. the observable trait expressed by an organism

   _a_ Phenotype   b. the genes for a trait present in an organism

   _b_ Genotype   b. the different varieties of a gene for a particular trait

   _e_ Homozygous   c. two genes traveling together on the same chromosome

   _d_ Linkage   d. the state of having two identical alleles for a particular trait

4. Having unattached earlobes is a dominant trait over attached earlobes. Calculate the probability of two parents who are heterozygous for unattached earlobes having a child with attached earlobes. Perform a Punnett square to check your answer *after* first using the probabilities method to answer this question.

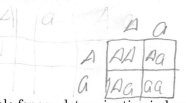

5. Who is responsible for sex determination in humans, the male or female parent?

   Male

   Explain.

   The Y chromosome contains the SRY determiner. If SRY is present, then the baby is male

6. Briefly explain sex linkage and linkage.

Sex-linkage is between disorders and sex chromosomes and can surface depending on the sex of a person. Normal linkage isn't determined by the sex

Why do X-linked genetic disorders occur more frequently in males?

Females have another X chromosome to mask the disorder.

7. In the space below, draw a pedigree chart for a biological family (yours or a classmate's). Assume that curly hair is the trait being considered. For simplicity's sake, completely darken boxes or circles for individuals with curly hair, darken half of the box or circle for wavy-haired family members, and leave the symbol unfilled for straight-haired people.

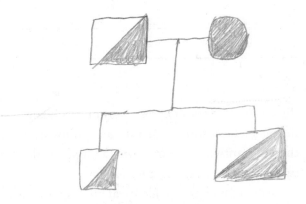

EXERCISE 21

DNA Technology and Genetic Engineering

Objectives

After completing this exercise, you should be able to

1. Explain the connection between nucleotide sequences, DNA structure, complementary basepairing, and DNA replication.
2. Extract DNA from plant cells and describe how it looks to the naked eye.
3. Explain how the structure of DNA allows us to identify individuals through DNA fingerprinting.
4. Perform the procedure for DNA fingerprinting.
5. Explain how genetically modified plants are created and explain their potential benefits and drawbacks.

Materials for Lab Preparation

Note to Instructors: Although we recommend the following kits, your laboratory setting may necessitate a different procedure. We recommend that you read through the activities in this exercise to ensure that you have all the needed supplies prior to starting the activity.

Equipment and Supplies

- ○ Transparent tape
- ○ Scissors
- ○ Blender
- ○ PowerPac power supply
- ○ Safety glasses
- ○ Heat-resistant vinyl gloves
- ○ Whole onions

- ○ Tissues (to wipe eyes)
- ○ Knife
- ○ 250-ml beakers
- ○ 100-ml graduated cylinder
- ○ 37°C and 65°C water baths or beaker and hot plate
- ○ Thermometer
- ○ Ice
- ○ Wooden splints
- ○ Cell lysis solution
- ○ DNA precipitation solution
- ○ Conical funnel and filter paper
- ○ Mini-Sub Cell GT
- ○ 1.5-ml micro test tubes
- ○ Test tubes and test tube racks
- ○ Gel staining trays
- ○ Gel support film
- ○ Micropipettes and tips
- ○ Computer with Internet access (optional)
- ○ Printer (optional)

Kits

- ○ Carolina Biological Supply Kit # WW-15-4704 (or equivalent)
- ○ Bio-Rad Laboratories Kit 1660007EDU (recommended)

Introduction

DNA (deoxyribonucleic acid) is found in all living cells, and it carries each cell's genetic information. **Genes** are the segments of DNA that carry the code for making a molecule or regulating a cellular function. When scientists discovered the structure and nature of DNA a little over six decades ago, it was an enormous leap forward in the field of biology. The growing knowledge of DNA function and advances in technology that allow manipulation of DNA molecules (DNA technology) have empowered biologists to manipulate the biochemistry of cells. In recent decades, the world has been greatly changed by the application of DNA technology to produce medical and commercial products. The ability to alter the genetics of organisms through DNA technology is called **genetic engineering.**

Chemical Characteristics of DNA

First, it is important to understand the basic structure of the DNA molecule. DNA is a double-stranded molecule composed of building-block molecules called **nucleotides.** Nucleotides are made up of a phosphate group, a five-carbon sugar molecule and a ringed nitrogen-containing molecule referred to as a **base** or **nitrogenous base** (Figure 21.1).

There are four varieties of nucleotides in DNA; each variety has a different base but they all have the same sugar and phosphate portions. Each nucleotide takes the name of the base that makes it unique: adenine, thymine, cytosine, or guanine. In a DNA molecule, the sugar of one nucleotide is covalently bonded to the phosphate of the next nucleotide, leaving the base poking inward. The opposing strand is set up similarly, and the inward-facing bases connect with each other through hydrogen bonding, giving DNA the "twisted ladder" shape referred to as a **double helix.** The sugar-phosphate backbones make up the sides, and the connected bases make up the rungs of the twisted ladder (Figure 21.2).

The bases of the nucleotides match up in the center in a specific pattern. Adenine (A) and thymine (T) nucleotides always pair with each other, as do guanine (G) and cytosine (C) nucleotides. This is called **complementary base pairing.** You will never find As or Ts matched with Gs or Cs in normal DNA molecules.

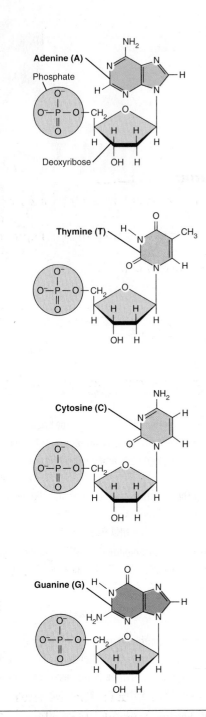

FIGURE 21.1 **The four nucleotides that comprise DNA.** The phosphate and sugar groups are identical in all four nucleotides.

ACTIVITY 1

Discovering the Structure of DNA

Materials for This Activity

Transparent tape

Scissors

Part 1: Structure of DNA

1. Cut out the nucleotide molecules on pages 271–272 by cutting along the dashed lines.
2. Connect four molecules in the sequence G-C-A-C (from top to bottom) by taping the rectangular sugar of one molecule to the plus-shaped phosphate of the next. This will be the *left strand* of your DNA molecule. Lay this out flat on your table with the G at the top and C at the bottom.

   With which nucleotide will you start when constructing the right side strand from the top?

   _____

3. To construct the right side of your DNA, take the cutout of the nucleotide you will place at the top, and try matching up the bonding "arms" of the nitrogenous base (which represent hydrogen bonding areas) with the bonding arms of the G nucleotide on the left strand. You should find that they don't seem to line up. What can you do to get those bonding arms to match up with each other? (*Hint:* Be creative, and leave no stone unturned when examining the possibilities!)

   _____

   _____

4. Once you have figured out the trick, line up (but do not tape) the nucleotides together at the central arms, and repeat this process for the remaining three nucleotides. Also remember the sugar-phosphate covalent bonds by lining up (but not taping) those molecules as you build the right strand.

   In this activity, you have essentially walked in the footsteps of James Watson, who was equally puzzled at first about the way these molecules could fit together.

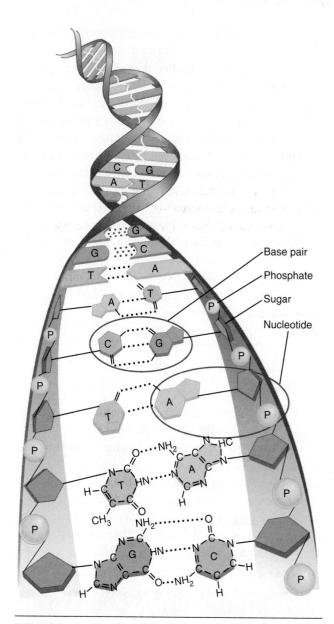

FIGURE 21.2   The double helical structure of DNA.

Part 2: Replication of DNA

1. Remove the nucleotides of your untaped right strand, leaving the left strand in place.
2. Now, reattach the loose nucleotides to the left strand in a new order, but be sure to maintain the appropriate bonds with the right strand.

   Can you do it? _____

   Why or why not?

   _____

   _____

This illustrates how the DNA molecule replicates itself: the strand splits, and new nucleotides are attached one at a time to create a complete strand again. Complementary base pairing allows each of the original "backbones" to produce an entirely new sister strand identical to its former partner. Enzymes help to split the strands and place the complementary nucleotides—just as your hands did in this activity. Compare your DNA molecule with that produced by other students in the lab. What does your attempt to rearrange the nucleotides suggest about the reliability of DNA replication?

_____

_____

_____ ■

Extracting DNA

DNA must be extracted from cells before it can be studied or manipulated. There are several ways to do this, but the process that typically works best in an educational laboratory setting is extraction using chemical techniques that break down the cell membrane and separate the DNA from other cell contents. In the following activity, we will use plant cells as our source of DNA, as they are somewhat easier to manipulate than human cells.

ACTIVITY 2

Extracting DNA from Onion Cells

Materials for This Activity

Carolina Biological Supply Kit # WW-15-4704 (or equivalent)

Cell lysis solution (e.g., 10% sodium dodecyl sulfate)

DNA precipitation solution (concentrated ethanol)

Test tubes

Filter paper

Wooden splints

Safety glasses

Heat-resistant vinyl gloves

Whole onions

Tissues (to wipe eyes)

Knife

Blender

250-ml beakers

100-ml graduated cylinder

65°C water bath or beaker and hot plate

Thermometer

Ice

Conical funnel

Note to Instructors: The directions that follow are provided for guidance; your kit or other source of supply may come with more detailed instructions.

1. Prepare a 65°C water bath. A beaker and hot plate may be used in place of a commercial laboratory water bath. Place a 250-ml beaker of DNA precipitation solution (concentrated ethanol) on ice to ensure it will be cold when needed.

2. Cut a medium onion (or half a large onion) into several pieces, and place in blender. Measure 65 ml of water in a graduated cylinder and add to the blender. Secure blender lid, blend at the *low* setting for 30 seconds, and then at the *liquify* setting for 10 seconds. Repeat 10 seconds of blending at the liquify setting until the onion is thoroughly liquified. Transfer the liquified onion into a clean beaker.

3. Add some of the liquified onion to a 15-ml test tube until it is almost half full. Add 7 ml of cell lysis solution and cap the tube. Invert and right the tube to mix contents (do not shake), and place in the 65°C water bath for 15 minutes.

4. After 15 minutes have elapsed, remove the tube from the water bath. Place a small conical funnel with filter paper in a 50-ml test tube, and pour the contents of your onion extract tube into the funnel. Allow the filtrate to drip for 10 to 15 minutes.

5. When most of the extract has filtered, remove the funnel and *slowly* add the cold ethanol while tilting the test tube about 45 degrees to minimize mixing and to allow the alcohol to float on top of the extract. Add an amount of ethanol about equal to the extract in the tube. If you have done this carefully, you should see a fairly clear layer of ethanol above a pinkish layer of extract.

6. The DNA is typically visible as a cloudy layer in between the two liquids (if not, add another 1–2 ml of cold ethanol). Slowly insert the wooden

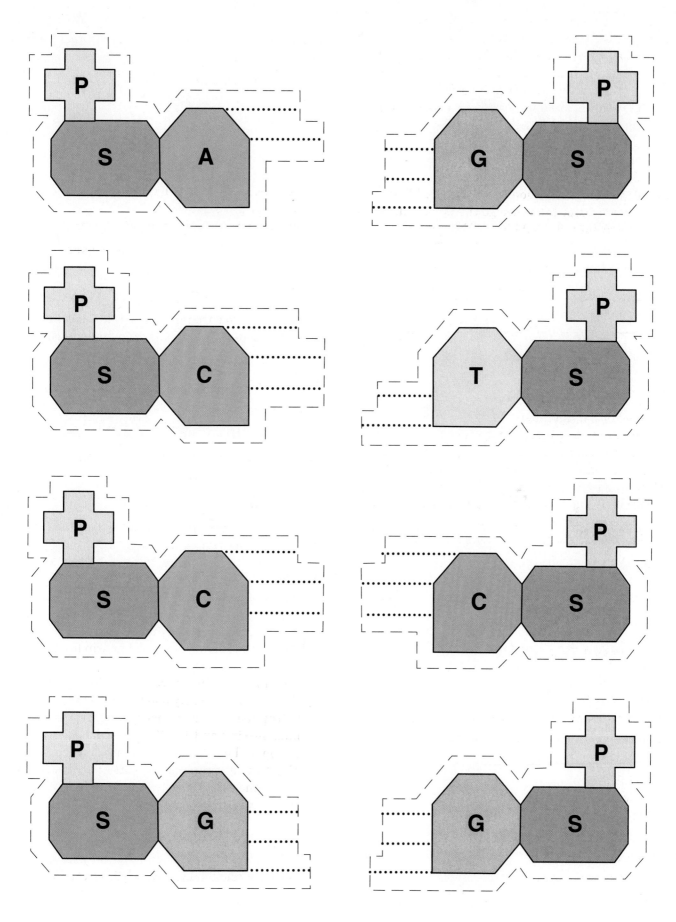

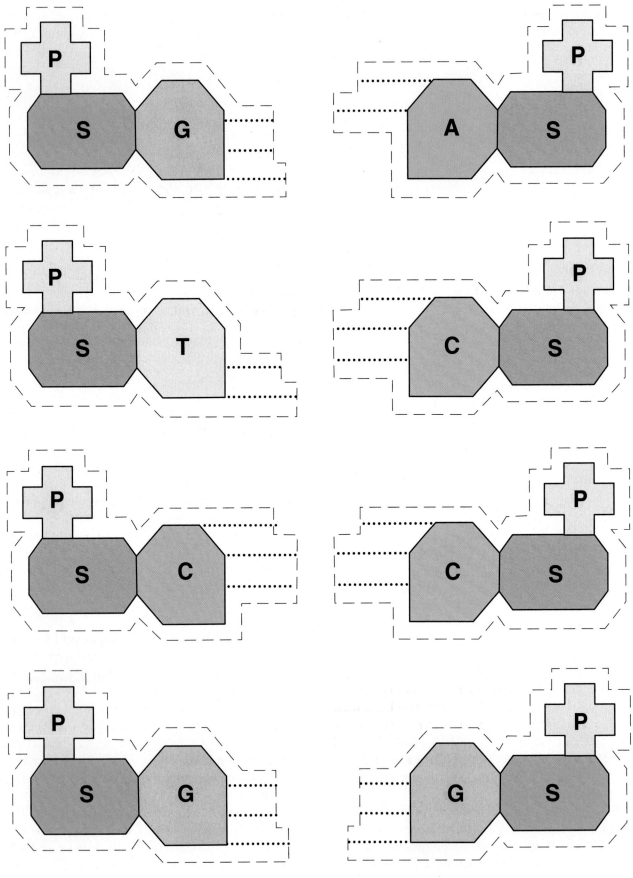

splint into the tube. Gently turn the splint as you slowly move it up and down to spool the DNA around it.

Describe the color and shape of the DNA on your splint.

_____

_____

Considering that the extraction/lysis solution was mostly water, speculate whether DNA is soluble in water:

_____

In ethanol:

_____

Test your answer for the aforementioned question by placing your splint with DNA into a small, clean test tube half filled with distilled water and stir. Does it dissolve?

_____

_____

Before testing this water with pH paper, predict whether DNA would be acidic, basic, or neutral.

_____

_____

Record your pH test results:

_____

_____

Check your results against your prediction. What chemical properties of DNA influence pH?

_____

_____

DNA Fingerprinting

A great deal of DNA is the same from person to person, assuring that each of us is usually born with one head, one four-chambered heart, and all the other things that make us human. But plenty of DNA varies greatly and can produce a reasonably unique pattern, or "fingerprint," when separated and analyzed.

DNA fingerprinting is accomplished by obtaining a sample of DNA and using **restriction enzymes** to cut up the DNA. Restriction enzymes are special enzymes that chop up DNA molecules

at certain locations (Figure 21.3a). Because even slight differences in DNA sequences may result in DNA that is broken up into very different fragments, each person's DNA sequences will chop up differently, and the pieces will exhibit different patterns when they are separated. The most common method of separating these chopped up pieces is **gel electrophoresis**, a process that separates DNA fragments by size and weight and represents them as bands (Figure 21.3b).

When the DNA technology of gel electrophoresis was first developed, the equipment was both somewhat difficult to use and prohibitively expensive for the average educational laboratory. Refinements in equipment design, and relatively easy-to-use kits for educational purposes have made it possible to include DNA fingerprinting in undergraduate laboratory activities.

Like most separation techniques, gel electrophoresis is based on the principle that molecules move through a medium at different speeds depending on physical and chemical characteristics. As a general rule, small molecules will be drawn through a medium faster than larger ones. In gel electrophoresis, an electrical current pulls the negatively charged DNA fragments from wells on one side of the apparatus toward the other positively charged side of the gel. When the process is terminated, the fragments are seen as bands at various distances from the well in which the sample was originally placed. The smaller fragments are observed as having moved farthest away from the well, while the larger fragments form bands a shorter distance from the well (Figure 21.3c). Identical DNA samples will be chopped up into the same sized fragments and will move through the gel in exactly the same way (Figure 21.3d).

With rare exception, the separated bands of DNA fragments in a gel produce patterns that are considered sufficiently unique that the odds of two unrelated samples matching would be extremely unlikely. And, unlike partial fingerprints, a very small sample of DNA is often sufficient to produce a very clear DNA fingerprint. For these reasons, DNA fingerprinting is often used in crime scene investigations where blood, skin cells, or bodily fluids have been found.

In the following activity, you will perform DNA fingerprinting on substitute blood samples in order to analyze which suspect was at a crime scene. DNA from each of several suspects is provided.

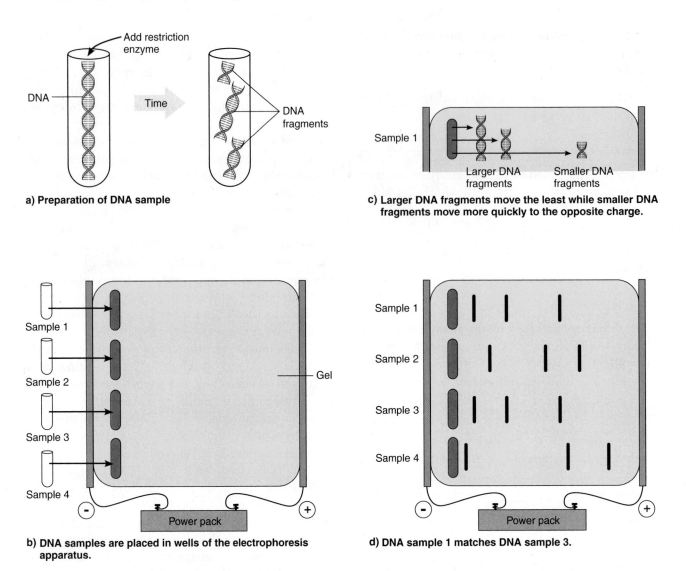

FIGURE 21.3 Gel electrophoresis. a) Preparation of DNA for gel electrophoresis. **b)** Setup for gel electrophoresis. **c)** Movement of DNA fragments across a gel plate. **d)** DNA fingerprints showing one set of identical fingerprints.

ACTIVITY 3

DNA Fingerprinting and Crime Solving

Materials for This Activity

Bio-Rad Laboratories Kit 1660007EDU (recommended)

DNA samples

DNA size standard

DNA sample loading buffer

Electrophoresis buffer

Agarose powder

DNA stain

Mini-Sub Cell GT

PowerPac power supply

1.5-ml micro test tubes

Test tube holders/racks

Gel staining trays

Gel support film

Micropipettes and tips

37°C water bath

Note to Instructor: Although we recommend the kit from Bio-Rad Laboratories, your laboratory setting and equipment may necessitate a different source of supply or procedure. The directions that follow are provided for guidance; your kit or other source of supply may come with more detailed instructions.

Part 1: Preparation of DNA Samples

1. Place the container of the restriction enzyme mix on ice.
2. Label the DNA sample test tubes, and add the DNA samples from the stock solutions according to the directions in your kit or as directed by your instructor. Use a new micropipette tip for each sample.
3. Mix the contents of each tube by flicking the bottom of the tube with a finger, and tap the bottom on your table top to collect the liquid at the bottom of the tube. Place the tubes in a 37°C water bath for 45 minutes. *Note:* If time does not permit completion of this activity in one class meeting, tubes may be placed in a refrigerator after this incubation period until the next lab period; or to finish in one period, the instructor or staff may prepare steps 1 through 3 in advance so samples are ready to process.

Part 2: Processing the Samples Using Gel Electrophoresis

1. Remove tubes and add loading dye solution to each tube according to the directions in your kit or as directed by your instructor.
2. Prepare your gel electrophoresis apparatus by adding the agarose gel and buffer, making sure the gel wells are near the black (negative) electrode and the base is near the red (positive) electrode.
3. Add a portion of the DNA samples from the tubes you prepared in Part 1 to each well according to the directions in your kit or as directed by your instructor. Use a new micropipette tip for each sample.
4. Place the lid on the electrophoresis chamber and plug it in to the power supply. Set the power at 100 V and allow to run for 30 minutes.
5. After 30 minutes, turn off the power and remove the lid of your electrophoresis apparatus. Carefully remove the gel and place in the staining tray. Stain and destain the gel according to the directions in your kit or as directed by your instructor (Figure 21.4).

Which suspect sample matches most closely to that found at the crime scene?

_____

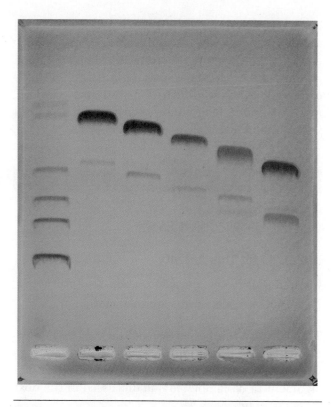

FIGURE 21.4 **Photograph of a gel plate after electrophoresis.**

What does this test suggest about the samples from the remaining suspects?

_____

_____ ■

Recombinant DNA and Transgenic Organisms

One of the ways that DNA technology has been useful to society is the placement of beneficial genes from one organism into the cells of other organisms. Organisms that have had foreign DNA segments or genes inserted into their cells—in other words, have undergone genetic engineering—are called **transgenic organisms.** Transgenic organisms can be created from bacteria or other single-celled organisms to produce specific chemicals. Hormones and medications have been made by this method. A somewhat controversial use of the process is the production of transgenic plants. These genetically altered plants can be made resistant to pests or infused with more nutrients and vitamins than they would normally produce.

Because transgenic plants can also be made pest resistant, they reduce or eliminate the need to be sprayed with potentially toxic pesticides. This can increase yields in areas of the world where people do not have access to farm chemicals. Food safety issues with transgenic plants, sometimes called **genetically modified organisms (GMOs),** are somewhat less clear, creating controversy within the industry.

ACTIVITY 4

Internet Research: Genetically Modified Foods

Materials for This Activity

Computer or mobile device to access the Internet
Notebook or printer to record information

This activity can be performed during the laboratory period if students have access to computers with Internet connections or as a homework assignment to be completed and discussed in the next class meeting.

1. Form groups of three or four students. Perform an Internet search for "Genetically Modified Food." Select several Web sites to investigate, choosing an equal number of those that seem to be for and against genetically modified foods. Attempt to find at least one site that seems to provide balanced coverage of risks and benefits.

   Using the information you find, summarize the process by which a GMO is produced.

   _____

   _____

   _____

2. Discuss the information you find with members of your group. Fill in Box 21.1, and discuss these risks and benefits. Attempt to reach a consensus about whether the risks outweigh the benefits or if the reverse seems true. Your instructor may wish to coordinate a class discussion during which each group presents its conclusions and rationale.

   Weighing these risks and benefits, do you feel genetically modified foods are relatively safe or relatively unsafe? Why?

   _____

   _____

   _____

Box 21.1 Benefits and Risks of Genetically Modified Food	
Benefits of Genetically Modified Food	**Risks of Genetically Modified Food**

DNA TECHNOLOGY AND GENETIC ENGINEERING
Critical Thinking and Review Questions

1. Summarize the relationship between the two opposite strands of the DNA molecule and the way in which DNA is copied.

   _____

   _____

2. Which portions of the DNA molecule are connected by covalent bonds?

   _____

   Which portions are connected by hydrogen bonds?

   _____

3. Draw the missing nucleotides on the right side of the DNA strand using colored pencils and the symbols shown in the key.

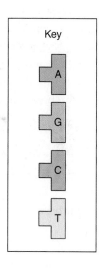

Key

4. What purpose was served by the blender and the cell lysis solution in the DNA extraction activity?

_____

_____

_____

5. Why did you add the cold ethanol solution to the liquefied onion in the DNA extraction activity?

_____

_____

_____

6. What are restriction enzymes, and what role do they play in DNA fingerprinting?

_____

_____

_____

7. Explain the term electrophoresis, and describe its role in DNA fingerprinting.

_____

_____

_____

8. What is the difference between genetically modified plants and hybrid plant varieties, which are produced by selective breeding?

_____

_____

_____

9. Summarize the risks and benefits of genetically modified plants.

_____

_____

_____

_____

_____

_____

EXERCISE

22

Evolution

Objectives

After completing this exercise, you should be able to

1. Define and understand the theory of evolution.
2. Describe the evidence for the theory of evolution.
3. Trace the scientific development of life on Earth.
4. Trace the scientific development of humans, from mammals to modern humans.
5. Define homologous, analogous, and vestigial structures.
6. Compare and contrast the forelimbs of human, whale, dog, and bird.
7. Explain natural selection.
8. Explain genetic drift and the founder effect.

Materials for Lab Preparation

- ❍ Skeletons of human arm, dog paw, whale flipper, and bird wing, if available.
- ❍ Colored pencils.
- ❍ Beads in 4 colors, 50 of each color
- ❍ Small paper bags
- ❍ Watch glasses
- ❍ Calculator
- ❍ Scratch paper
- ❍ Pencils

Introduction

Although the main focus of this book is the biology of humans, it is still important to consider where humans fit in with the rest of the living things on planet Earth. With all the amazing diversity of life around us, there is nonetheless clear evidence of shared biological underpinnings, and common ancestry. For example, essential organelles such as mitochondria are essentially the same in humans and houseflies. As different as we are, fungi and humans share many of the same DNA sequences. How do we reconcile the vast outward differences with the clear evidence of definite, even if more subtle, genetic connections?

The theory of evolution helps us to understand that question. DNA is **mutable** or changeable, and it logically follows that some changes would be bad for an organism, whereas others may good. This **genetic modification** causes changes over time that make later generations different from their ancestors, something biologists refer to as **descent over time.** What causes this change? Mostly, it is nature in the form of **natural selection.**

One of the great contributions of Charles Darwin to our understanding of evolution was providing a clear explanation of *how* it could happen. Natural selection provides a mechanism by which this descent over time can occur. Evolution by way of natural selection isn't "just a theory" as some of its detractors claim, it is a unifying concept in biology with a huge body of clear supporting evidence.

As you may recall from Exercise 1 on the scientific method, for something to become a theory it requires a number of rigidly applied steps. It must be based on unbiased, accurately obtained observations and information. The hypothesis must come from the body of information, and it must be tested. Not only has the theory of evolution followed the scientific method, but it has reached the level of a core principle or unifying concept in biology after many decades of being tested, revised, and modified accordingly.

ACTIVITY 1

Evolution and the Scientific Method

In Box 22.1, provide examples or explanations of how the concept of biological evolution fits with the scientific method. It may be helpful to reacquaint yourself with the steps of the scientific method from Exercise 1. You may wish to use your textbook and/or any other reference books your instructor may have available in the lab. Alternatively, your instructor may wish to assign this as a short library activity for you to do outside of class.

Box 22.1

Step of the Scientific Method	Summarize This Step	Evolution's Fit with This Step
1. Observation/Objectively acquired information		
2. Proposing a question or generalization		
3. Hypothesis testing		
4. Reject, revise, or confirm the hypothesis		
5. Theory		
6. Law or core principle		

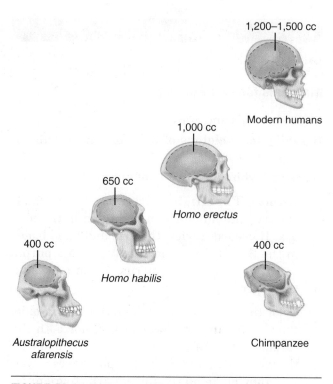

FIGURE 22.1 **Brain capacities of hominids and of modern chimpanzees.** Brain capacities are in units of cubic centimeters (cc).

Evidence for Evolution

Practically every scientific discipline that comes in contact with evolutionary biology has provided some strong evidence in support of it. Fossils, comparative biology, comparative biochemistry, genetics, and other scientific areas all contribute to the support of and understanding of evolution.

Fossils. First and foremost, the pattern of fossils we see upon superficial examination clearly fits with the concept of descent over time. Instead of a homogeneous mix of fossils one would expect if all life forms appeared at the same time, we see a clear overall pattern of older, sometimes simpler, and often extinct organisms in older, deeper rock layers. There are no fossils of more recently evolved things such as apple trees or other flowering plants until the newer rock layers closer to the surface are examined. Over the past few decades, more fossils have been found demonstrating descent of humans from a common hominid ancestor, with hips and feet modified for upright walking, and larger brain capacities (Figure 22.1).

Comparative Biochemistry

Whether we check the DNA directly, or the proteins it codes for, we see few differences between us and the primates predicted to be closely related to humans. More base sequences and amino acids are different in DNA and proteins when we look at animals further from us in the tree of life. An often-cited example, the molecule cytochrome c has great similarities among primates (human and chimpanzee cytochrome c molecules are identical) and increasing differences in its amino acids occur in more distantly related animals.

Comparative Anatomy and Embryology

Comparing the embryos and adult anatomical structures often provides insights regarding evolutionary relationships. As you can see in Figure 22.2, the

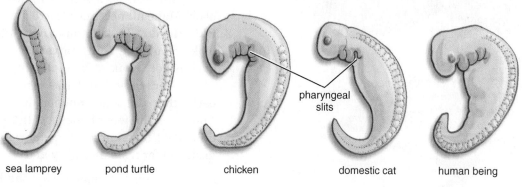

Pharyngeal slits exist in these five vertebrate animals . . .

pharyngeal slits

sea lamprey   pond turtle   chicken   domestic cat   human being

. . . evidence that all five evolved from a common ancestor.

FIGURE 22.2 **Note the similarities between vertebrate embryos, including development of rudimentary gill structures (pharyngeal slits), even in those vertebrates that do not actually produce gills.**

embryos of all vertebrates look somewhat similar, even producing rudimentary gill structures whether they are land or sea creatures. Rather than a complete "redesign" of appendages for different purposes, we see basic resemblances in skeletal structure that imply modification over time.

Comparative anatomy groups structures into three major types. **Homologous structures** are body structures that share many common features, which may imply a close evolutionary relationship. **Analogous structures** share common features but not a common origin. For example, the wings of birds and insects both evolved for flight, but from evolutionarily distant lineages (see Figure 22.3). **Vestigial structures** are structures that have little or no function in one organism while having significant functions in related organisms. For example, the arrector pili muscles you learned about in Exercise 7, The Integumentary System, thicken the fur coats of other mammals, but gives us nothing but goose bumps.

Let's compare four forelimbs: human, dog, whale, and bird. These forelimbs have very different functions such as grasping, running, swimming, and flying. Will an examination of the bone structure yield evidence to indicate that these are **homologous structures**—that is, that these four organisms may have shared a common ancestor? ■

Bird's wing

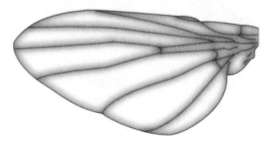

Insect's wing

FIGURE 22.3 Analogous structures.

ACTIVITY 2

Comparing Forelimbs

Materials for This Activity

Colored pencils and erasers

If available, skeletons of four forelimbs: human, dog, whale, bird

Black and white photocopies of Figure 22.4

1. **Human Forelimb.**   Examine Figure 22.4. Observe the human arm and hand in the first box. If needed, review the arm and hand bones in the Skeletal System chapter. Use a purple pencil to shade the humerus, which makes up the upper arm.

   Color the radius in orange and the ulna in blue; these make up the lower arm. Color both the carpals and metacarpals in green; they make up the wrist and palm. Color the phalanges in yellow; they make up the fingers, which are numbered 1 through 5.

   What main difference can you see between the human thumb and other fingers?

   _____

   _____

2. **Dog, Whale, and Bird Forelimbs.**   Closely examine the bones in the other three boxes. Which ones are most similar? Again, color the humerus in purple, the radius in orange, the ulna in blue, the carpals and metacarpals in green, and finally the phalanges in yellow.

   After you have colored in all of the bones, compare your work to a similar figure in your textbook. Make all necessary corrections.

3. **Skeletons.**   If skeletons are available, take a look at how the bones physically connect with each other. Look at bone density, length, and girth. Remember that the bones need to "fit" each other and that "function determines form." You'll see the wing bones are lighter and more delicate than human bones because they are designed for a smaller body and for flight.

   Speculate on the reasons for the heavier and bulkier whale bones.

   _____

   _____

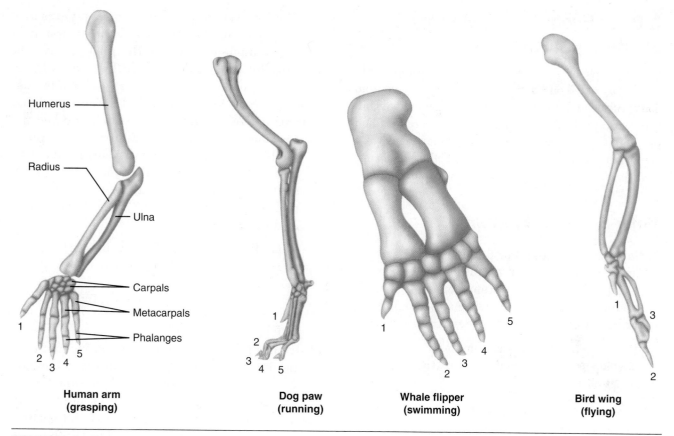

Humerus

Radius

Ulna

Carpals

Metacarpals

Phalanges

1 2 3 4 5

**Human arm
(grasping)**

1 2 3 4 5

**Dog paw
(running)**

1 2 3 4 5

**Whale flipper
(swimming)**

1 2 3

**Bird wing
(flying)**

FIGURE 22.4   **Four vertebrate forelimbs.**

4. **Whale.**   Compare the human and whale bones. What are the differences and similarities in structure and function?

_____

_____

5. **Bird.**   Compare the human and bird bones. What are the differences and similarities in structure and function?

_____

_____

6. **Dog.**   Compare the human and dog bones. What are the differences and similarities in structure and function?

_____

_____

7. **Similarity of Structure.**   What are the similarities between the forelimbs?

_____

_____

8. **Conclusion.**   Is there enough evidence to indicate that the four forelimbs are **homologous structures**?

_____

_____

9. The four animals you've been investigating in this activity are vertebrates, animals with a backbone. Vertebrates all evolved from a common ancestor, so you'll find their forelimbs are homologous structures. ■

A Brief Evolutionary History of Life on Earth

Humans have been on Earth a short time compared to the history of life on Earth. One-celled organisms from South Africa and Australia show life originated 3.8 billion years ago. Hominids have existed a mere 5 million years! In the next activity, we will compare the evolution of our human lineage with that of other life forms. First, however, you will need some basic information.

Major Evolutionary Events

Cells. An estimated 5 or 6 billion years ago our sun was born. About 4.5 billion years ago the planets in the solar system were first formed, including the planet Earth. At the beginning, Earth was too hot for life to develop.

What types of chemical compounds needed to develop before cells could appear? (Refer to your textbook.)

_____

_____

The first cells appeared 3.8 billion years ago. They are called **prokaryotes** (*pro* = first, *karyote* = nucleus) because the nucleus was not well-defined. About 1.82 billion years later, **eukaryotes** (*eu* = true, *karyote* = nucleus) appeared. They contained a true nucleus, with DNA, the genetic material, and became the dominant cell type (Figure 22.5). About a billion years ago eukaryotic organisms had diversified enough to allow for sexual reproduction, which led the way for rapid combinations of genetic material, by way of the **sperm and egg.**

What do you think are the advantages of sexual, as opposed to asexual, reproduction?

_____

_____

Multicellular life. **Invertebrates,** the first multicellular organism, appeared about 700 million years ago. They had specialized cells and tissues and complex forms including an outer shell, which are well preserved in the fossil records. Their descendants, such as the jellyfish, are

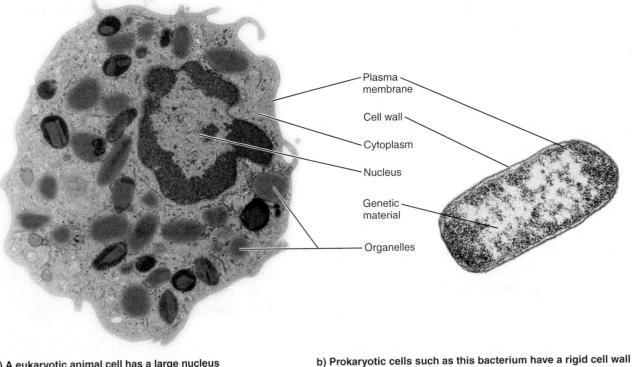

Plasma membrane
Cell wall
Cytoplasm
Nucleus
Genetic material
Organelles

a) **A eukaryotic animal cell has a large nucleus and numerous small organelles.** The cytoplasm is enclosed by a flexible plasma membrane.

b) **Prokaryotic cells such as this bacterium have a rigid cell wall surrounding the plasma membrane.** The genetic material is not surrounded by a membrane, and there are no organelles in the cell.

FIGURE 22.5 **Eukaryotes versus prokaryotes.**

usually found in the sea. The first **vertebrate** appeared about 500 million years ago. These ancient fishes had an internal skeleton with a backbone. They continued to diversify for several hundred million years, and eventually became amphibians, reptiles, birds, and mammals. **Dinosaurs** flourished about 200 million years ago, during the Jurassic period. They dominated the animal kingdom on Earth, with their huge bodies, sharp teeth, and ferocity. They reproduced by laying eggs.

What common animal is the dinosaur's descendant today?

_____

_____

Mammals. Small and nimble mammals first appeared about 130 million years ago. The **placental mammals** developed a more protective type of reproduction. The human fetus, for example, is nourished and protected inside the uterus for 9 months. After delivery, placental mammals continue to protect their young by producing milk for their infants and by actively teaching the young necessary life skills (Figure 22.6).

What are the pros and cons for this comparatively new type of reproduction?

_____

_____

Primates. About 65 million years ago, an asteroid hit Earth and initiated major climatic changes. In the new environment, the **dinosaurs** became **extinct**, creating an opportunity for mammalian radiation to occur. Five million years later, the first **primate** emerged. **Prosimians** ("pre-monkey") lived in trees, an environment that favored the development of grasping hands, forward-directed eyes, and a bigger brain. Their descendants today are the lemur and tree shrew.

What are the pros and cons humans face due to inheritance of these three traits?

_____

_____

Hominoids. Thirty-five million years ago, **arthropoids,** the next dominant primate, appeared. They descended from the prosimians, and included **monkeys** and **hominoids** (apes and hominids). Hominoids were larger, had bigger brains, lacked a tail, and had a more complex social structure. Apes also differed from monkeys because of their teeth and patterns of locomotion. Apes appeared 20 million years ago, and their descendants today are gibbons, orangutans, gorillas, and chimpanzees.

What are the advantages of a larger brain and lack of a tail?

_____

_____

FIGURE 22.6 **Aren't you glad that humans have mammary glands and the placenta instead?** It is useful to think of evolution as the way "Mother Nature" solves problems. One problem for most animals is how to nourish developing offspring until they are fully developed enough to feed themselves. This is a particularly serious issue for complex animals like mammals. Instead of developing for a few weeks inside an egg (which has a finite food supply), we have the placenta to provide nearly unlimited nutrition for months (with size of the fetus as the main limiting factor in how long it can grow within the human female). And of course, breastmilk seems far preferable to mom passing on predigested food like birds. (Hillary B. Price, Rhymes with Orange, October 4, 2010)

Hominids. The time line for human evolution began only a few million years ago. About 3.7 million years ago, the earliest **hominids** appeared. Named australopithecines ("southern ape"), they differed from apes because of a bipedal (upright) movement pattern and a larger brain. The bones of "Lucy," an *Australopithecus afarensis*, have been dated to 3.2 million years ago. She was 3 feet tall, walked upright, and had a heavy brow, low forehead, and forward-jutting jaw. *Homo habilis* ("handy man") was found in Tanzania and dated to about 1.6 million years ago. Found with him were primitive tools made from rocks and evidence that large animals were butchered. *Homo erectus* ("erect man") was found to date from 1.4 million years ago. He was 5 feet tall, with tools, weapons, and evidence of the use of fire. Modern humans, *Homo sapiens* ("wise man"), date back only 300,000 years. Over the course of time, the brain capacities of the hominids have continued to increase. Modern human brain capacities are 300% larger than both Lucy's and the chimpanzee's (Figure 22.7).

a) Gibbon.

b) Orangutan.

c) Gorilla.

d) Chimpanzee.

FIGURE 22.7 **The hominids.** Modern humans (not pictured) are also hominids, of the family *Hominidae*. Chimpanzees are our closest living relatives.

What is the relationship between the brain and manual dexterity (skill with your hands?). (*Hint:* Refer to Chapter 11 on the Nervous System in the text.)

_____

_____

ACTIVITY 3

Putting Human Evolution in Context

This activity is intended to provide a good grasp of the enormous time spans and major events involved in the evolutionary process. We will plot the historical development of the early life forms on Earth and see that humans are actually relative newcomers.

1. Complete Box 22.2, Major Evolutionary Organisms, filling in the time and major evolutionary features.

   What percent of time have the following cells or organisms been around (use 4.5 billion of years for Earth as the denominator): Prokaryotes, eukaryotes, invertebrates, vertebrates, mammals, primates, hominoids, and hominids.

   _____

   _____

   _____

   _____

2. Refer to Box 22.3, on page 288, which is a matching activity that features a time line on the left, and a series of life forms illustrated by icons on the right. Match the appropriate life form on the right with the evolutionary time period in which it appeared by connecting the dots. Refer to the section, "Major Evolutionary Events" for help. Answer the following questions based upon your completed time line.

   a. When did the earliest life evolve?

   _____

   b. What is the significance of the evolution of sexual reproduction?

   _____

   _____

   _____

Box 22.2	Major Evolutionary Organisms	
Organism	**mya (millions of years ago) or bya (billions of years ago)**	**Major Evolutionary Features**
a. Egg/sperm		
b. Arthropoid		
c. Eukaryotic cell		
d. Modern human		
e. Vertebrate—fish		
f. Hominoid		
g. Early human		
h. Invertebrate—jellyfish		
i. Early mammal		
j. Ape		
k. Prokaryotic cell		

c. What are features of the earliest vertebrates?

_____

_____

d. What distinguishes early mammals from other vertebrates?

_____

_____

_____

e. When did the dinosaurs become extinct? What is the importance of this event for the evolution of mammals?

_____

_____

_____

f. What distinguishes the earliest humans (hominids) from apes?

_____

_____

_____ ■

Natural Selection

As mentioned earlier, one of the great contributions of Charles Darwin to our understanding of evolution was providing a mechanism by which most evolution occurs. He was lead to the concept of **natural selection** by observing: 1) there was variation among individuals in most populations, 2) some traits would be of more survival value than others, and 3) many more offspring are produced each generation than will typically survive. Thus, those with the favored variations

Box 22.3   An Evolutionary Time Line for Life on Earth

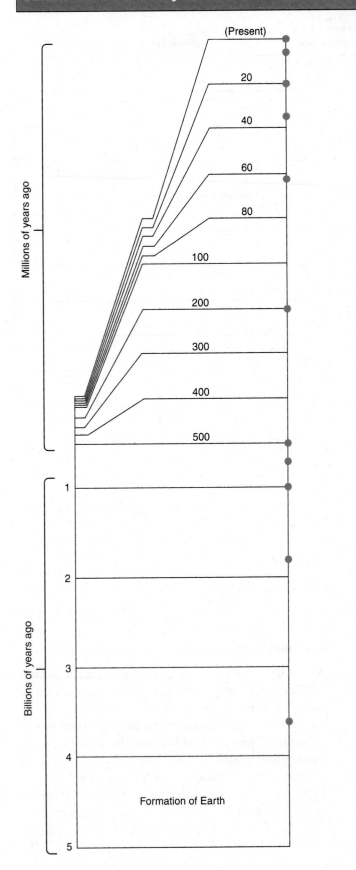

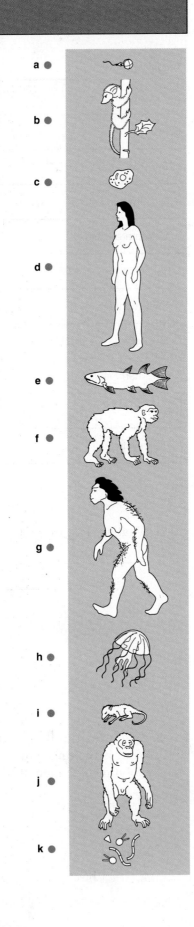

(the "fittest," as Darwin put it) are more likely to be the survivors who will pass on their traits to future generations.

Today, we have expanded on Darwin's work to add that evolution is not so much about survival of individuals, but of reproductive success. Although the two often do go together, a trait that makes an individual successful but leaves him sterile is useless in terms of evolution. We also know about DNA, genes, and how new material for new varieties (e.g., mutations) propel evolution in new directions. You learned about alleles or varieties of genes for a particular trait in Exercise 20, Genetics. Especially in evolving populations and/or changing environments, all alleles may not be of equal survival value.

ACTIVITY 4

Evolution and the Natural Selection

Materials for This Activity

Bag of 50 red and blue glass beads (or similar objects)

Two small watch glasses, petri dish bottoms, or similar small containers

Sticky notes (or similar items)

In this activity, you will observe how natural selection can change the gene frequencies of a population when alleles are not of equal survival value. The two colors represent two alleles or genes for a trait. Imagine that the blue gene's survival value is higher (e.g., it represents a trait such as larger brain size or better ability to walk upright), so that an individual with two blue alleles has an advantage over one with two red alleles.

Obtain a bag of 50 beads, 25 each of 2 different colors, preferably red and blue. What percentage of the total is each color bead in this bag? _____

Randomly select two beads representing an allele from each parent producing the offspring, and place them in the watch glass. If you obtain two red beads (= alleles), remove them to the second watch glass labeled with a note as "non-survivors." If you obtain two blue alleles, return them back to the bag (the "gene pool") as survivors. If you obtain a blue and a red allele, make a tally mark on another sticky note labeled "lesser fitness" and return the beads to the bag the first two times. At the third tally mark, add this mixed genotype to the "non-survivors" pool. Do the same for every third set of blue and red beads pulled from the bag.

1. After 10 offspring, count the number of remaining beads of each color in the bag, and record the percentage of each: Blue: _____; Red: _____. Also, note how many individuals (pairs of beads) you added to the non-survivors dish: _____

2. Repeat for another 10 offspring, count the number of remaining beads of each color in the bag, and record the percentage of each: Blue: _____; Red: _____. Also, note how many individuals (pairs of beads) you added to the non-survivors dish: _____.

3. Finally, repeat for another 10 offspring, count the number of remaining beads of each color in the bag, and record the percentage of each: Blue: _____; Red: _____. Also, note how many individuals (pairs of beads) you added to the non-survivors dish: _____.

In which trial, 1, 2, or 3, did you add the most individuals to the non-survivors dish? _____

In which trial, 1, 2, or 3, did you add the fewest individuals to the non-survivors dish? _____

This activity *loosely* approximated a few generations of natural selection at work. Speculate what the approximate percentage of the red allele might be like after 100 generations and why.

_____

_____

If this were a case of complete dominance and recessiveness, such that the heterozygous individuals were *not* at any disadvantage, would it still be correct procedure to remove every third individual with one blue and one red allele? _____

Propose an explanation of how this may partly account for why truly recessive disease alleles found in human populations are still present in small percentages.

_____

_____ ■

Genetic Drift

Most evolutionary change occurs through natural selection, but there are other ways that the variation in a population can change. For example, genetic drift is the term describing a situation where a small, potentially non representative portion of a population either moves or survives a population crash. These few individuals then produce a new population with different variation compared to the one they came from. The founder effect and population bottleneck are two examples of genetic drift.

In the **founder effect,** a small number of individuals from a population relocates and produces a new population in the new location. For example, imagine a bird scoops up and flies away with a yellow female fish from a pond that has members of her species that are yellow like her, but also about equal proportions of blue and red varieties. The fish is carrying a dozen fertilized eggs from a mating with another yellow member of her species. Then, before the bird can eat it, the bird accidently drops the fish in a new man-made pond a few hundred yards away. She and her dozen offspring will be the "founders" of a population of yellow fish in the new pond, very different from the variation in the original one they came from. Her yellow descendants will make up 100% of the individuals in the new location, compared to 33% in the original pond.

The **population bottleneck** is essentially the same concept, except it is the passage of time and some population-reducing event rather than a new location that causes the change in variation. For example, imagine the same situation with the yellow fish above. There is no bird picking up her and her eggs, but imagine instead that she is the last survivor in the last puddle of the pond during a long drought. Then the rains come, and it is she and her offspring that survive to repopulate the same pond.

ACTIVITY 5

The Founder Effect

Materials for This Activity

Bag of 40 different color glass beads or similar objects
A small watch glass, petri dish bottom, or similar small container

Obtain a bag of 40 beads, 10 each of 4 different colors: red, blue, yellow, and white. What percentage of the total is each color bead in this bag? _____

Randomly select four beads and place them in the watch glass. Multiply each bead by 25 to obtain a percentage, and report the results: Red: _____; Blue: _____; Yellow: _____; White: _____. Does this match the percentage of each color in the bag representing the original population? _____

Return the four beads to the bag, randomly select another four beads, and place them in the watch glass. Multiply each bead by 25 to obtain a percentage, and report the results: Red: _____; Blue: _____; Yellow: _____; White: _____. Does this match the percentage of each color in the bag representing the original population? _____

Finally, return the beads to the bag, randomly select another four beads, and place them in the watch glass. Multiply each bead by 25 to obtain a percentage, and report the results: Red: _____; Blue: _____; Yellow: _____; White: _____. Does this match the percentage of each color in the bag representing the original population? _____

In at least one of the above trials, did you obtain a percentage different from the original population? _____

Consider the first humans to migrate from Africa around 50,000 years ago. Based on this activity, propose a hypothesis about genetic variation in the populations moving into Asia and Europe compared to those that originated from Africa.

_____

_____

To expand on the last point, perform an Internet search (Google or other good search engine) of "out of Africa," "human variation," and "founder effect." As you start reading, you may come up with other key words to enter as well. Do this separately or in various search string combinations to obtain a good list of articles discussing genetic variation in human populations as they relate to the founder effect and human migration. Summarize your research next, connecting it to the validity of the hypothesis you proposed earlier. *Note:* Your instructor may wish to make this last item a homework assignment depending upon available time.

_____

_____

_____

_____

_____

_____

EVOLUTION
Critical Thinking and Review Questions

1. In your own words, what is the theory of evolution?

   _____

   _____

2. What are human traits, as compared to those of other animals discussed in this exercise?

   _____

   _____

3. Summarize four kinds of evidence for the theory of evolution.

   _____

   _____

4. What major events affected evolution in the first 4 billion years of Earth's history?

   _____

   _____

5. What major events affected evolution in the past million years of Earth's history?

   _____

   _____

6. What were the major events for human evolution, from mammals to modern humans?

   _____

   _____

7. How was the eukaryotic cell different from earlier life forms? In what way was this feature significant for human evolution?

   _____

   _____

8. How did the development of the hominoids progress?

   _____

   _____

9. How did the development of the hominids progress?

   _____

   _____

10. Compare homologous, analogous, and vestigial structures.

_____

_____

11. Which bones present in the human forelimb are present in the dog, whale, and bird forelimb? Why do you think these inclusions are significant?

_____

_____

12. What is the common ancestor for all four animals? Why is this important?

_____

_____

13. Define natural selection in your own words.

_____

_____

14. Why do you think natural selection does not produce perfect organisms?

_____

_____

15. Why doesn't evolutionary fitness mean "bigger and better"?

_____

_____

16. Why do you think inheritance and reproductive success are important in natural selection?

_____

_____

17. Define the founder effect. How is it different from the population bottleneck?

_____

_____

18. What is the relation of the founder effect to modern humans?

_____

_____

EXERCISE 23

Human Ecology

Objectives

After completing this exercise, you should be able to

1. Define human ecology.
2. Describe the 10 levels of biological organization.
3. Compare the lifestyle of hunter-gatherer, horticulturalist, and industrialized people.
4. Describe the impact of human lifestyles on the environment.
5. Compare human population growth trends in the last 2,000 years.
6. Perform a population estimate sampling technique for two organisms.

Materials for Lab Preparation

○ Ball of string

Introduction

In this last exercise, let's pause and briefly review what we have accomplished. Using a bottom-to-top approach, we have moved from smaller to larger biological units (Figure 23.1). In studying **human biology,** we systematically explored the different levels of organization in the human organism, starting with the chemicals, cells, tissues, organs, and organ systems. In **human evolution,** we looked at the effects of **evolving in time,** and found that all living organisms share a common origin, and that the rich diversity of life, including human beings, resulted from natural selection, descent over time, genetic modifications, and so forth. In this last exercise, on **human ecology,** we look at the effects of **living in the same space.** We will see that

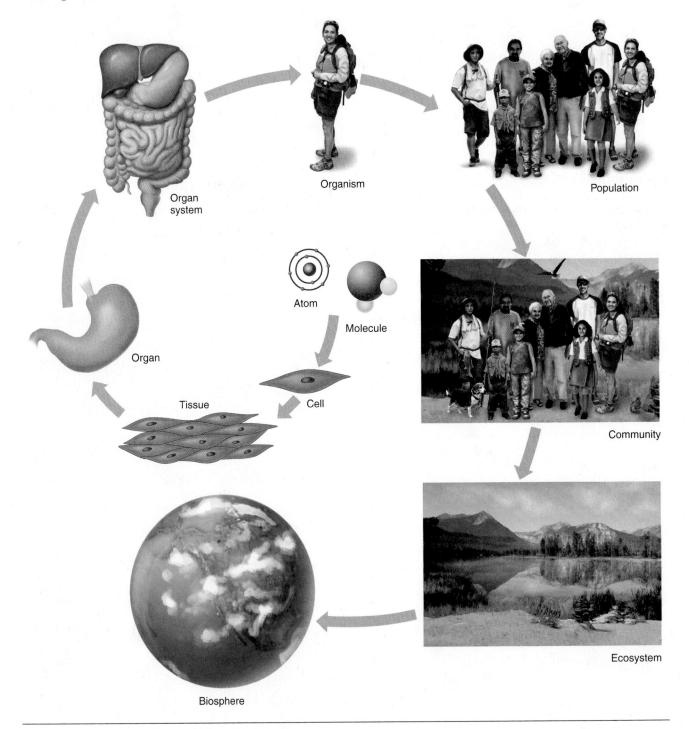

FIGURE 23.1 **Levels of organization in human biology.**

because we all share the same planet, the futures of all living organisms are inextricably linked.

Humans and Our Environment

Human ecology is the study of the relationship between people and their environment. Let's explore the concept of **ecological homeostasis,** or a state of dynamic balance between humans and our environment. When there is ecological homeostasis, most living organisms can obtain resources to grow, remain alive, reproduce, and so on. This balance is upset by **environmental crises,** such as pollution, global warming, diminished biodiversity, and depleted natural resources. As humans interact with the environment, what is the environmental impact of group size, human energy needs, and human use of resources? What is the balance between human and environmental needs?

From the past, we can learn that there is more than one approach to the way humans interact with the environment. To explore human impact on the environment, we will explore three different ways of human life: hunter-gatherer, horticulturalist, and industrialized people. What can we learn about the impact of human lifestyles on the environment? Are there benefits in nonindustrialized life styles that are lost in an industrialized society?

ACTIVITY 1

Human Lifestyles and Their Ecological Impacts

Your instructor will assign you one of three groups—hunter-gatherer, horticulturalist, or industrialized society. Use the detailed information about your assigned lifestyle to fill in Box 23.1. Afterward, each group will participate in a class discussion, using the group discussion questions at the end of this activity as a guide.

If you or your lab partner have a laptop or mobile device, you may also wish to do an Internet search of the specific people in your assignment (e.g., "Kung people," "Lese or Efe hunter-gatherer").

!Kung Hunter-Gatherers

Traditionally, the !Kung people lived in small groups (15–50) that changed according to seasonal resources. Their diet consisted of mostly plant food (60% of calories were derived from more than 200 species of plants) and lean animal meat, including small

animals such as rodents and reptiles and larger animals such as antelope (Figure 23.2). The !Kung used baskets, carrying bags, and digging sticks for gathering, and poison darts for hunting animals. The rich diversity of the diet meant that the !Kung had no observed vitamin or mineral deficiencies, with low levels of cholesterol and sodium intake. Food was shared with everyone in the group.

Anthropologists who studied the !Kung in the 1960s and 1970s were surprised to find that they worked only about 2.5 days per week to meet their subsistence needs. The rest of the time was spent socializing, eating, playing games, telling stories, and performing other activities. The !Kung had no concept of private property and had egalitarian relationships between men and women, with no defined leaders such as chiefs.

The age at menarche, or first menstruation, among the traditional !Kung was 16.5 years, on average. Women gave birth to their first child at age 19 and had an average of 4.7 children over

FIGURE 23.2 Hunter-gatherer lifestyle. Tribesmen gather around the fire.

Box 23.1   Characteristics of Three Human Lifestyles

	Hunter-Gatherers	Horticulturists	Industrialists
Group size			
Diet			
Tools/technology			
Hours worked per week			
Attitude toward private property			
Infant mortality			
Primary causes of death			
Age at menarche			
Age at first child's birth			
Number of children per woman (over a lifetime)			
Length of birth interval			
Approximate age at menopause			
Reproductive span*			
Number of years spent pregnant (over a lifetime)			
Number of years spent lactating			
Number of years of ovulatory/menstrual cycles			

*Reproductive span—age at menarche to age at menopause.

their lifetimes. Each child was breast-fed intensively (on demand, both day and night), and during this time ovulation was suppressed. Therefore, birth intervals were long, an average of 4.1 years. Age at menopause is more difficult to determine, but it seems that women reached menopause earlier than industrialized women, at about 40 years of age. This reproductive pattern means that !Kung women spent most of their lives either pregnant or lactating and not having ovulatory or menstrual cycles.

Infant mortality rates were high—as much as 20%—and consequently, overall life expectancy was short (about 30 years of age). However, individuals who did not succumb to accident or infectious disease in childhood could expect to live to old age. The primary causes of death were accident, complications of childbirth, or infectious disease.

Horticulturists

Horticulturists live in settlements that vary in size but may number in the thousands. Traditional horticulturists may grow more than one crop in their gardens so that they have some variety in their diets, but they have much less diversity of nutrients than hunter-gatherers (Figure 23.3). Typically, horticulturists may also eat meat, either from domesticated animals or from hunting or trading with hunting peoples. For example, the Lese horticulturists from the Ituri Forest in Africa trade plant food and other goods with the Efe hunter-gatherers (sometimes called Pygmies) for the meat that the Efe procure from the forest.

Horticulturists have simple metal tools such as hoes that they use to cultivate plants. They work more than hunter-gatherers, particularly in certain seasons of the year such as seed planting and harvest. They settle villages and have private family spaces and possessions.

Among the Lese women of the Ituri Forest who have been extensively studied by human ecologists, the age at menarche is around 16 years, and first birth typically occurs between age 18 and 19. Age at menopause is not well known. Lese women have

three children over their lifetimes, a lower number than what women bear in many simple agricultural societies. Women in such societies usually have four to eight children, with an average of six infants during their lives. Horticultural women typically breast-feed for shorter periods of time than the !Kung, and the interval between births is about 2 years. Infant mortality rates among traditional horticulturists are high—as much as 20–30%. Death is most often due to accident, complications of childbirth, or infectious disease.

Industrialists

Modern industrialized people may live in large groups (Figure 23.4); some cities have more than a million people. Rapid technological advances are taken for granted—think of the changes in technology over the 20th century when we went from the horse and buggy to rocket ships, jet planes, and reliance on computers. People in industrialized society have access to many goods but work hard to attain them. In the United States, a 40-hour workweek is the norm.

Women in industrialized societies reach menarche at an average of 12 to 12.5 years of age. Menopause occurs at about 50 years of age. A woman's age at the birth of her first child is often postponed with the use of contraceptives, and the size of the family may be planned. The average number of children over a woman's lifetime is 2. Length of lactation is extremely variable; some women do not breast-feed at all but feed their babies a prepared formula in

FIGURE 23.3 Horticultural Society. A farmer practicing sustainable agriculture with shade-grown crops.

FIGURE 23.4 Industrialized society. The El (elevated railway) crosses above a crowded street in Chicago, IL.

bottles. Even if women do choose to breast-feed, they are often separated from the infant while they work, so the frequency and quality of lactation are different from those of women in nonindustrialized societies.

Life expectancy in developed countries has risen to about 76 years of age, thanks to low infant mortality (about 96% of children survive to be teenagers) and well-developed medical technology. Industrialized people die of diseases that develop as a consequence of lifestyle—high-fat diets, use of tobacco and alcohol, and lack of vigorous physical activity contribute to heart disease (the leading cause of death), cancers, stroke, and emphysema.

Group Discussion Questions

1. What are the implications for energy resources extracted from the environment from the perspective of your group's lifestyle?

_____

2. Summarize the female reproduction pattern of your group.

_____

What are the implications for population growth?

_____

What are the implications for women's health?

_____

3. What are the dietary features of your group?

_____

How would this diet influence the health of the people?

_____

4. What is the major cause of death in your group, and how does it relate to lifestyle?

_____

5. How does the lifestyle of your group affect its specific environment?

_____

Human lifestyle has changed dramatically in the past 12,000 years, from the time that groups of people made the transition to reliance on domesticated plants and animals, to the breathtaking changes that have occurred in the past few hundred years

with industrialization. The study of human ecology, specifically of the different patterns of life of non-industrialized people such as hunter-gatherers and horticulturists, gives us insight into the impact of human behavior on the environment.

1. How have changes in human lifestyle contributed to the environmental challenges that we currently face?

_____

2. What are some of the ecological problems faced by people as we enter the twenty-first century?

_____

3. What can we learn from people with nonindustrialized ways of life that may offer solutions for some of our problems?

_____

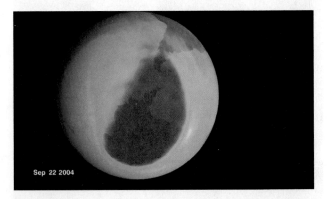

a) **This scientist is collecting insects from a tree-top in a tropical rainforest.** Studies such as this improve our understanding of the interrelationships of organisms in an ecosystem.

Sep 22 2004

b) **A satellite map documenting depletion of the ozone layer over Antarctica in 2004.** The area of greatest depletion appears dark blue.

FIGURE 23.5 Ecological studies.

World Human Population

In Figure 23.1, we study biological organization first at the organism level, examining individual living beings, and then at the level of groups of organisms. Ecologists are primarily interested in these last four **levels of biological organization—** population, communities, ecosystems, and biosphere. **Populations** make up **communities,** which make up **ecosystems,** which make up our **biosphere,** Earth. There are different ecological issues at each level. Population issues may involve rationing organs for transplants. Global warming and ozone destruction are problems at the biosphere level (Figure 23.5). What are the problems at the community and ecosystem level? (Refer to your textbook for examples of issues and controversies associated with the levels of human biology.)

_____

_____

The maximum rate of growth for any population under ideal conditions is called **biotic potential.** The opposing forces, called **environmental resistance,** are environmental factors such as limited nutrients, energy, and land. **Carrying capacity** represents the balance between biotic potential and environmental resistance; it is a steady level of population of a given species that the ecosystem can support at a given point in time.

FIGURE 23.6   Benefits and costs of industrialization. Modern farmers grow food crops much more efficiently than in the past, thanks to advances in such diverse fields as genetics, chemistry, and even the aerospace industry. Global positioning satellites and computers allow farmers to administer fertilizers precisely where needed, thereby eliminating waste, reducing environmental degradation, and improving yields.

What happens to the carrying capacity if environmental resistance is reduced? During the 1500s, the human carrying capacity was quite low. The world human population increased slowly because people lived in small, widely dispersed groups with limited ability to shape the environment. In the 1700s, the Industrial Revolution brought about enormous improvements in food, housing, transportation, medical care, education, and communication. The result was a steady, then exponential rate of growth in the world human population. Currently, there are about 6.4 billion people on this planet, and it is estimated that the population increases by a quarter of a million each day.

The initial effect of industrialization was to reduce the environmental resistance, thus allowing the carrying capacity for the human population to increase. Today, the opposite situation is beginning to develop. The unprecedented growth in the human population and the costs of industrialization have allowed environmental resistance to increase (Figure 23.6).

ACTIVITY 2

Population Explosion

Plot the changes in population over the past 2,000 years using the graph in Figure 23.7. The *x*-axis represents time (the number of years), while the *y*-axis represents population (numbers of people).

1. From the graph, approximately when did the growth curve increase the most? (*Hint:* Look for the area of the curve with the most slope ("rise over run").)

   _____

2. From the graph, if there was a major food problem in the years 1000, 1500, or 1900, which would have been the most devastating? Why?

   _____

   _____

3. What are the environmental implications of an explosive human population growth?

   _____

   _____ ■

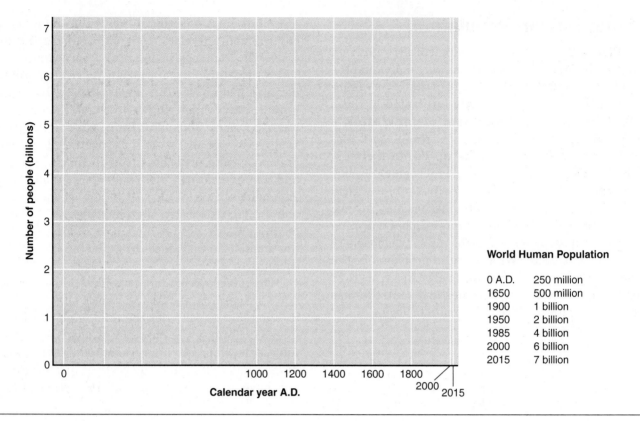

FIGURE 23.7 Growth curve.

Our population growth curve has a J-shape, which is very troubling because it resembles the exponential growth curve of an organism with a population growth rate tending toward the biotic potential. This suggests that there are unlimited environmental resources, which is clearly not the case. For all nations, both more-developed and less-developed, many medical, social, and environmental problems result from human overpopulation.

We would all like to see the achievement of major goals, such as ending waste and pollution. After all, preserving and protecting the planet's resources is just plain common sense. There are, however, smaller goals that are achievable by each of us. In the house, we can wrap insulation around the water heater, replace paper towels with cloth, wash our hands without running the water, turn off unused lights and appliances such as the TV and computer, and of course, recycle our trash. What suggestions do you have about driving, food preparation, and the use of the latest gadget?

_____

_____

_____

We are approaching our human biotic potential. From reading the text, what are some of the environmental changes that contributed to this population explosion?

_____

_____

_____

Speculate on Earth's carrying capacity for humans.

_____

_____

_____

Speculate on the effects of the human population on a major disaster such as the icebergs melting during climate change.

_____

_____

ACTIVITY 3

Population Estimates

How much territory does a group of living organisms need? How fast will a particular species be decimated by an ecological disaster? To answer these questions, ecologists need reliable information about that specific population, for how else can you compare the population before and after the change?

Remember, a **population** is defined as a group of individuals of the same species occupying the same geographic area and interacting with one another. **Population size** is a crucial piece of ecological information; for example, fluctuations in population size may indicate changes in the environment. Ecologists also need to know the size of a population in large areas, how and why it is distributed over a geographic area, and how and why its size changes over time.

Measuring a population by a **direct count** is usually not practical, especially over large areas

and with mobile organisms. A **population estimate** is used instead, with the understanding of a certain degree of error. In a very basic population estimation, an area is measured and divided into sections, then the specific population is counted and multiplied by the number of sections. Plant population sizes are generally easier to estimate because plants are immobile. Animals are harder, especially if they are numerous and in the air, such as birds, which must be estimated by counting the bird droppings.

Materials

Ball of string

1. The class is divided into groups of two to four students. Each group will perform the population estimate in the same general area and compare their final findings.
2. Identify an area for study near the campus, which is as uncultivated as possible (e.g., a nearby meadow). Each group should have an

Box 23.2 **Population Estimates Data**				
	Mobile Organism		**Immobile Organism**	
	Count	**Estimate**	**Count**	**Estimate**
Start Date: _____				
Plot 1				
Plot 2				
Total				
End Date: _____				
Plot 1				
Plot 2				
Total				

area of about 200 × 200 ft. Divide the area into 20 × 20 ft plots, for a total of 10 plots. Mark off the boundaries with string. Record the location so you can return to it a week later.

3. Decide on the two local (mobile and immobile) organisms. Select a plant for the immobile organism and animal droppings to represent the mobile organism (e.g., cloverleaf and bird droppings).

4a. Select the two plots that are the farthest from each other. Record the date in the Start Date section of Box 23.2. Count the number of plants, then animals, and record under the appropriate count. Multiply the counts by 10 and record under Estimate. Add Plot 1 and Plot 2 and record the total. Compare the number of mobile and immobile organisms between the two plots. Which population is larger? What might account for this difference?

_____

_____

_____

4b. Return one week later and record the date in the End Date section of Box 23.2. Repeat the procedure described in step 4a. Compare the total number of mobile and immobile organisms between the two dates. Which population is larger? What might account for this difference? How might weather factor in to your results? Did the populations increase or decrease? Why?

_____

_____

_____

5. What did you discover about the sampling method? Did you encounter any specific problems?

_____

_____

_____

6. Compare your data to those of the other students in your group. Are there variations in sampling data? Are the differences significant? Can you account for them?

_____

_____

_____

7. Preparing a total population line graph is a little more challenging than the foregoing; however, it will introduce you to the line graph, an important tool of scientists. Prepare line graphs of the total mobile and immobile population in your results on the start and end dates. The y-axis is the number of organisms, the x-axis is the number of days. Look at the slope, or angle of the line, between the start and end date population data. If the slope rises, it means the population has increased. The steeper the slope, the faster the population increase. If there was an increase over the 7 days, which population do the graphs say increased faster? Why do you think this was the result?

_____

_____

8. If you would like a bit more challenge, graph the population data for Plots 1 and 2 for the two dates in an individual plot population line graph. The x- and y-axes remain the same. The graphs will be a bit busier looking, as there are more data. If there was an increase over the 7 days time, which population do the graphs say increased faster? Why?

_____

_____

EXERCISE

23

HUMAN ECOLOGY

Critical Thinking and Review Questions

1. What are the main differences between the subject areas in human biology, human evolution, and human ecology? What are key similarities?

   _____

   _____

   _____

2. List the 10 levels of biological organization, from the smallest to the largest unit.

   _____

   _____

   _____

3. What do you think was the environmental impact of the hunter-gatherer, horticulturalist, and industrialized groups?

   _____

   _____

   _____

   _____

   _____

4. From studying nonindustrialized societies, what can we learn about environmental problems like pollution and global warming?

   _____

   _____

   _____

5. What are the main features of a hunter-gatherer lifestyle? What insight does the study of !Kung hunter-gatherers offer for understanding human biology?

   _____

   _____

   _____

   _____

   _____

6. How has the human lifestyle changed in the past 12,000 years? What impacts have the changes had on human biology? On the environment?

_____

_____

_____

_____

_____

7. Summarize human population growth over the past 2,000 years. Why has the human population increased so dramatically over the past century? What are some environmental effects of increasing population? Speculate on some of the consequences of the population increase for human biology.

_____

_____

_____

_____

_____

8. In the population estimate experiment, which population increased faster, the mobile or the immobile organisms? What might be the reasons? If there was a decrease, why do you think it occurred?

_____

_____

_____

_____

9. Speculate on the possible sources of errors in the sampling technique used in the population estimate experiment.

_____

_____

_____

THE METRIC SYSTEM

Measurement	Unit and Abbreviation	Metric Equivalent	Metric to English Conversion Factor	English to Metric Conversion Factor
Length	1 kilometer (km)	= 1,000 (10^3) meters	1 km = 0.62 mile	1 mile = 1.61 km
	1 meter (m)	= 100 (10^2) centimeters	1 m = 1.09 yards	1 yard = 0.914 m
		= 1,000 millimeters	1 m = 3.28 feet	1 foot = 0.305 m
			1 m = 39.37 inches	
	1 centimeter (cm)	= 0.01 (10^{-2}) meter	1 cm = 0.394 inch	1 foot = 30.5 cm
				1 inch = 2.54 cm
	1 millimeter (mm)	= 0.001 (10^{-3}) meter	1 mm = 0.039 inch	
	1 micrometer (μm) [formerly micron (μ)]	= 0.000001 (10^{-6}) meter		
	1 nanometer (nm) [formerly millimicron (mμ)]	= 0.000000001 (10^{-9}) meter		
	1 angstrom (Å)	= 0.0000000001 (10^{-10}) meter		
Area	1 square meter (m^2)	= 10,000 square centimeters	1 m^2 = 1.1960 square yards	1 square yard = 0.8361 m^2
			1 m^2 = 10.764 square feet	1 square foot = 0.0929 m^2
	1 square centimeter (cm^2)	= 100 square millimeters	1 cm^2 = 0.155 square inch	1 square inch = 6.4516 cm^2
Mass	1 metric ton (t)	= 1,000 kilograms	1 t = 1.103 ton	1 ton = 0.907 t
	1 kilogram (kg)	= 1,000 grams	1 kg = 2.205 pounds	1 pound = 0.4536 kg
	1 gram (g)	= 100 milligrams	1 g = 0.0353 ounce	1 ounce = 28.35 g
			1 g = 15.432 grains	
	1 milligram (mg)	= 0.001 gram	1 mg = approx. 0.015 grain	
	1 microgram (μg)	= 0.000001 gram		
Volume (solids)	1 cubic meter (m^3)	= 1,000,000 cubic centimeters	1 m^3 = 1.3080 cubic yards	1 cubic yard = 0.7646 m^3
			1 m^3 = 35.315 cubic feet	1 cubic foot = 0.0283 m^3
	1 cubic centimeter (cm³ or cc)	= 0.000001 cubic meter	1 cm^3 = 0.0610 cubic inch	1 cubic inch = 16.387 cm^3
		= 1 milliliter		
	1 cubic millimeter (mm³)	= 0.000000001 cubic meter		
Volume (liquids and gases)	1 kiloliter (kl or kL)	= 1,000 liters	1 kL = 264.17 gallons	1 gallon = 3.785 L
	1 liter (l or L)	= 1,000 milliliters	1 L = 0.264 gallons	1 quart = 0.946 L
			1 L = 1.057 quarts	
	1 milliliter (ml or mL)	= 0.001 liter	1 ml = 0.034 fluid ounce	1 quart = 946 ml
		= 1 cubic centimeter	1 ml = approx. ¼ teaspoon	1 pint = 473 ml
				1 fluid ounce = 29.57 ml
			1 ml = approx. 15–16 drops (gtt.)	1 teaspoon = approx. 5 ml
	1 microliter (μl or μL)	= 0.000001 liter		
Time	1 second (s)	= $\frac{1}{60}$ minute		
	1 millisecond (ms)	= 0.001 second		
Temperature	Degrees Celsius (°C)		°F = $\frac{9}{5}$ °C + 32	°C = $\frac{5}{9}$ (°F − 32)

ART, TEXT, AND PHOTO CREDITS

Exercise 1
Figure 1.1 Cinergi/Columbia/Tri-Star/Album/Newscom; **Figure 1.2** © 2010 Hilary B. Price. Distributed by King Features Syndicate, Inc.

Exercise 2
Figure 2.1 Charles D. Winters/Science Source; **Figure 2.2** © 2010 Hilary B. Price. Distributed by King Features Syndicate, Inc.

Exercise 3
Figure 3.3 Alan Bell, Pearson Education, Inc.; **Figure 3.4** Bert Atsma; **Figure 3.5** Science Source; **Figure 3.6** Ed Reschke/Peter Arnold/Getty Images.

Exercise 4
Figure 4.5 David M. Phillips/Science Source.

Exercise 5
Figure 5.1 PAL 3.0, Pearson Education, Inc.; **Figure 5.2** Alan Bell, Pearson Education, Inc.; **Figure 5.3** Alan Bell, Pearson Education, Inc.; **Figure 5.4** Alan Bell, Pearson Education, Inc.; **Figure 5.5** Bert Atsma; **Figure 5.6** Ed Reschke/Peter Arnold/Getty Images; **Figure 5.7** Ed Reschke/Peter Arnold/Getty Images; **Figure 5.8** Ed Reschke/Peter Arnold/Getty Images; **Figure 5.9** Pearson Education, Inc.; **Figure 5.10** Eric V. Grave/Science Source; **Figure 5.11** Ed Reschke; **Figure 5.12** Pearson Education, Inc.; **Figure 5.13** Ed Reschke/Peter Arnold/Getty Images.

Exercise 7
Figure 7.1 Ed Reschke/Peter Arnold/Getty Images; **Figure 7.2** SPL/Custom Medical Stock Photo; **Figure 7.3a–d** PAL 3.0, Pearson Education, Inc.; **Exercise 7.1** Ed Reschke/Peter Arnold/Getty Images; **Exercise 7.2** SPL/Custom Medical Stock Photo.

Exercise 8
Figure 8.1c Gene Cox/Science Source; **Figure 8.1d** Roseman/Custom Medical Stock Photo; **Figure 8.2** © 2009 Hilary B. Price. Distributed by King Features Syndicate, Inc.

Exercise 9
Figure 9.1b Science Source; **Figure 9.2c** Rosalind King/Science Source; **Figure 9.4b** Astrid & Hanns-Frieder Michler/Science Source.

Exercise 10
Figure 10.4 Victor Eroschenko, Pearson Education, Inc.

Exercise 11
Figure 11.5 Manfred Kage/Science Source; **Figure 11.8b–c** Sharon Cummings, University of California, Davis; **Figure 11.9b** Elena Dorfman, Pearson Education, Inc.; **Exercise 11.7** Sharon Cummings, University of California, Davis.

Exercise 12
Figure 12.9c Kresge Hearing Research Institute.

Exercise 13
Figure 13.4a–c Charles Venglarik, Pearson Education, Inc.

Exercise 14
Figure 14.6 John Scanlon.

Exercise 15
Figure 15.5a–c Wally Cush, Pearson Education, Inc.

Exercise 16
Figure 16.3c David M. Phillips/Science Source.

Exercise 17
Figure 17.5b Elena Dorfman, Pearson Education, Inc.

Exercise 19
Figure 19.8a Tomasz Trojanowski/Fotolia; **Figure 19.8b** Garo/Phanie/Alamy; **Figure 19.8c** Saturn Stills/Science Source; **Figure 19.8d** Xuejun li/Fotolia; **Figure 19.8e** Debi Treloar/Jules Selmes/Dorling Kindersley, Ltd.; **Figure 19.8f** A. Wilson/Custom Medical Stock Photo/Newscom;

INDEX

Note: A *b* following page numbers refers to boxes, an *f* refers to figures, and a *t* refers to tables.

A

Abdominopelvic body cavity, 65
ABO blood typing, 171, 174*f*
  characteristics of, 173*f*
Accessory organs, 210, 214
  of digestive system, 211*f*
Acid, lactic, 111
ACTH. *See* Adrenocorticotropic hormone
Action potentials, 116
ADH. *See* Antidiuretic hormone
Adipose tissues, 51–52
  microscopic observation of, 52*b*
  from subcutaneous layer under skin, 51*f*
Adrenal cortex, 156
Adrenal gland, 156, 159
Adrenal medulla, 156
Adrenocorticotropic hormone (ACTH), 154
Afterbirth, 240
Agglutination, 173, 174*f*
AIDS, 245
Air forced to move into lungs, 198
Air pressure in airways decreases, 198
Air-blood barrier, 197
Airflow into and out of lungs, 198–199
Airways decreases, air pressure in, 198
Aldosterone, 156
Allele, 258
Alveoli, 196
  gas exchange between blood, 197*f*
Ampulla, 242
Analogous structures, 282, 282*f*
Anaphase, 254
Anatomical landmarks and reference
  points, 59
Anatomical planes, 64*f*
Anatomical position. *See* Standard
  anatomical position (SAP)
Androgens, 156
Animal cell, generalized, 28*f*
Ankle jerk reflex, 121
Anterior and posterior body cavity, 65
Anterior and posterior surface regions,
  62–63, 63*f*
Antibody, 171, 172*f*
Antidiuretic hormone (ADH), 154, 229
Antigen, 171
Antigen-antibody complex, 171–173
Aorta, 178
Aortic arch, 179
Aortic arteries, 179
Aortic semilunar valve, 180
Appendicular skeleton, 89, 93
Aqueous humor, 142
Arachnoid, 126
Arachnoid mater, 126
Areolar connective tissues, 50
  loose, 50*f*
  microscopic observation, 50*b*

Arthropoids, 285
Asthma, 202
Astrocytes, 117
Atria, 178, 182
Auditory canal, 145
Auditory tube, 145
Auricles, 182
Autonomic nervous system, 131,
  133–134
  parasympathetic division, 131
  pulse rate data, 134*b*
  schematic drawing, 134*f*
  sympathetic division, 131
Avascular, 46
Axial skeleton, 89–90
Axon, 105, 117

B

Balance receptors, 139
Basal metabolic rate (BMR), 218
Basal metabolism, 218
Basement membrane, 46
Bell jar model, lungs, 198*f*
Bicuspid valve, 180
Binocular vision, 144–145
Biochemistry, comparative, 281
Biological organization, levels of, 299
Biosphere, 299
Biotic potential, 299
Bipolar cells, 142
Birth control methods, 246*f*
Bladder, 228*f*
Blood, 166–174
  components of, 166–169, 166*t*
  gas exchange between alveoli, 197*f*
Blood cells, 169–171, 171*f*
  sketches, 170*b*
Blood flow in heart, 181*f*
Blood flow through heart, tracing,
  181*b*
Blood pressure, 185–188
  measurements, 187*b*
  and relationship to cardiac cycle,
  185–188
  systolic and diastolic, 188*t*
Blood types, 171–174
  results, 174*b*
Blood vessels, 78, 174–185
  heart and, 178–182
  pulmonary and systemic circuits,
  186*f*
BMR. *See* Basal metabolic rate
Body, 92
Body cavities, 65–66, 65*f*
  abdominopelvic, 65
  anterior, 65
  cranial, 65
  orientation to, 65–66

  posterior, 65
  thoracic, 65
  vertebral canal, 65
Body regions, 62–63
Body sections, identifying, 64
Bone fusion, 94
Bones, 86–88
  anatomy, 86–88
  compact, 88
  coxal, 94
  cranial, 90
  facial, 90
  frontal, 90
  nasal, 91
  occipital, 91
  parietal, 90
  spongy, 87
  structures, 86–88
  temporal, 90
  tissues, 88
  zygomatic, 91
Bones of pelvic girdle, leg, and foot, 95*f*
Bones of right side of pectoral girdle, arm,
  and hand, 94*f*
Boyle's law, 198
Brain, 116
  functional divisions of, 129–130
  human, 129*f*
  intact sheep, 132*f*
  stem, 129
Brain, sheep, 131
  midsagittal section of sheep, 133*f*
Branch of science, 5
Bronchi, 196
Bronchioles, 196
Bulbourethral glands, 243

C

Calcitonin, 156
Caloric content, foods and estimated, 219*t*
Calorie, 218
Cardiac cycle, 185
Cardiac muscle, 104
Cardiac tissues, 52–53
Cardiovascular system, 166–188
  blood, 166–174
  heart and blood vessels, 178–188
Carriers, 260
Carrying capacity, 299
Cartilage, hyaline, 51
  microscopic observation, 51*b*
  from trachea, 51*f*
Cartilaginous joints, 96
Cauda equina, 128
Cecum, 214
Cell body, 117
Cell physiology, 36–41
  osmosis-part one, 38–39

osmosis-part two, 39–41
plasma membranes, 36–37
selectively permeable membranes, 36–38, 38*f*
Cells, 284
definition, 28
generalized animal, 28*f*
muscle, 104–106, 104*f*
structures and organelles, 29*t*
tissues, and organs-link between, 46
Cells, anatomy and diversity of, 28–32
diversity of human cells, 30–32
organelles, 28–29
Cells, diversity of human cells, 30–32
observing cellular diversity, 30–31
observing cellular diversity within single organ, 32
Cellular diversity, 31*b*
observing, 30–31
within single organ, 32
Central nervous system (CNS), 116
Cerebrospinal fluid (CSF), 126
Cerebrum, 129, 130*f*
Cervical curve, 91
Cervical enlargement, 128
Cervical vertebrae, 91
Cervix, 240
Chemoreceptors, 138
Chlamydia, 245
Chondrocytes, 51
Chordae tendineae, 180, 182
Choroid, 141–142
Chromatids, sister, 254
Chromosomes, 254
Chronic bronchitis, 202
Ciliary muscle, 141
Ciliated columnar epithelium, pseudostratified, 49
lining human trachea, 49*f*
microscopic observation, 49*b*
CNS. *See* Central nervous system
Coagulation, 167–168
Coccyx, 92
Cochlea, 145
Coin-data, toss, 7*b*
Coin-toss, 259
procedure, 7*f*
Punnett square, 259*b*
worksheet, 259*b*
Cold, common, 5–7
Collecting duct, 229
Colon, 214
Color blindness Punnett square, 264*b*
Columnar epithelium, simple, 48–49
microscopic observation, 49*b*
of stomach mucosa, 48*f*
Communities, 299
Compact bone, 88
Comparative anatomy, embryology and, 281–283
Comparative biochemistry, 281
Complementary base pairing, 268
Cones, 142
Connective tissues, 46
areolar, 50
dense white fibrous, 50–51
Contractile proteins, 52
Contraction, 185
skeletal muscles, 110–111
sliding filament mechanism, 110*f*

Control group, 6
Control variables, 6
Core principle, 5
Cornea, 141
Coronal planes, 64
Corpus callosum, 130
Corpus luteum, 236
Cortex, 226
Cow eye, 143
Cowper's glands, 243
Coxal bones, 94
Cranial body cavity, 65
Cranial bone, 90
Cranial nerves, 131
Crenation, 41
CSF. *See* Cerebrospinal fluid (CSF)
Cuboidal epithelium, simple, 48
in kidney tubules, 48*f*
microscopic observation, 48*b*
Cytokinesis, 254

D
Dendrites, 117
Dense white fibrous connective tissues, 50–51, 50*f*
microscopic observation, 51*b*
Deoxyribonucleic acid (DNA), 268
chemical characteristics of, 268–270
double helical structure of, 269*f*
extracting, 270–273
fingerprinting, 273–275
four nucleotides that comprise, 268*f*
recombinant, 275–276
technology and genetic engineering, 268–276
Dermis, 73
Dermis, structure of, 76–78
blood vessels, 78
collagen fibers, 76
elastic fibers, 76
hair follicle cells, 76
reticular fibers, 76
sebaceous glands, 76
sensory nerve endings, 78
smooth muscles, 76
sweat glands, 76, 78
Descent over time, 280
Diaphysis, 86
Diastolic pressure, 187, 188*t*
Diencephalon, 129
Diffusion, 36
through nonliving selectively permeable membrane, 38*f*
Digestive organs of fetal pig, 215*f*
Digestive system, 210–216
anatomy and basic function, 210–216
organs and accessory organs of, 211*f*
Digestive system and nutrition, 210–220
digestive system anatomy and basic function, 210–216
enzymes, 216–218
metabolism and nutrition, 218–220
Dinosaurs, 285
Direct count, 301
Directional terms, 59–61
orientation and, 61*t*
using, 60
Direction of movement, 149
Disorders, respiratory, 202–204
Dissection protocols, review of, 10–12

Dissection tools, 10–11
commonly used, 11*f*
and their purposes, 10*t*
Distal tubule, 229
DNA. *See* Deoxyribonucleic acid
Dominance and recessiveness, principle of, 257
Dorsal root, 127
Dorsal root ganglion, 128
Double helix, 268
Ductus deferens, 244
Duodenum, 214
Dura mater, 126
Dynamic equilibrium, 148

E
Ear, 145–149
and balance, 148–149
examining structure of, 147
and hearing, 145–147
middle and inner, 146*f*
structure of human, 145*f*
Ear drum, 145
Earth, brief evolutionary history of life, 284–290
major evolutionary events, 284–287
Earth, evolutionary time line for life, 288*b*
Ecological homeostasis, 295
Ecological studies, 298*f*
Ecosystems, 299
Egg, sperm and, 284
Ejaculatory duct, 243
Electrophoresis, 275
photograph of a gel plate, 275*f*
Embryology, comparative anatomy and, 281–283
Embryos, vertebrate, 281*f*
Emphysema, 202
Endocardium, 178
Endocrine glands, 157*f*
of fetal pig, 160*f*
in fetal pig and models, 156–160
miscellaneous, 156–161
Endocrine system, 154–161
hypothalamus and pituitary gland, 154–155
miscellaneous endocrine glands, 156–161
Endometrium, 238, 239
Endosteum, 87
Environmental crises, 295
Environmental resistance, 299
Enzymes, 216–218
activity as measured by color change, 217*b*
restriction, 273
Epicardium, 178
Epidermis, 73
layers, identifying, 74–75
showing five epidermal strata, 74, 75*f*
Epidermis, structure of, 74–76
keratinocytes, 74
melanocytes, 74
stratum basale, 74
stratum corneum, 74
stratum granulosum, 74
stratum lucidum, 74
stratum spinosum, 74
thick skin, 74
thin skin, 74

Epididymis, 245
Epiglottis, 195
Epinephrine, 156
Epiphyseal (growth) plate, 86
Epiphyses, 86
Epithelial tissues, 46–49
Equator, 254
Equilibrium, 148–149
   dynamic, 148
   maintenance of, 149
   static, 148
   testing, 149
Error/BIAS, more practice identifying, 3
Erythrocytes, 170
Esophagus, 179, 212
   cross section, trachea and, 195*f*
Estrogen, 156, 236, 239
Eukaryotes, 284
   versus prokaryotes, 284*f*
Eustachian tube, 145
Evidence, for evolution, 281
Evolution, 281–290
   brief evolutionary history of life on
      Earth, 284–290
   evidence for, 281
   human, 294
Evolutionary history of life on Earth,
      brief, 284–290
   genetic drift, 289–290
   major evolutionary events, 284–287
   natural selection, 287–289
Evolutionary organisms, major, 286*b*
Evolutionary time line for life on Earth,
      288*b*
Evolving in time, 294
Exhale, 198
Experimental error, math and, 7
   size and experimental error, 7–8
Experimental group, 6
External respiration, 197–198
External urethral sphincter, 226
Extinct, 285
Extracting DNA, 270–273
Eye, 141–145
   cow, 143
   examining structure of, 143
   models, 144
   sheep, 143
   structure of, 142*f*
   and vision, 141–145

F
Facial bone, 90
Fallopian tube, 239
False ribs, 92
Fascicles, 106, 118
Fatigue, muscle, 111
Female reproductive organs, 240*f*
Female reproductive system, 236–241
Femur, 94
Fertilization, 239
Fertilized egg, 239
Fetal pig, 156–160
   digestive organs of, 215*f*
   dissection, 158*f*
   endocrine glands in, 156–160
   endocrine glands of, 160*f*
   jejuno-ileum, 214
   male and female urinary system, 228*f*
   reproductive system, 243–245, 244*f*

Fibroblasts, 50
Fibrous joints, 96
Fimbriae, 239
Fingerprinting, 273–275
First heart sound, 187
Flat bone, 86
Floating ribs, 92
Follicle, 236
Follicle-stimulating hormone (FSH), 154
Follicular cells, 161
Foods, 276
   and estimated caloric content, 219*t*
   genetically modified, benefits and risks,
      276*b*
Foramen magnum, 91
Forebrain, 129–130
Forelimbs, vertebrate, 283*f*
Fossils, 281
Founder effect, 290
Fovea centralis, 142
Fractions, converting percentages to, 10
Frontal bone, 90
Frontal lobe, 130
Frontal planes, 64
FSH. *See* Follicle-stimulating hormone
Fungus, 245

G
Gallbladder, 214
Gametes, 254
Ganglion cells, 142
Gas exchange, 197
Gastrointestinal (GI) tract, 210, 210*f*
   layers of, 210, 212
Gel electrophoresis, 273, 274*f*
   photograph of a gel plate after, 275*f*
General sensation, 138
Genes, 258, 268
Genetically modified foods, 276*b*
Genetically modified organisms (GMOs),
      276
Genetic drift, 289–290
Genetic engineering, 268
Genetic modification, 280
Genetics, 254–264
   meiosis, 254–257, 256*f*
   mitosis, 254, 255*f*
   probabilities, 260–261
   sex determination, linkage, and
      pedigree charts, 261–264
   using Punnett square, 258–260
Genital herpes, 245
Genital warts, 245
Genotype, 258
GH. *See* Growth hormone (GH)
Glands, 156–161
   adrenal, 156, 159
   bulbourethral, 243
   Cowper's, 243
   endocrine, 156–161, 157*f*
   pituitary, 154, 155*f*
   prostate, 243, 244
   salivary, 214
   sebaceous, 76
   sweat, 76, 78
   thyroid, 156
Glomerular capsule, 228
Glomerular filtrate, 228
Glomerulus, 228
Glucagon, 156

Glucocorticoids, 156
GMOs. *See* Genetically modified
      organisms
Gonads, 156
Gonorrhea, 245
Graafian follicle, 236
Gray matter, 127
Growth hormone (GH), 154

H
Hair cells, 146
Hair follicle cells, 76
Hairy hands Punnett square 1, 259*b*
Hairy hands Punnett square 2, 260*b*
Hand, dorsal and ventral outlines, 79*b*
Hearing, 147
   conduction pathways, 147
   localization, 147
   receptors, 139
Heart, 178–182
   blood flow in, 181*f*
   blood vessels, 174–185
   gross anatomy of, 60*f*
   internal view of, 180*f*
   sheep, 183*f*
   as situated in thoracic cavity, 179*f*
   tracing blood flow through, 181*b*
Heart rates, 182–184
   measurements, 184*b*
Hematocrit calculations, 169*b*
Hematocrit tests, 166, 168
   results, 168*f*
   steps in performing, 167*f*
Hemolysis, 40
Heparin, 167–168
Hepatitis B, 245
Heterozygous, 258
High power lens, 16
Hindbrain, 129
Hominids, 286, 286*f*
Hominoids, 285
Homologous structures, 282
Homozygous, 258
Hormones, 154
Horticultural society, 297*f*
Horticulturists, 297
Human biology, 294
   levels of organization, 294*f*
Human body, orientations to, 58–66
   body cavities, 65–66, 65*f*
   directional terms, 59–61
   planes, 63–64
   standard anatomical position, 58–59
   surface regions, 62–63
Human cells, 30–32
   diversity in, 28*f*
   diversity of, 30–32
Human ecology, 294–302
   humans and our environment, 295–298
   world human population, 299–302
Human population, world, 299–302
   growth curve, 300*f*
Human skin, photomicrographs of, 77*f*
Humans, 294–298
   evolution, 294
   and our environment, 295–298
   respiratory system, 194*f*
   urinary system, 227*f*
Hunter-gatherer lifestyle, 295*f*
Hunter-gatherers, !Kung, 295–297

Hyaline cartilage, 51
  microscopic observation, 51*b*
  from trachea, 51*f*
Hypertonic fluid, 38
Hypodermis, 73
Hypothalamus, 129, 155*f*
  pituitary gland and, 154
Hypothesis, 3
Hypotonic fluid, 38

I

Ileum, 214
Immature sperm, 242
Incubation period, 6
Incus, 145
Industrialists, 297–298
Industrialization, benefits and costs of, 299*f*
Industrialized society, 297*f*
Inferior vena cava, 179–180, 182
Inhale, 198
Inner ear, 145, 146*f*
Insulin, 156
Integration centers, 117
Integumentary system, 72–80
Intercalated discs, 53
Internal urethral sphincter, 226
Interneurons, 117
Interphase, 254
Interventricular septum, 179
Intervertebral disks, 92
Invertebrates, 284
Iris, 142
Irregular bone, 86
Isotonic fluid, 38

J

Jejuno-ileum, 214
Jejunum, 214
Joint movements, 96*b*
Joints, 96–98

K

Keratinocytes, 74
Kidney, 226
Knee jerk reflex, 120–121

L

Labor, 240
Laboratory protocol, scientific method, 1–12
  converting percentages to fractions, 10
  math and experimental error, 7–8
  metric measurements, 8–10
  scientific method of discovery, 2–7
Laboratory safety, 10–12
  equipment, 12*f*
  protocols, 11
  review of dissection protocols, 10–12
  review of lab safety protocols, 11
Lactic acid, 111
Lacunae, 51
Large intestine, 213, 213*f*
Large Small Right (LSR), 9
Larynx, 196
Left atrioventricular (LV) valve, 180, 182
Legs, feet, and pelvic girdle, 94–95
Lens, 141
Leukocytes, 170

LH. *See* Luteinizing hormone (LH)
Life on Earth, 284–290
  brief evolutionary history of, 284–290
  evolutionary timeline for, 288*b*
Lifestyle, 295–298
  characteristics of three human, 296*b*
  hunter-gatherer, 295*f*
Liver, 214
Living in same space, 294
Lobes, 130
Localizing sounds, 146
Locomotion, 86
Long bone, 86, 87*f*
  anatomy, 86–88
Loop of Henle, 229
Loose areolar connective tissues, 50*f*
Lower respiratory tract, 194
Low power lens, 16
LSR. *See* Large Small Right
Lumbar, 92
  curve, 91
  enlargement, 128
  nerves, 118–119
  vertebrae, 91
Lungs, 178
  airflow into and out of, 198–199
  air to move into, 198
  bell jar model, 198*f*
  volumes, 200
Luteinizing hormone (LH), 154

M

Macula lutea, 142
Male reproductive system, 241–243, 242*f*
Malleus, 145
Mammals, 285
Mandible, 91
Manubrium, 92
Math and experimental error, 7
  size and experimental error, 7–8
Matrix, 49
Maxilla, 91
Mechanoreceptors, 138
Median (midsagittal) planes, 63–64
Medulla, 226
Medulla oblongata, 129
Meiosis, 254–257, 256*f*
  stages of, 257*b*
Melanocytes, 74
Membranes, 36–38
  plasma, 36–37
  respiratory, 197
  selectively permeable, 36–38, 38*f*
Membranous organelles, 28, 29*t*
Meninges, 126
Menstrual period, 238
Menstrual phase, 238
Menstruation, 238
Menu and calorie totals, proposed, 220*b*
Metabolism and nutrition, 218–220
Metaphase, 254
Metric measurements, 8–10, 8*f*
  conversions, 9–10, 9*t*
  standard references, 8–9
Metric prefixes and conversions, 9*t*
Metric system, 8
Microglial cells, 117
Microscopes, 16–24
  applying magnification and field size, 22–23

calculating total magnifications for, 17*b*
care of, 16
check cells at different magnifications, 24*b*
checklist for proper microscope storage, 24*b*
cleaning and caring for, 19
cleanup and storage of equipment, 24
determining depth of field, 21
drawing of letter "e" using scanning power, 20*b*
focusing, 19–21
letter "e" at different magnifications, 20*b*
and magnifications, 16–17
parts of compound light, 18*f*
preparing wet mount slide, 23–24, 23*f*
understanding size of field, 22
Microscopes, identifying parts of, 17–19
  arm, 17
  base, 19
  condenser, 17–18
  condenser-adjustment knob, 18
  course-adjustment knob, 18
  fine-adjustment knob, 18–19
  iris diaphragm, 18
  iris diaphragm wheel, 18
  lamp, 19
  nosepiece, 17
  objective lens, 17
  ocular lens, 17
  slide, 17
  stage, 17
  stage controls, 17
Midbrain, 129
Middle ear, 145, 146*f*
Minute respiratory volume (MRV), 202*f*
Mitosis, 254, 255*f*
  stages of, 255*b*
Mitotic spindle, 254
Monkeys, 285
Motor division, 116
  autonomic division, 116
  somatic division, 116
Motor neuron, 105
Motor unit, 105
Mouth, 195, 212
MRV. *See* Minute respiratory volume
Mucosa, 210
Multicellular life, 284–285
Multipolar neuron, 30*f*, 54*f*
  microscopic observation, 54*b*
Muscle tissues, 46
  cardiac muscle, 52–53
  smooth muscle, 53
Muscles, 104–106
  cardiac, 104
  cells, 104–106, 104*f*
  fatigue, 111
  organization of whole, 106–107, 107*f*
  papillary, 180, 182
  skeletal, 30*f*
  sphincter, 212
  spindles, 138*f*
  tissues, 52
  uterine, 239
Muscles, skeletal, 30*f*, 52*f*, 104
  microscopic observation, 52*b*
Muscles, smooth, 53, 53*f*, 104
  dermis structure, 76
  microscopic observation, 53*b*

Muscular system, 104–111
  contraction of skeletal muscles, 110–111
  energy use by skeletal muscles, 111
  gross anatomy of skeletal muscles, 107–109
  skeletal muscle cell microstructure, 104–105
  stimulation of skeletal muscle cells, 105–106
  whole skeletal muscle organization, 106–107, 107*f*
Muscularis, 210
Mutable DNA, 280
Myelin, 117
Myocardium, 178
Myofibrils, 104
  structures of, 104*f*
Myofilaments, 104
Myometrium, 239
Myosin, 104
Myosin heads, 110

N
Nails, 80
  cuticle, 80
  eponychium, 80
  free edge, 80
  lunula, 80
  mail matrix, 80
  nail bed, 80
  nail body, 80
  nail root, 80
  structures of, 80*f*
Nasal bone, 91
Nasal cavity, 195
Natural selection, 280, 287–289
Nephron, 226
  tubular structure of, 229*f*
  and urine formation, 228–232
Nerves, 116
  cervical, 118
  cranial, 118
  lumbar, 118–119
  peripheral, 119*f*
  PNS and, 118–119
  sacral, 119
  spinal, 118, 118*f*
  thoracic, 118
Nervous system, 116–121, 126–134
  organization, neurons, nervous tissue, and spinal reflexes, 116–121
  organization of, 116, 116*f*
  spinal cord, brain, and autonomic nervous system, 126–134
Nervous tissues, 46, 54
Neuroglia, 54, 117
Neuromuscular junction, 105, 105*f*
  drawing and photomicrograph of, 106*f*
Neurons, 54, 117
  sensory, 117
  types and components of, 117*f*
Neurons, motor, 105, 117
  drawing and photomicrograph of, 106*f*
Neurons, multipolar, 30*f*, 54, 54*f*
  microscopic observations of, 54*b*
Neurotransmitters, 105
Nitrogenous base, 268
Nodes of Ranvier, 118
Nonkeratinized structures, 212

Nonmembranous organelles, 28, 29*t*
Norepinephrine, 156
Normal respiratory rate, 201
Nose, 195
Nucleotides, 268
Nucleus, 28
Nutrition, digestive system and, 210–220
  digestive system anatomy and basic function, 210–216
  enzymes, 216–218
  metabolism and nutrition, 218–220

O
Occipital bone, 91
Occipital lobe, 130
Oil immersion, 16
Oligodendrocytes, 117
Oocytes, 236, 239
  path of, 241*f*
Optic disk, 143
Organelles, 28–29
  cell structures and, 29*t*
  identifying, 29
Organs, 46
  of digestive system, 211*f*
  link between cells, tissues, and, 46
  observing cellular diversity within single, 32
  tissues, and functions, 46*b*
Osmosis, 38–41
  definition, 38
  observing in potato cubes, 38–39
  observing in red blood cells, 40–41
  part one, 38–39
  part two, 39–41
  in potato cubes, 39*f*
  in red blood cells, 41*b*
  weight in potato cubes, 39*b*
Osteoblasts, 87
Osteoclasts, 87
Outer ear, 145
Oval window, 145
Ovarian cycle, 236–238, 237*f*, 238*f*
Ovaries, 159, 239, 243
Oviduct, 239
Ovulation, 236
Oxytocin, 154

P
Packed cell volume (PCV), 166, 169
Pain receptors, 138
Pancreas, 156, 159, 214, 216
Papillary muscles, 180, 182
Parafollicular cells, 161
Parasympathetic division, 131
Parfocal, 20
Parietal bone, 90
Parietal lobe, 130
Passive transport , 36
  forms of, 36*f*
PCV. *See* Packed cell volume
Pectoral girdle, 93–94
  arm and hand, 94
Pedigree charts, 261–264
  blank, 263*f*
  displaying X-linked disorders, 262*f*
  sex determination, linkage, and, 261–264

Pelvic girdle, 93–95
  legs and feet, 94–95
Penis, 243
Percentages to fractions, converting, 10
Pericardium, 178, 182
Periosteum, 87
Peripheral nervous system (PNS), 116
Pharynx, 195, 212
Phenotype, 258
Phospholipid, 36
Phospholipid bilayer, 36
Photoreceptors, 138
Pia mater, 126
Pig, 156–160
  digestive organs of fetal, 215*f*
  endocrine glands in fetal, 156, 158–160, 160*f*
  urinary systems, male and female, fetal, 228*f*
Pinna, 145
Pituitary gland, 155*f*
  hypothalamus and, 154
Placenta, 239, 240
Placental mammals, 285
Planes, 63–64
  anatomical, 64*f*
  coronal, 64
  cross section, 63
  frontal, 64
  longitudinal section, 63
  median (midsagittal), 63–64
  sagittal, 63
  transverse, 64
Plasma membrane, 28
  definition, 36
  measuring diffusion rates of dye through agar gel, 36–37
PNS. *See* Peripheral nervous system
Pons, 129
Population, 299, 301
  bottleneck, 290
  estimate, 301
  estimates data, 301*b*
  size, 301
Population, world human, 299–302
  growth curve, 300*f*
Populations, communities, and ecosystems, 299
Power of lens, 16
Pressure gradient, 198
Pressure receptors, 138
Primary curvatures, 92
Primates, 285
PRL. *See* Prolactin
Probabilities, 260–261
Progesterone, 156, 239
Prokaryotes, 284
  versus eukaryotes, 284*f*
Prolactin (PRL), 154
Proliferative phase, 238
Proprioceptors, 140
Prosimians, 285
Prostate gland, 243, 244
Proximal tubule, 228
Pseudostratified ciliated columnar epithelium, 49
  lining human trachea, 49*f*
  microscopic observation, 49*b*
Pulmonary arteries, 182
  right and left, 180

Pulmonary circuit, 180, 185
  blood vessels of, 186*f*
Pulmonary semilunar valve, 180
Pulmonary trunk, 179, 182
Pulmonary veins, right and left, 180
Punnett square, 258–260
  coin-toss, 259*b*
  color blindness, 264*b*
  demonstrating sex determination by the X and Y chromosomes, 262*t*
  demonstrating two individuals who are heterozygous for freckles, 258*t*
  setting up, 258*b*
  using, 258–260
Punnett square 1, hairy hands, 259*b*
Punnett square 2, hairy hands, 260*b*
Pupil, 142

R
Rain shadow phenomenon, 5*f*
Ratio, 258
RBS. *See* Red blood cells
Receptors, 138–141
  adaptation, 139
  balance, 139
  chemo, 138
  hearing, 139
  mechano, 138
  pain, 138
  photo, 138
  pressure, 138
  sensory, 138*f*
  stretch, 138
  tendon, 138*f*
  thermo, 138
  touch/tactile, 138
Recessiveness, principle of dominance and, 257
Recombinant deoxyribonucleic acid, 275–276
  transgenic organisms and, 275
Rectum, 214
Red blood cells (RBS), 40*f*
  observing osmosis in, 40–41
  results of osmosis in, 41*b*
Reflexes, 119–121
  ankle jerk, 121
  knee jerk, 120–121
  spinal, 119–121
Relaxation, 185
Renal pelvis, 226
Reproductive system, 236–247
  female reproductive system, 236–241
  fetal pig reproductive system, 243–245, 244*f*
  male reproductive system, 241–243, 242*f*
  preventing STDs and unintended pregnancies, 245–247
  sexually transmitted diseases, 245
Respiration, 194
Respiratory anatomy, 194–197
Respiratory disorders, 202–204
Respiratory membrane, 197
Respiratory system, 194–204
  breathing, 198–200, 199*f*
  external respiration, 197–198
  human, 194*f*
  lung capacity, 200–202
  respiratory anatomy, 194–197
  respiratory disorders, 202–204

Respiratory tree, 196
Restriction enzymes, 273
Retina, 142–143
  structure of, 143*f*
Rh+, 172
Rib cage, 90, 92–93, 93*f*, 179
Ribs, 92
Right atrioventricular (AV) valve, 180, 182
Rods, 142

S
Sacral curve, 91
Sacrum, 91
Sagittal planes, 63
Salivary glands, 214
SAP. *See* Standard anatomical position
Sarcomere, 105
Scanning power lens, 16
Scapula, 92
Schwann cells, 117
Science, branch of, 5
Scientific method, 2
Scientific method, developing, 2–5
  observation in collecting information, 2–3
  proposing question, 3–4
  reject, revise, or confirm hypothesis, 4
  some hypotheses become theories, 4–5
  some theories become laws, 5
  testing hypothesis, 4
Scientific method and laboratory protocol, 1–12
  converting percentages to fractions, 10
  math and experimental error, 7–8
  metric measurements, 8–10
  scientific method of discovery, 2–7
Scientific method of discovery, 2–7
  theory building-common cold, 5–7
  analyze data and modify hypothesis, 6–7
  design experiment, 6
  formulate hypothesis, 5–6
  formulate valid test criteria and ideas, 6
  observe and generalize, 5
Sclera, 141
Scrotum, 242, 244
Sebaceous glands, 76
Secondary curvatures, 92
Second heart sound, 187
Secretory phase, 239
Segregation, principle of, 258
Selective perception, 2
Selectively permeable membrane, 36–38
  diffusion through nonliving, 38*f*
Self-fulfilling prophecy, 3
Semen, 243
Semicircular canals, 145, 148
  vestibular apparatus, and balance, 148*f*
Seminal vesicle, 243, 244
Seminiferous tubules, 242
Senses, 138–149
  ear and balance, 148–149
  ear and hearing, 145–147

  eye and vision, 141–145
  receptors, 138–141
Sensory adaptation, 79
Sensory division, 116
Sensory receptors, 138*f*
Septum, 178
  interventricular, 179
Serosa, 210
Sex cells, 254
Sex determination, linkage, and pedigree charts, 261–264
Sex-linked, 261
Sexually transmitted diseases (STDs), 245
  preventing and unintended pregnancies, 245
Sheep eye, 143
Sheep heart, 183*f*
Short bone, 86
Simple columnar epithelium, 48–49
  microscopic observation, 49*b*
  of stomach mucosa, 48*f*
Simple cuboidal epithelium, 48
  kidney tubules, 30*f*
  in kidney tubules, 48*f*
  microscopic observation, 48*b*
Simple squamous epithelium, 47
  microscopic observation, 47*b*
  in normal lung alveoli, 47*f*
Single organ, observing cellular diversity within, 32
Sister chromatids, 254
Skeletal muscle cells, 104–106
  microstructures, 104–105
  stimulation of, 105–106
Skeletal muscles, 30*f*, 52*f*
  contraction of, 110–111
  energy use by, 111
  gross anatomy of, 107–109
  groups and their functions, 108*f*
  microscopic observation, 52*b*
  selected, 109*t*
Skeletal system, 86–98
  bone structure, 86–88
  bone tissue, 88–89
  joints, 96–98
  organization of, 89–95
  whole body growth, 95, 95*f*
Skeleton, 89–90, 93
  appendicular, 89, 93
  axial, 89–90
  human, 89*f*
Skin, 72, 72*f*
  functions of, 72
  layers and organization, identifying, 73–74
  photomicrographs of human, 77*f*
  structure of, 72–74
  types, 78–79
Skull, 89, 90*f*
  examining, 90–91
Sliding filament mechanism, 110, 110*f*
Small intestine, 212–213
  walls of, 213*f*
Smoking, 202
Smooth muscles, 53, 53*f*
  dermis structure, 76
  microscopic observation, 53*b*
Somatic sensation, 138
Space, living in same, 294

Special senses, 138
Sperm, 284
  cells from sperm smear, 30*f*
  egg and, 284
  mature, 28*f*
Spermatic cord, 244
Spermatogenesis, 242
Spermatogonia, 242
S-phase, 254
Sphenoid, 91
Sphincter muscles, 212
Sphygmomanometer, 185, 188*f*
Spinal cord, 116, 126–129
  cross section, 126*f*
  longitudinal view of, 128*f*
  meninges and, 126*f*
  photomicrograph of, 128*f*
Spinal curvatures, 92
  abnormal, 92*f*
Spinal reflexes, 119–121
  ankle jerk reflex, 121
  knee jerk reflex, 120–121
  pain withdrawal reflex, 120*f*
Spirometers, 201*f*
Spongy bone, 87
Squamous epithelium, simple
  microscopic observation, 47*b*, 47
  in normal lung alveoli, 47*f*
Squamous epithelium, stratified, 47–48
  lining of esophagus, 48*f*
  microscopic observation, 48*b*
Standard anatomical position (SAP),
    58–59
Standard references, 8–9
Stapes, 145
Starch test, 38
Static equilibrium, 148
STDs. *See* Sexually transmitted diseases
Sternum, 92
Stomach, 212
Stratified squamous epithelium, 47–48
  lining of esophagus, 48*f*
  microscopic observation, 48*b*
Stretch receptors, 138
Striations, 52
Subclinical cases of cold, 6
Submucosa, 210
Sugar test, 38
Superior vena cava, 179, 180, 182
Surface landmarks, 62–63
  for identification, 63*f*
  standard anatomical position, 58*f*
Surface regions, 62–63
Sweat glands, 76, 78
Sympathetic division, 131
Synaptic cleft, 105
Synovial joints, 96
  movements, 97, 97*b*, 98*f*
Syphilis, 245
Systemic circuit, 179, 185
  blood vessels of, 186*f*
Systolic pressure, 187, 188*t*

T
Tapetum lucidum, 143
Taste buds, 141*f*
Telophase, 254
Temporal bone, 90
Temporal lobe, 130
Tendon, 107
Tendon receptors, 138*f*
Testes, 159, 242
Testis, 245
Testosterone, 156, 241
TH. *See* Thyroid hormone
Thalamus, 129
Thermoreceptors, 138
Thoracic body cavity, 65
Thoracic cavity, 178, 198
  heart in, 179*f*
Thoracic vertebrae, 91
Thoracic volume, 198
Thymus, 156, 159
Thyroid follicles, 161
Thyroid gland, 156
Thyroid hormone (TH), 156
Thyroid stimulating hormone (TSH), 154
Time, 37*b*
  descent over, 280
  evolving in, 294
Tissues, 46–54
  adipose, 51–52
  areolar connective, 50
  cardiac, 52–53
  connective, 49–52
  dense white fibrous connective, 50–51
  epithelial, 46–49
  muscle, 52–53
  nervous, 54
  organs, and functions, 46*b*
  and organs, link between cells, 46
  smooth, 53
TLC. *See* Total lung capacity
TM. *See* Total magnification
Tonicity, 38
Total lung capacity (TLC), 200
  components of, 200*f*
Total magnification (TM), 17
Touch/tactile receptors, 138
Trachea, 179, 196
  and esophagus cross section, 195*f*
Transgenic organisms, 275–276
Transport, passive, 36, 36*f*
Transverse planes, 64
Tricuspid valve, 180
Triglyceride, 216
Tropic hormones, 154
True ribs, 92
TSH. *See* Thyroid stimulating hormone
Two-point discrimination test, 139
Tympanic membrane, 145

U
Universal donor, 171
Upper respiratory tract, 194

Ureter, 226
Urethra, 226, 228*f*, 243, 244
Urinary bladder, 226
Urinary system, 226–232
  human, 227*f*
  male and female fetal pig, 228*f*
  nephron and urine formation, 228–232
  organs of, 226–228
Urine formation and nephron, 228–232
Urine test data, 231*b*
Urogenital sinus, 243
Uterine cycle, 238–241, 238*f*
Uterine muscles, 239
Uterine tube, 239, 243
Uterus, 239, 243
  full-term fetus inside, 241*f*

V
Vagina, 240, 243
Vaginal opening, 240
Variables, 6
Vas deferens, 242
VC. *See* Vital capacity
Ventral root, 128
Ventricles, 126, 178, 182
Vertebral canal, 65
Vertebral column, 91, 91*f*
Vertebral regions and spinal curves,
    91–92
Vertebrate, 285
  embryos, 281*f*
  forelimbs, 283*f*
Vestibule, 145, 148
  in balance, 148*f*
Vestigial structures, 282
Visual acuity, 144
Vital capacity (VC), 201
Vitreous humor, 142
Vocal cords, 196
Voluntary motor division, 116

W
Wet mount slide, preparing, 23–24, 23*f*
White matter, 127
Whole body growth, 95, 95*f*
Working distance, 20
World human population, 299–302
  growth curve, 300*f*

X
X and Y chromosomes, 262*t*
Xiphoid process, 92
X-linked disorders, pedigree chart
    displaying, 262*f*

Y
Yeast infections, 245

Z
Z-line, 105
Zygomatic bone, 91
Zygous, 258